Water and Sanitation Services

Water and Sanitation Services

Public Policy and Management

Edited by José Esteban Castro and Léo Heller

earthscan
publishing for a sustainable future

London • Sterling, VA

First published by Earthscan in the UK and USA in 2009

ISBN: 978-1-84407-656-7

Typeset by JS Typesetting Ltd, Porthcawl, Mid Glamorgan
Cover design by Dan Bramall
Cover images: Aalok Hiti stone sprout, Lafitpur (Nepal), December 2006. © Photographer: Laia Domènech Pretus, Institute of Environmental Science and Technology (ICTA), Autonomous University of Barcelona, Spain
Sewer main pipe in Bangkok, Thailand © Loic Bernard/istoclephoto.com
Structure for Multipurpose Water Use © African Development Bank funded RWSS I Project, Sao Tome (2008)

For a full list of publications please contact:

Earthscan
Dunstan House
14a St Cross St
London, EC1N 8XA, UK
Tel: +44 (0)20 7841 1930
Fax: +44 (0)20 7242 1474
Email: earthinfo@earthscan.co.uk
Web: **www.earthscan.co.uk**

22883 Quicksilver Drive, Sterling, VA 20166-2012, USA

Earthscan publishes in association with the International Institute for Environment and Development

A catalogue record for this book is available from the British Library

Library of Congress Cataloging-in-Publication Data
Water and sanitation services : public policy and management / edited by José Esteban Castro and Léo Heller.
p. cm.
Includes bibliographical references and index.
ISBN 978-1-84407-656-7 (hardback)
1. Water treatment plants–Management. 2. Water-supply–Government policy. 3. Sewage disposal plants–Management. I. Castro, José Esteban. II. Heller, Léo.
TD434.W355 2009
363.6'1–dc22
2008044547

At Earthscan we strive to minimize our environmental impacts and carbon footprint through reducing waste, recycling and offsetting our CO_2 emissions, including those created through publication of this book. For more details of our environmental policy, see www.earthscan.co.uk.
This book was printed in the UK by Antony Rowe, Chippenham.
The paper used is FSC certified and the inks are vegetable based.

FSC
Mixed Sources
Product group from well-managed forests and other controlled sources
Cert no. SGS-COC-2953
www.fsc.org
© 1996 Forest Stewardship Council

Contents

Part II Country and Regional Experiences

List of Figures, Tables and Boxes

FIGURES

TABLES

BOXES

List of Contributors

Susanne Balslev Nielsen graduated as a civil engineer in 1994 and is now an associate professor in urban technology and management at DTU Management at the Technical University of Denmark. She has specialized in the planning and management of transformation processes towards a sustainable built environment. Since January 2008, she has been vice chairman of the Centre of Facilities Management, Realdania Research, at the Technical University.

Bernard Barraqué is a professor at the National Institute for Water and Forestry Management and Engineering (ENGREF) and a research director of the National Centre for Scientific Research (CNRS) in France. He is based at the International Centre for Research on Environment and Development (CIRED), where he studies the water policies of European Union member states. He is particularly interested in the comparative development of water and sanitation services, and their articulation with regional (or basin) water resources management policies. He is also the coordinator of the United Nations Educational, Scientific and Cultural Organization (UNESCO) Urban Water Conflicts Task Force in Paris.

Okke Braadbaart is professor of the urban environment at Wageningen University and Research Centre, The Netherlands. He is based at the Centre of Excellence for Sustainable Water Technology (WETSUS) in Leeuwarden. His main research interests are the comparative performance of water companies and the development of urban hydro-spatial networks.

José Esteban Castro is professor of sociology at the School of Geography, Politics and Sociology, Newcastle University, United Kingdom (UK). He has a DPhil in politics from the University of Oxford, and has done extensive research on the sociology of water conflicts, water and citizenship, and changing power configurations in relation to the control and management of water and water services in Europe and Latin America. He has recently published *Water Power and Citizenship: Social Struggle in the Basin of Mexico* (Palgrave-Macmillan, 2006).

Laia Domènech is a PhD candidate at the Institute of Environmental Science and Technology (ICTA) of the Autonomous University of Barcelona (UAB). She is currently conducting research in the field of water supply and sanitation in Spain and South Asia. Her main research topics are community participation and the use of alternative water sources (i.e. rainwater harvesting, grey water and reclaimed water reuse).

Lili Gan is a PhD student at the Department of Environmental Science and Engineering, Tsinghua University, Beijing, China. His main research topic is urban storm-water control and sewerage system optimization.

Maria J. Gunnarsdóttir has been working in the water sector for 16 years. She has a BSc in civil and infrastructure engineering and an MSc in environmental science (2005). She has done research on the quality of drinking water and water resources protection in Iceland, and has been responsible for developing and implementing the country's Water Safety Planning in Water Supply. She is currently a PhD student at the University of Iceland and her doctoral dissertation focuses on water safety planning and the safety of drinking water.

David Hall is director of the Public Services International Research Unit (PSIRU) in the Business School, University of Greenwich, London. He researches and teaches the history, politics and economics of public services and privatization, with special reference to water and energy (and waste management and healthcare); European Union and World Bank policies on public services; multinational companies; and corruption. He has published numerous articles and reports and two books, and his recent publications cover sewerage, private equity and water privatization in the UK, most of which are available at www.psiru.org. He is currently working on re-nationalizations and water and energy in Southern Africa.

Léo Heller is professor of sanitary and environmental engineering at the Federal University of Minas Gerais (UFMG), Brazil. He has a BA in civil engineering, an MSc in sanitary engineering and a PhD in epidemiology. He recently carried out post-doctoral research at the University of Oxford, focusing on the theoretical aspects of public policy and management in the water supply and sanitation sector. He has extensive experience of research in the fields of environmental health and public policy in water and sanitation services, has coordinated several national and international research projects on these topics, and has a track record of scientific publications on water supply and sanitation, health and environmental sanitation. He coordinates the interdisciplinary academic network Ethics, Politics and Management in Water and Sanitation (POLISAN), which includes European and Brazilian scholars.

Peder Hjorth is senior lecturer at the Department of Water Resources Engineering, Lund Institute of Technology, Lund University, Sweden. Dr Hjorth has a strong interest in interdisciplinary approaches to management problems related to water, natural resources and the environment. He works from a systems analysis perspective, and his current research aims at developing the holistic view called for in Agenda 21 in both developed and developing countries.

Jarmo J. Hukka is adjunct professor in futures research in the water sector at Tampere University of Technology, Finland. He currently works in the African Development Bank in Tunis. His research interests include governance, asset management and strategic development of water services. He has authored several publications.

Blanca Jiménez Cisneros is a professor at the Institute of Engineering, National Autonomous University of Mexico (UNAM), Mexico. Her main field of interest is water quality and reuse. She has recently edited *Water Reuse: An International Survey – Contrasts, Issues and Needs Around the World* (jointly with Takashi Asano; IWA Publishing, 2007) and *Urban Water Security: Managing Risks, UNESCO–IHP* (jointly with Joan Rose; Taylor and Francis, 2008).

Arne Kaijser is professor of history of technology at the Royal Institute of Technology (KTH), Stockholm. He worked as a civil servant during the 1970s and 1980s on futures studies, funding of energy research and development aid. Since 1991 he has been based at KTH. His main research interest is comparative studies of the historical development of large technical systems. His latest book in English is (co-edited with Erik van der Vleuten) *Networking Europe: Transnational Infrastructures and the Shaping of Europe, 1850–2000* (Science History Publications, 2006).

Tapio S. Katko is a sanitary engineer, senior research fellow and adjunct professor at Tampere University of Technology, Finland. His main research interests are the long-term development of water and related infrastructure, as well as institutional, management and policy issues of water and sanitation services. He is the author of several books and papers.

Emanuele Lobina is senior research fellow at the Public Services International Research Unit (PSIRU), University of Greenwich, London, UK. Emanuele joined PSIRU in 1998, where he specialized in water. His work is on the institutional issues revolving around the reform of water supply and sanitation in developed, transition and developing countries. His interests include private-sector participation (PSP) and the in-house restructuring of public water operations. Emanuele has contributed to most of PSIRU's work on the subject, writing and co-authoring refereed articles and reports.

Gordon McGranahan directs the Human Settlements Group at the International Institute for Environment and Development. Dr McGranahan works primarily on urban environment and poverty issues. He has been the co-author/editor for Earthscan Publications of *Citizens at Risk* (2001), *Air Pollution and Health in Rapidly Developing Countries* (2003), *Scaling Urban Environmental Challenges* (2007) and *The New Global Frontier* (2008).

Martin Mulenga is a senior researcher in the Human Settlements Group at the International Institute for Environment and Development (IIED) in London. He works on water and sanitation projects in low-income urban settlements in developing countries. He has a PhD in civil and environmental engineering from the University of Southampton in the UK. He is the main author of *Building Links for Improved Sanitation in Poor Urban Settlements – Recommendations from Research in Southern Africa* (Institute of Irrigation and Development Studies, University of Southampton, 2004).

Roldan Muradian is a senior research fellow at the Centre for International Development Issues Nijmegen (CIDIN), Radboud University, Nijmegen, The Netherlands. His main research interests include economic instruments for ecosystem management; globalization; trade and the environment; integrated assessment of global commodity chains; and community participation in the provision of water supply and sanitation in poor urban areas. He has published extensively in the fields of ecological and development economics. He is currently a member of the board of the European Society for Ecological Economics.

David Nilsson is an associate researcher at the Royal Institute of Technology in Stockholm, Sweden, and is also employed at the Swedish International Development Cooperation Agency (Sida). Dr Nilsson is currently based in Nairobi, Kenya, working in the field of development cooperation on water resources and environment in East Africa. His main research focus is on long-term development of the water sector in East African countries, with a specific focus on urban water supply systems.

Jorge Olcina is professor of regional geography at the University of Alicante, Spain. His main research topics are climatology, flood and drought management, and land-use planning. He has participated in several European projects and has authored books and papers on regional climatology, natural hazards and land-use planning, particularly in relation to water.

Christelle Pezon is a Programme Officer at the International Water and Sanitation Centre (IRC) in Delft, The Netherlands, where she is in charge of the West Africa Programme and of the Financing thematic area. She was previously a lecturer in economics and management at the National Institute for Water and Forestry Management and Engineering (ENGREF), France, where she also directed the Water Services and Sanitation Management Research Unit from 2003 until August 2008.

Pekka Pietilä is a senior researcher at the Institute of Environmental Engineering and Biotechnology (IEEB), Tampere University of Technology, Finland. He has 30 years of experience in water and environmental engineering both in the private and public sectors. Dr Pietilä's experience includes consulting engineering tasks in Finland and overseas, development cooperation tasks in Africa, public-sector administration and management assignments in Finland and the Baltic Region, and research and teaching duties at universities in Finland and overseas.

Antonio Rico is professor of regional geography at the University of Alicante, Spain. His main research topics are agricultural geography, water resources management and land-use change. He has worked as a consultant for the Spanish government on the Water and Urban Development project in the Jucar River Basin, as well as for several irrigation communities. He has authored several books on water management in eastern Spain.

Mark W. Rosenberg received his PhD from the London School of Economics and Political Science (LSE), UK. He is currently a professor of geography and cross-appointed as a professor in the Department of Community Health and Epidemiology at Queen's University, Canada. From 2000 to 2008, Professor Rosenberg was the chairperson of the International Geographical Union's (IGU's) Commission on Health and the Environment. He is currently the co-chairperson of the Scientific Committee of the Earth System Science Partnership on Global Environmental Change and Human Health.

Michael Rouse is a distinguished research associate at the University of Oxford, UK. He was formerly head of the Drinking Water Inspectorate in London and managing director of the Water Research Centre. He is also a past president of the International Water Association and an Honorary Member of the American Waterworks Association. He is a visiting professor at Tsinghua University in Beijing and at the Shanghai Academy of Social Science. In 2000 he was awarded a CBE (Commander of the British Empire) for his professional services.

Andrés Sanz Mulas is programme director at the Institute for Fiscal Studies in Madrid, Spain. He was previously associate professor in public policy analysis at the School of Political Science and Sociology, Complutense University in Madrid, and coordinator of the Public Services Observatory based at the same university. He has participated in several research projects on public management of water services and public finance, and has co-authored several books and articles on these topics. In addition to his academic activities, he is also active as a practitioner in the field of public policy control and evaluation and budgetary programmes through his work as a civil servant in several national and regional audit institutions in Spain.

David Saurí is professor of human geography at the Autonomous University of Barcelona (UAB), Spain. Dr Saurí has conducted research on natural hazards, water management and urban development in the Barcelona area, as well as on land-use changes in northern Spain.

Osmo Seppälä is currently the managing director of Espoo Water, the second biggest water utility in Finland. He also has considerable experience as a consultant in developing water and sanitation services in Africa and Asia. Dr Seppälä's research interests and experience are mainly related to the strategic and visionary management of water services.

Abu Jafar Shamsuddin is an engineer trained at Bangladesh University of Engineering and Technology (BUET), Bangladesh, and has an MSc in water and waste engineering from Loughborough University, UK. He works on the administrative and management aspects of water supply and environmental sanitation and is involved in participatory institutional development and capacity-building in the Bangladeshi water and sanitation services (WSS) sector. He is also active in programme formulation, appraisal, evaluation and impact assessment, has numerous publications to his name related to the field, and has acquired substantial experience in technological and organizational aspects of WSS through his work with a number of national and international organizations both in Bangladesh and abroad.

Vijay P. Singh holds the Caroline and William N. Lehrer Distinguished Chair in Water Engineering, and is professor of biological and agricultural engineering, and civil and environmental engineering at Texas A&M University, Kingsville, Texas, United States (US). His interests are environmental and water engineering. He has been president of the American Institute of Hydrology, and serves as editor-in-chief of the *Journal of Hydrologic Engineering* and Springer's *Water Science and Technology* book series. He has authored 14 textbooks, 70 book chapters, 445 journal articles and 230 conference proceedings papers, and has edited 44 reference books. He has received more than 45 national and international awards.

Erik Swyngedouw is professor of geography in the School of Environment and Development at Manchester University, UK. He was previously professor of human geography at Oxford University, UK. He has written extensively on the political ecology of water, urban governance and the political economy of uneven development. Recent books include *Social Power and the Urbanisation of Nature* (Oxford University Press, 2004) and *In the Nature of Cities* (Routledge, 2006).

Bishwa Nath Tiwari is associate professor of economics in Tribhuvan University, Kathmandu, Nepal. His main research topics are poverty and human development; food insecurity and vulnerability; drinking water and sanitation; and monitoring and evaluation, including monitoring of the Millennium Development Goals. He holds an MSc in environmental and resource economics from University of London, UK, and a PhD from Rajasthan University, India.

María Luisa Torregrosa Armentia has a PhD in social sciences with a major in sociology from El Colegio de México, and is a professor of sociology at the Latin American Faculty of Social Sciences (FLACSO), Mexico. Her main research and teaching interests focus on the social aspects of natural resources management, particularly land and water, and on the institutional and social transformations driven by the liberalization and deregulation of land and water in Mexico since the 1980s. She has participated in several international research projects related to these issues, including the Private Participation in Water and Sanitation Services (PRINWASS) project (www.prinwass.org).

Venkatesh Uddameri is an associate professor of environmental engineering at Texas A&M University, Kingsville, Texas, US. His research and teaching interests are in the broad area of integrated water resources management, with a special emphasis on addressing groundwater availability and water quality issues using combined simulation-optimization and fuzzy set theoretic techniques. He serves on the editorial board of the *Clean Technologies and Environmental Policy Journal* and is the associate editor (groundwater section) for the *Journal of Spatial Hydrology*.

Jiane Zuo is professor of environmental engineering at the Department of Environmental Science and Engineering, Tsinghua University, Beijing, China. His research fields cover anaerobic wastewater treatment processes, anaerobic biomass waste digestion, anaerobic sludge digestion, microbial fuel cell (MFC), urban storm-water control, and sewerage system optimization. He is a leading Chinese authority on anaerobic processes, particularly anaerobic ammonia oxidation (ANAMMOX) and psychrophilic anaerobic processes. He is the main author of *The Principle and Technology of Anaerobic Biological Treatment* (China Architecture and Building Press, 2004). He has published more than 100 papers.

List of Acronyms and Abbreviations

A/A/O	anaerobic–anoxic–oxic
ADB	Asian Development Bank
AdP	Aguas de Portugal
AFLRA	Association of Finnish Local and Regional Authorities
AGBAR	Aigües de Barcelona (Spain)
AIM	Asian Institute of Management (the Philippines)
ANAMMOX	anaerobic ammonia oxidation
ANA	National Water Agency (Brazil)
A/O	anaerobic–oxic
ARBAN	Association for Meeting Basic Needs (Bangladesh)
ASCE	American Society of Civil Engineers
ASTM	American Society for Testing and Materials
ATO	optimal territorial unit (Italy)
BAF	biological aeration filtration
BANOBRAS	National Works and Public Services Bank (Mexico)
BC	Before Christ
BLM	Bureau of Land Management (US)
BOD	biochemical oxygen demand
BOT	build–operate–transfer
BS	basic sanitation
BUET	Bangladesh University of Engineering and Technology
CA	community of agglomeration (*communauté d'agglomération*) (France)
CADWES	capacity development in water and environmental services (Finland)
CAG	Castalia Advisory Group
CAS	conventional activated sludge
CBC	Canadian Broadcasting Corporation
CBE	Commander of the British Empire
CBO	Congressional Budget Office (US)
CC	community of communes (*communauté de communes*) (France)
CCW	Consumer Council for Water (UK)

CEDECON	Central Department of Economics, Tribhuvan University (Nepal)
CEO	chief executive officer
CEPR	Centre for Economic Policy Research (UK)
CESB	provincial company for water supply and sanitation (Brazil)
CIDIN	Centre for International Development Issues Nijmegen (The Netherlands)
CIRED	International Centre for Research on Environment and Development (France)
CLTS	community-led total sanitation
CMH	Commission on Macroeconomics and Health (of the WHO)
CNA	National Water Commission (Mexico)
CNRS	National Centre for Scientific Research (France)
COD	chemical oxygen demand
CSS	combined sewer system
CU	urban community (France)
CWA	Clean Water Act (US)
CWEF	Colorado Water Education Foundation (United States)
DALY	disability-adjusted life year
DAM	Dhaka Ahsania Mission (Bangladesh)
DANVA	Danish Water and Waste Water Association
DBP	disinfection by-product
DDC	District Development Committee (Nepal)
DFID	Department for International Development (UK)
DHSP	District Health Strengthening Programme (Nepal)
DMAE	Municipal Department of Water and Sewerage (Porto Alegre, Brazil)
DNA	deoxyribonucleic acid
DPHE	Department of Public Health and Engineering (Bangladesh)
DSK	Dushtha Shasthya Kendra (Bangladeshi NGO)
DSP	delegation of public services (France)
DU	urban district (France)
DWASA	Dhaka Water Supply and Sewerage Authority (Bangladesh)
DWSS	Department of Water Supply and Sewerage (Nepal)
ECLAC	United Nations Economic Commission for Latin America and the Caribbean
ECOSAN	ecological sanitation
ED	endocrine disruptor
EEA	European Economic Area
EMASESA	Empresa Metropolitana de Abastecimiento y Saneamiento de Aguas de Sevilla (Spain)
ENGREF	National Institute for Water and Forestry Management and Engineering (France)
ENPHO	Environment and Public Health Organization (Nepal)
EPCI	inter-communal cooperation organism (France)
ESA	European Space Agency

ESCAP	United Nations Economic and Social Commission for Asia and the Pacific
EU	European Union
FCM	Federation of Canadian Municipalities
FDI	foreign direct investment
FEI	Finnish Environment Institute
FEMA	Federal Emergency Management Agency (US)
FGTS	Employment Guarantee Fund (Brazil)
FLACSO	Latin American Faculty of Social Sciences
FNDAE	National Fund for the Development of Water Sources (France)
GDP	gross domestic product
GIS	geographic information system
GNI	gross national income
GNP	gross national product
GOBACIT	Research Network on Governance, Citizenship, Water Management and Environmental Health in Developing Countries (www.gobacit.org)
GON	Government of Nepal
GWP	Global Water Partnership
HDI	Human Development Index
hm^3	cubic hectometres
HSWA	Hazardous and Solid Waste Amendments (US)
IATP	Institute for Agriculture and Trade Policy (US)
IBGE	Brazilian Institute for Geography and Statistics
ICSID	International Centre for the Settlement of Investment Disputes
ICTA	Institute of Environmental Science and Technology (Spain)
ICWE	International Conference on Water and the Environment
IDA	international development agency
IDB	Inter-American Development Bank
IEEB	Institute of Environmental Engineering and Biotechnology (Finland)
IEEE	Institute of Electrical and Electronics Engineers, Inc (US)
IFI	international financial institution
IGU	International Geographical Union
IHE	International Institute for Infrastructural, Hydraulic and Environmental Engineering (Delft, The Netherlands)
IIED	International Institute for Environment and Development (UK)
IIED–AL	International Institute for Environment and Development – America Latina (Argentina)
IMF	International Monetary Fund
IRBM	integrated river basin management
IRC	International Water and Sanitation Centre (Delft, The Netherlands)
IT	information technology
ITN-BUET	International Training Network Center – Bangladesh University of Engineering and Technology
IWA	International Water Association

IWRM	integrated water resources management
JBIC	Japan Bank for International Cooperation
JMP	Joint Monitoring Programme for Water Supply and Sanitation (World Health Organization and United Nations Children's Fund)
KIWASCO	Kenyan water utility operator
km	kilometre
KTH	Royal Institute of Technology (Sweden)
l	litres
lcd	litres per capita per day
LGB	local government body (Bangladesh)
LGED	local government engineering department (Bangladesh)
lpd	litres per day
LPM	local public management (France)
LPUPAP	Local Partnerships for Urban Poverty Alleviation Project (Bangladesh)
LSE	London School of Economics and Political Science
LVSWSB	Lake Victoria South Water Services Board (Kenya)
m^3	cubic metres
m^3/d	cubic metres per day
m^3/s	cubic metres per second
MAF	Ministry of Agriculture and Forestry (Finland)
MAG	Ministry of the Attorney General (Ontario, Canada)
MBR	membrane bioreactor
MCL	maximum contaminant level
MDG	Millennium Development Goal
MEP	Ministry of Environmental Protection (China)
MFC	microbial fuel cell
mg/l	milligrams per litre
MHURD	Ministry of Housing and Urban–Rural Development (China)
MM	Mahila Milan (India)
mm	millimetre
MOE	Ministry of the Environment (Canada)
MWR	Ministry of Water Resources (China)
MWSS	Metropolitan Waterworks and Sewerage System (the Philippines)
NAFTA	North American Free Trade Agreement
NEWAH	Nepal Water for Health
NGO	non-governmental organization
NIMBY	not in my backyard
NMFS	National Marine Fisheries Service (US)
NPDWR	National Primary Drinking Water Regulation (US)
NPS	National Park Service (US)
NSDF	National Slum Dwellers Federation (India)
NSDWR	National Secondary Drinking Water Regulation (US)
NWC	New Water Culture (Spain)
NWS	National Weather Service (US)

NWSC	National Water and Sewerage Corporation (Uganda)
NWSC	Nepal Water Supply Corporation
O&M	operation and maintenance
OECD	Organisation for Economic Co-operation and Development
OFWAT	Office of Water Services (England and Wales)
OPP	Orangi Pilot Project (Pakistan)
OPP–RTI	Orangi Pilot Project – Research and Training Institute (Pakistan)
PAC	powdered activated carbon
PAC	Programme for Accelerating Growth (Brazil)
PCA	principal components analysis
pcpy	per capita per year
p.e.	person equivalent
PESTEL	political and policy, economic, social, technological, environmental and legislative (aspects of water and sanitation services)
PLANASA	National Basic Sanitation Plan (Brazil)
PLC	public limited company
POLISAN	Ethics, Politics and Management in Water and Sanitation (Brazil)
POP	persistent organic pollutant
PNSB	*National Survey of Basic Sanitation* (Brazil)
PPI	private participation in infrastructure
PRINWASS	Private Participation in Water and Sanitation Services project (www.prinwass.org)
PROMAGUA	Programme for the Modernization of Water and Sanitation Utilities (Mexico)
PSIRU	Public Services International Research Unit (UK)
PSP	private-sector participation
PSTC	Population Services and Training Centre (Bangladesh)
PUC	Public Utilities Commission (Canada)
PUP	public–public partnership
PWA	Palestinian Water Authority
PWO	People Welfare Organization (Nepal)
PWS	public water supply
PYLL	potential years of life lost
QMRA	quantitative microbial risk assessment
QUANGO	quasi non-governmental organization (England and Wales)
RCRA	Resource Conservation and Recovery Act (US)
REPDA	Public Water Rights Registry (Mexico)
RMB	Chinese yuan renminbi
RMHO	Regional Medical Health Office (Canada)
RO	reversed osmosis
RWA	Regional Water Authority (England and Wales)
SAHOP	Ministry of Human Settlements and Public Works (Mexico)
SAN	new agglomeration syndicate (France)
SARH	Ministry of Agriculture and Hydraulic Resources (Mexico)

SDWA	Safe Drinking Water Act (US)
SEDUE	Ministry of Urban Development and the Environment (Mexico)
SEMARNAT	Ministry of Environment and Natural Resources (Mexico)
SEP	Ministry of Public Education (Mexico)
SEPA	Swedish Environmental Protection Agency
SGAB	Sociedad General de Aguas de Barcelona (Spain)
Sida	Swedish International Development Cooperation Agency
SISS	Superintendent of Sanitary Services (Chile)
SIVOM	multi-purpose inter-communal syndicate (*syndicat intercommunal à vocation multiple*) (France)
SIVU	single-purpose inter-communal syndicate (*syndicat intercommunal à vocation unique*) (France)
SM	mixed syndicate (France)
SNSA	National Department of Environmental Sanitation (Brazil)
SPARC	Society for the Promotion of Area Resource Centres (India)
SRH	Ministry of Hydraulic Resources (Mexico)
SS	suspended solid(s)
SSA	Infrastructure Commission of the Ministry of Health (Mexico)
SSS	separate sewer system
TMDL	total maximum daily load (of pollutants)
TN	total nitrogen
TP	total phosphorus
TSD	treatment, storage and disposal
TSS	total suspended solids
UAB	Autonomous University of Barcelona (Spain)
UDC	Urban Development Centre (Bangladesh)
UF	ultra-filtration
UFMG	Federal University of Minas Gerais (Brazil)
UK	United Kingdom
UN	United Nations
UNAM	National Autonomous University of Mexico
UNCED	United Nations Conference on Environment and Development
UNDP	United Nations Development Programme
UNESA	United Nations Economic and Social Affairs
UNESCO	United Nations Educational, Scientific and Cultural Organization
UN-Habitat	United Nations Human Settlements Programme
UNICEF	United Nations Children's Fund
UNMP	United Nations Millennium Project
US	United States
USACE	US Army Corps of Engineers
USAID	US Agency for International Development
USBR	US Bureau of Reclamation
USEPA	US Environmental Protection Agency
USFWS	US Fish and Wildlife Service

USGS	US Geological Survey
USPHS	US Public Health Service
UV	ultraviolet (radiation)
VDC	Village Development Committee (Nepal)
VERC	Village Education and Resource Center (Bangladesh)
WASREB	Water Services Regulatory Board (Kenya)
WATERTIME	Research Project on Water and Sanitation in European Cities (www.watertime.net)
WEDC	Water Engineering Development Centre (UK)
WEF	Water Environment Federation (US)
WETSUS	Centre of Excellence for Sustainable Water Technology (The Netherlands)
WFD	Water Framework Directive (Europe)
WHO	World Health Organization
WPI	Water Poverty Index
WRI	World Resources Institute
WSB	water service board (Kenya)
WSP	waste stabilization pond
WSP	Water and Sanitation Programme (World Bank)
WSP	water service provider (Kenya)
WSS	water and sanitation services
WSUC	Water Supply Users Committee (Nepal)
WWDR	*World Water Development Report* (UNESCO)
WWTP	wastewater treatment plant

Introduction

José Esteban Castro and Léo Heller

The adequate provision of water and sanitation services (WSS)[1] continues to be an essential requirement for protecting public health and maintaining basic living conditions, and the sustainable universalization[2] of these services at a global scale constitutes one of the major challenges of the early 21st century. Although in the last few decades the pace of technological progress in the WSS sector has been enormous, access to the benefits of this progress continues to be precluded to a significant share of the world's population. The protracted inequalities in the access to essential WSS continue to pose one of the most significant ethical dilemmas facing the international community.

In simple words, a very large proportion of the world's population continues to lack access to even a few daily litres of clean water and to the basic sanitary conditions required for survival, and there is no consensus at the international level that access to such essential basic services should be considered an inalienable human right (see, for instance, Amnesty International, 2003). The prospects for overcoming this situation look grim in the light of the continued failure of the international community in achieving its targets. During the 1980s, the United Nations Water Decade aimed at providing universal access to essential volumes (around 40 litres) of safe drinking water by 1990 (UN, 1980; UNDP, 1990); but still in the year 2000, around 1.1 billion people (17 per cent of the world's population) lacked access to safe drinking water, while around 2.4 billion, or 40 per cent, had no adequate sanitation (EC, 2002, 2003). The commitment adopted by the international community through the Millennium Development Goals (MDGs) seeks to halve the proportion of the world population without access to these services by 2015 (UN, 2000, 2002); but despite significant progress in some countries, the evidence shows that the MDGs may be unattainable.

In addition, recent reports on the progress made towards the MDGs suggest that it is unlikely that the global target for sanitation will be met and that some regions will also fail to attain the drinking water target (WHO, 2005, pp27, 71; WHO and UNICEF, 2006, p6). An up-to-date assessment concludes that 'the world is not on track to meet the MDG sanitation target... At the current rate, the world will miss the MDG sanitation target by over 700 million people' (UNICEF and WHO, 2008, p8). The reports predict that not only will the MDGs for WSS be unobtainable, but that regions such as sub-Saharan Africa and parts of Asia and Oceania will also experience a significant increase in the absolute number of un-served people in the face of rapid population growth (UNMP, 2005, p21; ADB et al, 2006; WHO and UNICEF, 2006). Moreover, even those countries that are expected to meet their MDG targets will continue to experience significant difficulties given that the MDGs merely aim at halving the proportion of un-served people, and the remaining population happens to be mostly the very poor (see, for instance, ADB et al, 2006). In addition, the scenario depicted above is based on the quantitative aspects of the problem but if we focus on the qualitative dimension of WSS, it is evident that the figures used to report progress in meeting the MDGs need to be carefully scrutinized. For instance, although some reports show that by 2004 Brazil had already achieved its MDG target for drinking water and Mexico would have met its sanitation target too (see WHO and UNICEF 2006, pp8, 17), as shown in this volume (see Chapters 20 and 21), the actual situation is far worse than is portrayed in the official reports. The lessons from this analysis can be extended by analogy to most countries, including the developed North.

In this regard, the fact that much of the discussion on the MDGs is concentrated on the problems facing the global South may obscure the existence of significant challenges for the provision of WSS in developed countries, as illustrated in Part II of this volume. Among other issues affecting WSS in the developed North are the escalating costs of infrastructure renewal, the impact of old and emergent water-related health hazards, and the effects of climate change and widespread pollution on the availability of water sources (WHO-Europe, 2006). Although WSS, in principle, became universally available in developed countries during the 20th century, a number of factors also continue to preclude large numbers of people from access to these essential services. For instance, recent reports show that 16 per cent of the European population (140 million people) still lack in-house drinking water, 10 per cent (85 million people) do not have access to improved sanitation and 5 per cent (41 million people) lack safe drinking water (WHO-Europe 2006, pp4–5). As a result, some European countries are subject to 'an epidemic of morbidity from water-related diseases', including hepatitis A, diarrhoeal infections and typhoid fever (UNECE, 2007). Although much of this problem affects some of the poorer European countries recently incorporated within the European Union, wealthier countries are also facing significant problems in guaranteeing universal access to WSS. For instance, in England and Wales, between 2 million and 4 million families live in 'water poverty' (Fitch and Price, 2002; Klein, 2003; UKP, 2003; NCC, 2005). This problem is intimately connected with the fact that between 15 and 20 per cent of WSS users are failing to pay their bills (Defra, 2004; OFWAT, 2004; CCW, 2006; Fitch, 2006), which has a significant impact upon the WSS and casts shadows on the financial future of the privatized water and sanitation companies (Water UK, 2005; Hall and Lobina, 2007). Another example comes from the US, where, although nominally there is universal coverage for WSS, ethnic minorities and poor neighbourhoods continue to face enormous obstacles in accessing these services owing to structural social inequalities (Bath et al, 1998; Berry, 1998; Whiteford and Cortez Lara, 2005). These are just some examples to illustrate the fact that the ethical dilemmas concerning the access to essential WSS have a global dimension and are not restricted to poor developing countries.

We are convinced that these and other crucial challenges facing the water and sanitation sector worldwide are neither the consequence of a shortage of technological solutions to the problems nor the result of physical–natural constraints, such as water 'scarcity'. As acknowledged in the 2006 United Nations world water report, the global water crisis, including the crisis of WSS, is primarily 'a crisis of governance' (UNESCO–WWAP, 2006, p1), a fundamentally ethical challenge that we face in the 21st century (Davis and McGinn, 2001; Delli Priscoli et al, 2004; EUWATER Network, 2005). However, despite this increasing recognition of the (broadly speaking) social and political nature of the crisis, the key activities connected with the organization of WSS, including the production of scientific knowledge, to a large extent continue to be monopolized by the techno-sciences. Borrowing from John Dryzek, the prevailing understanding in relation to environmental management processes, including the management of WSS, is that they constitute a preserve of the experts – that is, of the technical experts (Dryzek, 1997).

In this connection, we firmly believe that overcoming the unacceptable state of affairs characterizing the crisis of WSS will require significantly greater synergy between the techno-scientific and social science domains. Necessarily, achieving the sustainable universalization of essential WSS requires adequate technological approaches and techniques for the proper planning, design, construction and operation of service units and systems that meet the increasingly higher standards driven by the goals of ecological sustainability and by rising user expectations. However, WSS are also in the sphere of public policy that involves activities of planning, assessment, institutional organization and political decision-making, often in the interfaces between WSS and, among other, water resources management, social policy and public health – activities that respond to very different criteria from those prevalent in the technological sphere. This book aims to bridge the gap between the public policy and management aspects of WSS and to overcome the entrenched disciplinary and professional divides that, in our perspective, constitute one of the main obstacles for successfully tackling the crisis of WSS.

In this regard, the need for an integrated approach that brings together the technological and the public policy dimensions of water and sanitation services has received relatively scant attention in the specialized literature, both in terms of conceptual development and intellectual debate. The lack of a consistent and in-depth treatment of the interweaving between technology and public policy in WSS is particularly noticeable at the level of individual disciplines. Much of the existing work on WSS approaches the topic from the perspectives of a wide range of disparate disciplines, including administration, demography, economics, engineering, geography, history, urban planning, political science and sociology. This is perhaps a reflection of the multidimensional character of the activities involved in managing and governing these services, which tends to prompt *ad hoc* responses from a range of disciplinary approaches. However, it can also be argued that the slow development of a cross-disciplinary body of literature on WSS suggests that there is a need for a more systematic effort geared at organizing the field in conceptual and methodological terms. We believe that the contributions included in this volume have the potential to greatly stimulate the development of more coherent theoretical and methodological approaches to the topic, including work oriented towards the evaluation of empirical experiences. We also hope that the book will provide support for policy design and planning in the interfaces between WSS and other interlinked areas of activity, such as public health and water resources management. The main goal is to trigger the development of new models of public policy and management of WSS that are primarily focused on meeting the needs of the users through enhancing the effectiveness, efficacy and efficiency of WSS interventions.

The premises and rationale of our book

The book is structured on the basis of several assumptions that provide a common framework to the different contributions, although these presuppositions are the responsibility of the editors and they are not fully shared by all of the participating authors. First, we assume that WSS policies must be grounded on the principle that these services constitute a social right of citizenship – that is, it is a state obligation to guarantee their universal access. This conceptualization is underpinned by the principles of universality and equity, according to which all citizens, independently of their social class, gender, ethnic origin or any other factors of social differentiation, have the unrestricted right to those goods and services deemed to be essential for sustaining life in a civilized society. The countries in the developed North were able to achieve the universalization of essential WSS during the 20th century thanks to the adoption of these principles, which were defended by a broad range of social and political forces. There is no reason to believe that the universalization of these services at a global scale could take place in the absence of a similar commitment. However, we are aware that by adopting this assumption we are distancing ourselves from rival positions that argue for the organization of WSS on the basis of market principles, where these services must be considered a private good, a commodity and no longer a public or social good (e.g. WSP and PPIAF, 2002; UNESCO–WWAP, 2006, p409). These positions are defended, among others, by multilateral institutions, international organizations, academics and practitioners, who emphasize market efficiency as the primary objective in the management of WSS over the principles of universality, equity and social efficacy. We strongly believe that the organization and delivery of WSS cannot be subordinated to market interests, should not be subject to market criteria as their main organizing principle, and that the pursuit of sound and sustainable management of WSS' economic, financial and techno-infrastructural aspects does not require treating these essential services as private goods. The access to essential WSS constitutes a fundamental right that cannot be subject to the payment capacity of individual users and must be guaranteed by the state.

Second, we assume that public policy and management in the field of WSS depend upon the historically specific configurations of physical–natural, socio-economic, political and cultural conditions characterizing different countries and regions. In particular, there exist fundamental disparities between and within countries and regions with regard to the socio-economic conditions that provide the basis for the organization and delivery of WSS and other essential public services. For instance, the bulk of the world population without access to WSS is concentrated in poor developing countries (UN-HABITAT, 2003). A recent report suggests that many of these countries will not be able to achieve the MDGs, including the WSS targets, because they have 'fragile states ... with weak governance and institutions' (WHO, 2005, pp27, 71). Institutional fragility and weakness are the expression of structural socio-economic and political processes, but conventional WSS policies often adopt a technocratic, supposedly neutral, approach that fails to incorporate these processes within the analysis. Therefore, the implementation of models of public policy and management must consider the existence of *systemic conditions* – that is, factors and processes that are external to the specific field and internal logic of essential public services, but that shape and even determine the ways in which these services are organized and delivered. These include a wide range of issues such as the socio-economic characteristics of the population, hydro-geological constraints, demographic patterns, ethnic and cultural cleavages, the prevailing model of development, political processes, and so on. For instance, while in developed countries WSS policies may be designed and implemented in a given context of priorities (e.g. the renewal of WSS infrastructures inherited from the 19th century) and trends (e.g. stable or declining domestic water consumption), in developing countries the requirements and conditions for these policies may be radically different. Thus, in most developing countries WSS policies must consider the need for expansion and continued improvement of the services in conditions that are seldom amenable to policy initiatives designed in the context of developed countries, such as full-cost recovery or the financial self-sufficiency of the utilities. In these cases, the structural social inequalities that underlie the conditions of widespread poverty and extreme poverty affecting a large proportion of the population in the global South often constitute intractable systemic constraints for the implementation of policies that presume the existence of a willing-to-pay consumer base for commodified and full-cost-recovered WSS – a consumer base that is often non-existent or is, at best, limited to a relatively small fraction of the population. Unfortunately, the prevailing policy-making traditions in the WSS sector tend to neglect the importance of these and other systemic conditions that influence and structure the actual working of these services on the ground, and our book makes a contribution towards their incorporation in both analysis and practice.

In addition to these implications that follow from our initial assumptions, the structure of the book reflects the fact that WSS need to be examined in two different dimensions:

1 the public policy dimension, comprising both the process of establishing and enacting political, legal and institutional frameworks and ensuring accountability and responsibility in public-decision making, as well as policy processes such as agenda-setting and the selection of policy alternatives followed by their implementation and evaluation; and
2 the management dimension that refers to the actual organization of the services, which (depending upon the particular model adopted) may be implemented at the local, regional or national level.

In this regard, the history of WSS is a history of institutional and organizational diversity, which our book reflects through examining the inter-linkages between public policy and management in a range of different contexts. On the one hand, it explores what are arguably the main paradigms for the organization of these services – the models organized around, broadly speaking:

- public-sector principles, and especially forms that emphasize the role of local authorities,
- private-sector principles;
- co-operative approaches; and
- community-driven approaches.

On the other hand, the book examines a number of cases from developed countries and the global South to offer elements for a comparative perspective of this historical diversity as it has been documented in Africa, Asia, Europe and the Americas.

Finally, an important aim of the book is to emphasize the need to deepen and consolidate the principles of substantive, not merely rhetorical, democracy for achieving the goals of universalization, equity and sustainability in WSS. In this regard, as already stated, WSS are not (could not be) insulated from the wider world, as it seems to be sometimes understood, but are, instead, subject to multidimensional systemic conditions and constraints, from physical–natural factors to cultural and socio-historical processes that have significant influence in shaping and even determining how these services are organized and run. For instance, WSS public policies are part and parcel of the process of governance, which includes the development and enforcement of legal-administrative and institutional frameworks and mechanisms, but also the more crucial social and political processes informing the design, implementation and monitoring of public policies. In this conception, governance is not merely an instrumental political device for the execution of decisions taken by technical experts and professional politicians, but is, rather, a process driven by the pursuit of rival, even incompatible, social and political ends and means (Hanf and Jansen, 1998), which is therefore characterized by ongoing social and political struggle. Thus, the process of governance in relation to WSS involves decisions about how these services should be organized, financed and governed, by whom, on what principles, to serve what interests, and so on.

In consequence, the editors adopt a critical perspective towards the mainstream WSS policies that have been implemented worldwide since the 1980s, a perspective shared by most co-authors. Not only have these mainstream policies failed to deliver the expected results, as shown, for instance, in Chapters 2 and 6, but they have also set in motion processes whose inertial forces will continue to shape the way in which WSS are organized in many countries for years to come. We are strongly critical of these policies but keep a constructive and positive approach because we are persuaded that the heterogeneous and often fragmented forces defending the principles of social justice and substantive democracy in relation to water and life-sustaining water services are making considerable progress. However, the fact remains that achieving the universalization of WSS (and other essential services) will require significant changes in the prevailing socio-economic, policy-institutional and political arrangements, and the evidence suggests that these changes are slow and may not take place in time to meet the targets set by the international community where it matters most. There is little reason for complacency, and the book aims at stimulating new thinking and practical action to successfully tackle the WSS crisis.

THE CONTENT

The collection is divided into two main sections. Part I is composed of ten chapters presenting the 'theoretical and conceptual dimensions' of the book, while the 11 chapters of Part II cover a wide range of country and regional experiences that explore the diversity of institutional and policy options characterizing WSS, as well as the main challenges facing the sustainable and equitable universalization of these services worldwide. As already explained, although the book is structured along key assumptions and premises, these are not shared to the same extent by all authors, and this is particularly clear in Part I. In this regard, there are important undercurrents of tension and contradiction in the chapters of the first section, among which it is worth highlighting the controversy between public-, market- and community-led WSS systems, including the question of funding sources. In our view, these unresolved contradictions that run through our chapters reflect the existence of substantive disagreements and even irreconcilable positions in the debate about which policies and management models are better suited for developing the fully sustainable, efficient and socially just organization and delivery of WSS.

Chapters 1 and 2 offer a broad analysis of some of the main forces that, to a large extent, shape and drive the organization and running of WSS. In Chapter 1, José Esteban Castro explores the significance of systemic conditions, mostly external to the WSS sector, on the development, organization and functioning of these essential services, with emphasis on the social dimension. In particular, the chapter pays attention to structural social inequalities that often pose major constraints to the universalization of WSS, and examines the inertial forces unleashed by mainstream WSS policies implemented worldwide since the 1980s. In this connection, in Chapter 2, Erik Swyngedouw scrutinizes the particular contradictions arising from the corporatization of WSS promoted by mainstream public policies, given the essentially collective nature of these services, especially in urban areas. The chapter contends that the corporatized market-oriented model of WSS being promoted at the global level is socially and environmentally unsustainable, and that achieving the WSS targets envisaged in the MDGs will not be possible without a proper embedding of these services within a public or social financial and regulatory infrastructure backed by massive national and international public funding.

Chapters 3 to 7 look at different facets of the relationship between public policy and management in WSS. In Chapter 3, Andrés Sanz Mulas provides an introduction to a public-policy approach for the analysis of WSS grounded on the premises that:

- these services have a multidimensional character and therefore require inter- and multidisciplinary efforts; and

- universal access to these services constitutes a social right of citizenship and cannot be subject to market criteria.

The chapter pays particular attention to budgetary policy as a crucial tool in what the author believes is an essentially public-sector duty and responsibility, and focuses on the policy processes that concern the design and enforcement of political, legal and institutional aspects of WSS. Okke Braadbart adopts a historical perspective in Chapter 4 to examine how the eventual predominance of the technological paradigm associated with the development of networked domestic WSS in Europe and the US during the 19th century was closely interwoven with the consolidation of public, mainly municipal, forms of management and administration of these services. Braadbart challenges established notions that the early development of urban piped WSS in Europe was mainly the outcome of rising concerns with public health, and discusses the evidence emerging from recent research suggesting that responding to the challenges of fire fighting and water scarcity in urban areas, complemented by the requirements of environmental protection and the pursuit of private profit, were much more important driving forces than public health in this development. The chapter also explores the process of municipalization of WSS utilities during the late 19th century in Europe and the US and offers a critical, somewhat pessimistic, overview of the transfer of the 'piped water and wastewater paradigm' to developing countries.

From a different angle, in Chapter 5, Osmo Seppälä and Tapio S. Katko elaborate upon the results of recent research on water resources and services management in European cities, and highlight the complex institutional and organizational diversity that characterizes this sector of activity. They argue that WSS initiatives must consider the long-term experiences and nature of these services, take into account the fundamental impact of local conditions, and avoid 'one-type-fits-all' solutions that tend to neglect the historical lessons of successful WSS management practices. Chapter 6 by David Hall and Emanuele Lobina aims at establishing a balance in a debate often centred on water supply to the neglect of sanitation and, especially, sewerage services. They argue that despite the well-known positive impact of sewerage services on public health, and particularly child health, for the most part current debates on sanitation tend to marginalize sewerage, often on the assumption that it would be unaffordable for poor countries. Hall and Lobina strongly reject such claims by examining how sewerage systems have been developed in the North, argue that sewerage systems would be actually affordable for the countries where the mass of un-served people is concentrated, and claim that public sector action, and principally tax-based public finance, is the key factor to secure the universalization of these services. In close relation to their argument about WSS and public health, Léo Heller contends in Chapter 7 that public policy and management of WSS should be reoriented by making public health their central objective. This would require a radical change to break with the hegemonic WSS paradigm that prioritizes technological development as the main driver of WSS management and focus on what should be the main goal of these essential services: the enhancement of the living conditions of the population. Such a radical strategy would require the adoption of genuinely interdisciplinary approaches in the field of knowledge production and cross-sector public policies, particularly in the interfaces between water resources management, environment policy, public health, social policy, urban and regional planning, and public services.

The last three chapters of Part I focus in more detail on some of the main paradigms for the organization of WSS, although important aspects of public-led models are also covered in Chapters 3 to 7. This is perhaps the section in the book where the disagreements between the co-authors are more evident, including some positions that are largely at odds with the editors' main line of argumentation. In this regard, in Chapter 8, Michael Rouse claims that the most successful strategy for running WSS is the introduction of market forces as the key organizing principle, whether through outright privatization or by creating the conditions to make public companies operate on commercial principles. Rouse dismisses the municipal model of WSS and re-states the key principles promoted by mainstream water policies since the 1980s – in particular, the de-politicization of operational management, the cancellation of general public subsidies and their replacement by full-cost recovery policies (though keeping targeted subsidies for the poor), comparative (bench-marking) competition, transparent procurement of goods and services, and independent regulation. He discusses a number of examples to back up his claims in favour of mainstream WSS policies and to explain why some market-led WSS projects such as in Atlanta in the US, La Paz-El Alto in Bolivia, and Chengdu in China have failed.

Jarmo J. Hukka and Tapio S. Katko start from a very different position in Chapter 9 to examine what they term 'complementary paradigms' for the organization of WSS. They explore five different models identified in their research on Europe and the Americas: the customer/citizen-responsive paradigm, the multiple-stakeholder paradigm, the local government paradigm, the co-operative paradigm, and the multiple producers paradigm. Hukka and Katko emphasize that these models are not mutually exclusive, but rather complementary, and argue that policy-makers and implementers should analyse the strengths and weaknesses of the various paradigms in relation to local requirements and preferences, avoiding the mechanical adoption of one-fits-all solutions. They also argue that there is a need to persuade decision-makers that the main goal of WSS is to secure the satisfaction of the essential needs of human communities, which requires superseding the techno-centred management approaches still dominating the sector. Finally, in Chapter 10, Gordon McGranahan and Martin Mulenga stage a strong defence of community involvement in WSS. The authors criticize what they call idealized management models, whether state, market or community led, and warn against the belief in 'perfect' solutions to the challenging problems facing WSS, especially in the poorest countries. The chapter discusses successful examples of improvements in rural sanitation and in urban WSS, where the authors argue that the key factor was community organization and involvement. McGranahan and Mulenga go to great lengths to emphasize that the solution in their view is not to replace market- or government-led WSS by community-driven approaches, but rather that community participation is often the crucial factor in successful projects.

Part II features 11 chapters offering a wide diversity of country and regional experiences that provide excellent empirical material to complement and put in perspective the previous discussion. We present here experiences from developed countries, including

European and North American cases, followed by analyses of the situation in the global South, focusing on African, Asian and Latin American examples.

In Chapter 11, Christelle Pezon opens Part II with an in-depth historical examination of the development of WSS in France. She argues that key explanatory factors of the French tradition of WSS are the country's high territorial fragmentation, which made local authorities central players in the organization and delivery of these services, and the particular ways in which private-sector management was developed through public finance and investment. A crucial lesson is that the universalization of WSS in France was made possible through fiscal finance (local taxes in urban centres, and a national funding mechanism in rural areas), and cost-recovery policies for urban users (rural users are still subsidized) were only implemented two or three generations after the universalization of the services. In turn, David Saurí, Jorge Olcina and Antonio Rico explore the Spanish situation in Chapter 12, with special attention on the Mediterranean coast of the country. The Mediterranean coast has recently been the scene of heated debates around the management of water and WSS, driven by higher demands posed by rapid urban and touristic development in conditions of climatic and hydrological constraints. In this context, the authors examine the recent trends towards privatization of WSS fuelled by a combination of financial policies, local authority initiatives in a context of relative autonomy and the expansion of private capital. The chapter also discusses a number of water conflicts such as the Barcelona 'water war', a popular revolt against a new government tax aimed to fund water infrastructures in the 1990s, and the new frontiers opened by large-scale desalinization projects as a response to the rapidly increasing water demand. In the words of the authors, water management in Spain is dominated by 'technological fixes' – that is, large-scale technological interventions, ranging from traditional infrastructures such as dams and river transfers to the most sophisticated desalinization and water reuse technologies.

Chapter 13 takes the reader to a very different set of examples grouped under the broad theme of Nordic countries' experiences, looking at the cases of Denmark, Finland, Iceland, Norway and Sweden. Pekka Pietilä, Maria J. Gunnarsdóttir, Peder Hjorth and Susanne Balslev Nielsen explain how despite their diverse geographic-natural features, the organization of WSS in these countries share certain common characteristics, such as a decentralized public administration, the central role of local authorities, institutional stability and universal access. Thus, despite important policy reforms in recent years, WSS continue to be considered as essential public services that must be kept under public control and are the responsibility of local authorities. Nevertheless, and illustrating some of the examples discussed in Chapters 5 and 9, the provision of WSS in Nordic countries is characterized by institutional diversity, wide participation of different actors, including a very active private sector, and the rejection of one-size-fits-all approaches.

To close the European case studies, in Chapter 14 Bernard Barraqué offers an overview of key lessons to be learned from the European experience with WSS. Among other arguments, he contends that, as a general rule, the optimal size of WSS utilities (local, regional, etc.) is not only dependent upon physical-natural or demo-geographic conditions such as rainfall patterns, catchment extensions or population densities, but even more importantly on the level of social trust in the relevant institutional arrangements. One lesson from the European experience is that, for different reasons and in different forms, European countries were able to universalize their WSS as a result of the development of a social structure, where the access to essential goods and services such as WSS came to be considered a fundamental right and duty of citizens. This, argues Barraqué, has important consequences for developing countries. For instance, it may well be that improving WSS will require the adoption of different strategies for wealthier and poorer sectors of the population, such as networked services in established urban areas and other forms such as community distribution of bulk water in the periphery. The author is aware of the potential controversies arising from this argument, but claims that such strategies may be the only practical solution, in many cases, where the basic conditions for universal provision are poorly developed or entirely lacking, such as the achievement of a 'common feeling of citizenship' and solidarity among the different social sectors in societies that are highly fragmented on socio-economic or ethnic grounds.

Chapters 15 and 16 introduce the cases of Canada and the US, respectively. In Chapter 15, Mark W. Rosenberg scrutinizes the striking contradictions characterizing WSS in Canada, a country favoured with abundant water resources where many Canadians and aboriginal peoples continue to be at risk from preventable water-related diseases and threats. Rosenberg argues aboriginal peoples, in particular, are paying a very high price for policy failures that continue to prevent them from accessing safe water supply services and adequate sanitation systems. The author examines two high-profile incidents that took place in Walkerton, Ontario, in 2000, and in the Kashechewan First Nations reserve in 2005, which exposed the risk of nationwide failures in managing WSS owing to a deadly combination of aging infrastructures, declining interest in WSS by federal and provincial governments, and far-reaching neoconservative reforms such as deregulation and privatization of crucial aspects of essential public services. In Chapter 16, Venkatesh Uddameri and Vijay P. Singh address the case of the US, a country whose historical ascendancy as a developed nation, they argue, has been clearly connected with its capacity for the appropriate organization of water and water services, including sanitation and waste management. However, while the country provides safe WSS to around 95 per cent of the population and the services are fully universalized in urban centres, the impact of very large urban agglomerations such as Los Angeles or San Antonio upon the aquatic environment has been significant. Uddameri and Singh state that these and other problems, such as health risks derived from inadequate waste management, have prompted the development of a 'comprehensive socio-political and technological framework' for environmental management, including the management of water sources and WSS. Still, the future maintenance of universal coverage in conditions of socio-economic and environmental sustainability constitutes an enormous challenge for the country since it will require massive investments running into the hundreds of billions of dollars in infrastructure renewal and further expansion to attend un-served communities.

The final group of chapters, Chapters 17 to 21, present experiences from the global South, focusing on cases from Africa, Asia and Latin America. In Chapter 17, David Nilsson and Arne Kaijser delve into the historical roots of current developments in the WSS of East African towns. The authors argue that the colonial WSS built in Kenya, Uganda and Tanzania during the early 20th century followed the paradigm of large-scale technological development – a model of WSS that was preserved after independence and that has been characterized by the systematic discrimination of the urban poor. They also examine recent reforms introduced since the 1990s, which have been centred mainly on cost-recovery policies rather than on improving and expanding the services to the un-served poor population located mostly in informal settlements. Nilsson and Kaijser propose the adoption of a 'nested systems approach' to the organization of WSS, where the existing large-scale infrastructure provides the backbone of the system; but it is then complemented by small distribution networks that can be managed by local agents. This potential solution, however, requires that both techno-infrastructural and socio-institutional aspects be designed, taking into account the local circumstances in order to sustainably secure universal access to WSS.

Chapter 18 by Roldan Muradian, Bishwa Nath Tiwari, Abu Jafar Shamsuddin and Laia Domènech looks at a number of successful experiences of community participation in improving WSS in poor areas of Bangladesh and Nepal, and explores the factors that may facilitate or inhibit community involvement. The authors found that through enhanced local involvement, poor communities can successfully implement cost-shared schemes for improved WSS, and that the key factors behind their success are flexibility and adaptation of the schemes to meet local needs and circumstances. In addition, the chapter argues that successful WSS initiatives driven by community involvement often trigger additional beneficial impacts, such as the strengthening of local capacities and leadership. However, the authors also found evidence of the well-known limitations of participatory processes at the community level, such as institutional capture by interest groups, the lawful segregation of socially marginalized groups through democratic majority decisions, or the outright exclusion of the very poor from decision-making bodies. In balance, though, the chapter ends on a very positive note regarding the potential of WSS interventions based on strong community participation to improve the quality and access of this service in poor urban areas.

In Chapter 19, Jiane Zuo and Lili Gan provide an introductory overview of the current situation and future challenges affecting WSS in China in a context of very rapid economic development, urbanization and rising user expectations regarding service quality and access. The chapter shows that despite the significant improvements achieved by the country since the creation of the People's Republic of China in 1949, there exist striking asymmetries in the access to safe water supply and adequate sanitation systems that particularly affect the rural areas concentrating the bulk of the population, but also a large number of urban centres. Zuo and Gan argue that while the country is making substantial progress in developing or adapting technologies for enhanced WSS management, some of the main obstacles are policy-institutional and organizational, especially the overlap and duplication of administrative and monitoring structures. Although the far-reaching reforms introduced in China since the late 1970s have also modified the institutional and organizational structure of WSS – for instance, opening up the possibility for private-sector involvement and the adoption of commercial principles for the running of WSS utilities – the responsibility and control for these essential services remain a state preserve and the authors argue that this will continue to be the case for the foreseeable future.

Last, but not least, the book is closed with two chapters on Latin America, focusing on the cases of Brazil and Mexico. Léo Heller considers the Brazilian experience in Chapter 20, exploring the development of WSS in the country during the 20th century with particular emphasis on the last two decades. Heller discusses the progressive replacement of the notion of water and sanitation in Brazil by a more comprehensive definition, 'basic sanitation', that includes not just water supply and sewage collection and disposal, but also a whole range of policy interventions, including the adequate management of solid waste and storm water, and vectors control. Heller argues that the development of WSS since the 1970s has been characterized by a crucial contradiction between the continuous improvements in service quality achieved through technological development and the increasing social inequalities in the access to the resulting benefits. Although, since 2003, the government of President Lula da Silva has introduced important reforms that have the potential to reverse these trends, Heller warns that systemic conditions mainly external to the WSS sector, such as the country's traditional power configurations and the prevailing development policies, still present formidable obstacles to progress. He argues for the adoption of interdisciplinary and cross-sector approaches that bring together the technological and social-science dimensions of WSS policy and management as the key strategy for achieving the sustainable universalization of these services.

In Chapter 21, María Luisa Torregrosa Armentia and Blanca Jiménez Cisneros address the case of WSS in Mexico. In a way not too dissimilar from the Brazilian situation, Mexico has experienced significant progress in the technological and infrastructural dimensions of WSS since the 1970s in a context of rapid economic development and urbanization. However, there exist enormous asymmetries in the access to the benefits of improved WSS, and the authors show that Mexico compares very poorly with countries with a lower level of economic development that have achieved much better results in the expansion of service coverage, particularly in the case of rural sanitation. The chapter examines the sequence of legal, institutional and administrative reforms introduced since the mid-1970s in the WSS sector, which included decentralization of responsibilities for service organization and delivery, and the introduction of commercial principles and private-sector participation. The authors argue that these policies have failed to achieve their main objectives because, among other reasons, of the protracted social inequalities characterizing Mexican politics where access to WSS continues to be the object of authoritarian and clientelist practices. They contend that WSS must continue to be a state responsibility, but that there is a need for more comprehensive and long-term planning strategies involving the whole spectrum of actors and the development of truly interdisciplinary approaches to tackle the complex social, environmental and technological challenges facing the goal of sustainable universalization of these services in Mexico.

CONCLUSIONS

It is apparent from the above discussion that, although the book has unambiguous objectives and structuring assumptions, it is also a collective product that reflects both the rich diversity characterizing WSS experiences worldwide and the deep-rooted contradictions and rival understandings about how these services should be organized and run. In this regard, the book will offer readers an opportunity to examine the challenges facing public policy and management activities in the WSS sector from a range of different perspectives and in the spirit of open academic debate. We hope that the book will make a positive contribution towards a better understanding of these challenges and towards the development of higher levels of interdisciplinary coordination in the production of knowledge about water and essential water services.

In this connection, it is apposite to mention that the book is the result of the joint initiative of a sanitary engineer, Léo Heller, and a sociologist, José Esteban Castro, and it is the product of an ongoing interdisciplinary exchange in the context of the research network on Governance, Citizenship, Water Management and Environmental Health in Developing Countries (GOBACIT) (www.gobacit.org), which focuses on issues of governance and citizenship in the interface between water management, WSS and public health. Many of the book's co-authors are partners in the GOBACIT network, while others have kindly agreed to contribute with their expert knowledge to our enterprise. This has been an interdisciplinary enterprise that brought together, among others, economists, engineers, geographers, historians, hydrologists, political scientists and sociologists, and we have a strong feeling that despite the difficulties involved, this has been a very fruitful endeavour for the group.

We also want to thank Professor Leslie Bethell, former director of the Oxford Centre for Brazilian Studies at Oxford University, who generously accepted to fund the international workshop that helped us to put together the project and the core group of contributors. The workshop focused on Public Policy in Water Supply and Sanitation: Theoretical Perspectives and a Review of Experiences from Brazil and European Countries, and took place in Oxford on 5 May 2006. We are also very grateful for the support provided by Earthscan during the preparation of the book, and particularly to the commissioning editor, Tim Hardwick. Many thanks also to Alison Kuznets, who oversaw the design of a wonderful cover for the book, and to Andrew Miller for his work on the publicity aspects. Finally, we want to praise Camille Bramall for her very careful editing of the text, which has helped to correct many mistakes while also substantially enhancing the overall readability of the collection.

We know that owing to space restrictions we have left unacknowledged a great deal of help received in many forms and from many people, which has made this work possible. We wish to partly compensate for that omission by dedicating the book to the workers, intellectual and manual, whose daily and mostly anonymous work in running essential WSS helps to make life possible for most human beings.

NOTES

1 We refer in the text to 'water supply and sanitation', 'water and sanitation', 'water and sewerage' and similar forms as equivalents, unless there is an explicit definition to the contrary. In some chapters, these definitions will be further specified, given that the concrete services included under the umbrella term 'water and sanitation' vary in different organizational models and

cultures (e.g. in Brazil the concept of integrated sanitation includes urban drainage, solid waste collection and disposal, vector controls, etc.).

2 Here 'sustainable' has a multidimensional meaning and refers to the sustainability of WSS in social, economic-financial, infrastructural, political-institutional and ecological terms. In other words, we refer to the comprehensive and long-term viability of universally available WSS.

REFERENCES

ADB, UNDP, WHO and ESCAP (Asian Development Bank, United Nations Development Programme, World Health Organization and United Nations Economic and Social Commission for Asia and the Pacific) (2006) *Asia Water Watch 2015: Are Countries in Asia on Track to Meet Target 10 of the Millennium Development Goals?*, ADB, WHO, ESCAP and UNDP, Manila, Geneva, Bangkok and New York

Amnesty International (2003) *Human Right to Water*, Amnesty International, London

Bath, R. C., J. M. Tanski and R. Villarreal (1998) 'The failure to provide basic services to the colonias of El Paso County: A case of environmental racism?', in D. E. Camacho (ed) *Environmental Injustices, Political Struggles: Race, Class and the Environment*, Duke University Press, Durham and London, pp125–137

Berry, K. A. (1998) 'Race for water? Native Americans, eurocentrism, and Western water policy', in D. E. Camacho (ed) *Environmental Injustices, Political Struggles: Race, Class and the Environment*, Duke University Press, Durham and London, pp101–124

CCW (Consumer Council for Water) (2006) *Rise in Consumer Debt Calls for Swift Action*, CCW, Birmingham

Davis, C. K. and R. E. McGinn (eds) (2001) *International Perspectives on Ethical Dilemmas in the Water Industry*, American Water Works Association, Denver, CO

Defra (Department for Environment, Food and Rural Affairs) (2004) *Cross-Government Review of Water Affordability Report*, Defra, London

Delli Priscoli, J., J. Dooge and R. Llamas (2004) 'Water and ethics: Overview', in *Water and Ethics*, United Nations Educational Scientific and Cultural Organization (UNESCO), International Hydrological Programme and World Commission on the Ethics of Scientific Knowledge and Technology, Paris

Dryzek, J. S. (1997) *The Politics of the Earth: Environmental Discourses*, Oxford University Press, Oxford

EC (European Commission) (2002) *Water Management in Developing Countries: Policy and Priorities for EU Development Cooperation*, Communication from the Commission to the Council and the European Parliament, EC, Brussels

EC (2003) *EU Water Initiative: International Cooperation – From Knowledge to Action*, EC, Brussels

EUWATER Network (2005) *European Declaration for a New Water Culture*, New Water Culture Foundation, Saragossa

Fitch, M. (2006) *Fair and Affordable Water*, UNISON, London

Fitch, M. and H. Price (2002) *Water Poverty in England and Wales*, Centre for Utility Consumer Law and Chartered Institute of Environmental Health, London

Hall, D. and E. Lobina (2007) *From a Private Past to a Public Future? The Problems of Water in England and Wales*, Public Services International Research Unit (PSIRU), University of Greenwich, Greenwich

Hanf, K. and A.-I. Jansen (1998) 'Environmental policy: The outcome of strategic action and institutional characteristics', in *Governance and Environment in Western Europe: Politics, Policy and Administration*, Addison Wesley Longman, Harlow, pp1–15

Klein, G. (2003) *Life Lines: The NCC's Agenda for Affordable Energy, Water, and Telephone Services*, National Consumer Council, London

NCC (National Consumer Council) (2005) *Affordability Problems in the Privatised WSS Sector in England and Wales*, NCC, London

OFWAT (Office of Water Services) (2004) *Annual Report 2003–2004*, OFWAT, Birmingham

UKP (United Kingdom Parliament) (2003) *Memorandum Submitted by the Public Utilities Access Forum*, UKP, London

UN (United Nations) (1980) *International Drinking Water Supply and Sanitation Decade: Report of the Secretary General*, UN, New York

UN (2000) *Millennium Declaration*, UN, New York

UN (2002) 'Key commitments, targets and timetables from the Johannesburg Plan of Implementation', in *World Summit on Sustainable Development*, UN, Johannesburg

UNDP (United Nations Development Programme) (1990) *The New Delhi Statement: Some for All rather than More for Some: Global Consultation on Safe Water and Sanitation for the 1990s*, UNDP and the Government of India, New Delhi

UNECE (United Nations Economic Commission for Europe) (2007) *After Historic Ratification, European Countries Meet for the First Time to Improve Water Management and Curb Water-Related Diseases*, UNECE, Geneva

UNESCO–WWAP (United Nations Educational Scientific and Cultural Organization–World Water Assessment Programme) (2006) *Water, a Shared Responsibility: The United Nations World Water Report 2*, UNESCO and Berghahn Books, Paris, New York

UN-Habitat (2003) *Water and Sanitation in the World's Cities: Local Action for Global Goals*, Earthscan, London

UNICEF and WHO (United Nations Children's Fund and World Health Organization) (2008) *Progress on Drinking Water and Sanitation: Special Focus on Sanitation*, UNICEF and WHO's Joint Monitoring Programme for Water Supply and Sanitation (JMP), New York and Geneva

UNMP (United Nations Millennium Project) (2005) *Final Report on Health, Dignity and Development: What Will It Take?*, Task Force on Water and Sanitation, Swedish Water House, Stockholm International Water Institute, Stockholm

Water UK (2005) *Household Debt in the Water Industry: A Water UK Briefing Paper*, Water UK, London

Whiteford, S. and A. Cortez Lara (2005) 'Good to the last drop: The political ecology of water and health on the border', in L. Whiteford and S. Whiteford (eds) *Globalization, Water and Health: Resource Management in Times of Scarcity*, School of American Research Press and James Currey, Santa Fe, CA, and Oxford, pp231–254

WHO (World Health Organization) (2005) *Health and the Millennium Development Goals*, WHO, Geneva

WHO-Europe (World Health Organization Europe Office) (2006) *The Protocol on Water and Health: Making a Difference*, WHO-Europe, Rome

WHO and UNICEF (World Health Organization and United Nations Children's Fund) (2006) *Meeting the MDG Drinking Water and Sanitation Target: The Urban and Rural Challenge of the Decade*, WHO and UNICEF, Geneva and New York

WSP and PPIAF (Water and Sanitation Programme and Public–Private Infrastructure Advisory Facility) (2002) *New Designs for Water and Sanitation Transactions: Making Private Sector Participation Work for the Poor*, World Bank, Washington, DC

Part I

Theoretical and Conceptual Dimensions

1

Systemic Conditions and Public Policy in the Water and Sanitation Sector

José Esteban Castro

Introduction

This chapter addresses the import of systemic conditions on the development, organization and functioning of water and sanitation services (WSS), with particular emphasis on the social conditions and constraints that are mostly neglected or even ignored in traditional public policy and management practices in this sector of activity. The analysis of the influence of systemic conditions has a long tradition in the social sciences, particularly in relation to the intertwining of structural and agent-driven factors in almost all spheres of social interaction. It has also been deployed in a number of fields, from the analysis of the barriers and constraints facing the introduction of preventive health public policies (Berkeley and Springett, 2006) to business studies examining the constraints posed by market systems on economic agents who have to make choices between the pursuit of profit maximization or 'more morally preferable alternatives' such as environmental sustainability (Alexander, 2007).

In recent decades, the analysis of systemic conditions in the WSS sector has been dominated by studies seeking to overcome the obstacles to the implementation of deregulation, commercialization, liberalization and privatization policies, such as regulatory barriers, legal and institutional constraints, or cultural resistance to the expansion of these initiatives (see, for instance, Richard and Triche, 1994; Farley, 1997; Brook Cowen and Cowen, 1998; GWP, 2003; Moss et al, 2003). However, much less has been said about the impact that socio-economic and political conditions have on such issues as the lack of access to reliable WSS affecting a large section of the world population, and this chapter makes a contribution towards addressing this gap. Given the limited space available, the chapter focuses mainly on two aspects: structural social inequalities and mainstream WSS policies as systemic conditions affecting the universalization of these essential services.

THE SIGNIFICANCE OF SYSTEMIC CONDITIONS FOR WATER AND SANITATION SERVICES (WSS) POLICY

I focus here on those factors and processes that, although being external to WSS-related activities, have significant influence in shaping and directing their organization and functioning and, in some circumstances, even in their original development. For instance, in some countries the historical origin of urban WSS was not the outcome of intentionally planned policies but the by-product, often accidental, of such activities as the development of the railways (see, for instance, Chapter 17) or of fire-fighting systems (e.g. Katko, 1997; see also Chapters 4 and 20). Clearly, certain systemic conditions have traditionally been part and parcel of policy-making and management activities in the WSS sector, such as physical-natural factors (e.g. type of available water sources, hydro-geological and geographical constraints on infrastructural and technological solutions, etc.), economic-financial aspects (e.g. determinants of the financial viability of building piped water supply networks in low-density areas or the steep increase in investment requirements driven by stricter environmental and public health quality standards) or demo-geographic processes (e.g. the challenges posed by rapid urban growth and highly differentiated spatial patterns of urbanization on the organization of WSS). However, perhaps because of the traditional dominance of techno-centric approaches in the management of WSS, a range of systemic conditions that have significant influence on the running of these services have been historically downplayed, if not entirely neglected. These concern factors and processes that belong, broadly speaking, to the social dimension.

For example, systemic conditions tend to influence, shape and even determine consumption trends and practices. Thus, while African countries, on average, withdraw 31m^3 of water per capita per year for domestic use, in Asia and the Pacific the figure is 49m^3, in Europe 86m^3, in Latin America and the Caribbean 98m^3, and in North America the withdrawal reaches 221m^3 per capita per year (ESCAP, 2007, p165). These differences cannot be explained merely by physical-natural or technological factors, as clearly suggested by the striking differences in domestic water withdrawal between the similarly developed regions of Europe and North America. The analysis becomes more complex when we consider differences in water consumption between regions, cities or social groups, where factors such as structural social inequalities are often more significant in explaining asymmetries in service coverage and quality or differences in consumption practices than, for instance, climatic or technological constraints (e.g. Bath et al, 1998; Berry, 1998; Castro, 2004a; Crenzel and Forte, 2004; Heller, 2007; Laurie, 2007).

Likewise, the high diversity found worldwide in terms of institutional structures, policy options or even technological choices (see, for instance, Juuti and Katko, 2005) also lends support to the argument that long-term social factors and processes have significant influence in shaping and even determining how WSS services are organized. The fact, for example, that WSS in England are currently run as privately owned and profit-oriented businesses, while in the US or in the Nordic European countries these services are owned and managed mainly by public municipal organizations, is not a necessary or intrinsic characteristic of WSS, but is rather the product of historical processes where social factors have paramount influence (see, for instance, Chapters 9 and 13), a fact that continues to be largely neglected. For instance, since the 1980s, public policies promoted by the international financial institutions (IFIs) to reform WSS in developing countries, such as deregulation, liberalization, commercialization and privatization, have often been designed and implemented as one-fit-all 'tool kits' that are supposed to work in any condition, from global metropolises such as London to middle-sized cities in 'emerging economies' or small urban areas in very poor developing countries (e.g. OECD, 1994, 1995; *Economic Perspectives*, 1997; Brook Cowen and Cowen, 1998; World Bank, 1998; PPIAF and World Bank, 2006), a public policy approach that, in most cases, has led to disastrous results for the countries involved and, in some cases, also for the private companies (see Chapters 2 and 6).[1]

Learning from the lessons has been very slow, even if, increasingly, certain aspects of the social dimension previously excluded from consideration in water policy are now given more attention, at least rhetorically (see, for instance, GWP, 2003; Moss et al, 2003; ADB et al, 2006; UNESCO–WWAP, 2006). Nevertheless, techno-centric, including techno-social,[2] approaches continue to prevail in this field of activity to the neglect, for the most part, of significant systemic conditions such as structural social inequalities.

Structural social inequalities as systemic conditions

Let us consider, for instance, recent assessments of the progress made towards meeting the WSS targets set for the Millennium Development Goals (MDGs), which aim to reduce by half the proportion of the world's un-served population by 2015 (UN, 2000, 2002a). Successive reports have confirmed that these targets will not be met by many countries, particularly the poorest. In fact, as shown in Table 1.1, recent estimates suggest that in certain regions such as sub-Saharan Africa, Oceania and Western Asia, the actual number of un-served people is increasing rather than decreasing given that the expansion of service coverage cannot match the rate of population growth (see also UN, 2007, pp21–22).

The problem is particularly severe in the case of the sanitation target. As a recent report from the United Nations Children's Fund and the World Health Organization states (UNICEF and WHO, 2008, p8):

> *The world is not on track to meet the MDG sanitation target. Between 1990 and 2006, the proportion of people without improved sanitation decreased by only 8 percentage points. Without an immediate acceleration in progress, the world will not achieve even half the MDG sanitation target by 2015.*

Table 1.1 *Estimates of the progress to be achieved between 2005 and 2015 in meeting the Millennium Development Goals (MDGs) targets for water and sanitation (by world region)*

Region	*Population without water by 2015 (millions)*	*Population without improved sanitation by 2015 (millions)*
Sub-Saharan Africa	+47	+91
Oceania	+1	+1
Northern Africa	–1	–11
Western Asia	–5	+3
South-Eastern Asia	–5	–55
Latin America and the Caribbean	–25	–24
Eastern Asia	–30	–157
Southern Asia	–139	–69
World total	–150	–221

Source: adapted from WHO and UNICEF (2006, pp10, 18)

The monitoring reports have argued that political, institutional and financial factors are the key explanatory factors for this failure. For example, the final report from the United Nations (UN) task force that monitored the MDGs between 2000 and 2005 highlighted a number of issues, such as lack of political commitment, weak institutional frameworks and the sluggish pace of reforms, political manipulation and patronage, and financial constraints, particularly in the poorest countries where governments lack the resources to invest and the services are unaffordable for very large sections of the population (UNMP, 2005, pp26–32; see also ADB et al, 2006, pp37–40). The relevance of these factors for explaining the slow progress or even the regression experienced in many countries is confirmed by a wealth of recent research (see, for instance, GWP, 2003; UN-Habitat, 2003; Hall et al, 2004; UN-Habitat, 2006). Moreover, the reports already warn that in Asia, which houses the largest, un-served population:

> *... [even] achieving the MDG targets is still not the end of the story, including the end of the need for significant investments in water supply and sanitation... In most countries, many millions of people will not have their needs met despite the MDG targets being achieved. Given that it is likely to be the poorest people, often living in the most challenging locations and whose needs are unmet, the challenges of providing improved water supply and sanitation for all of the people of Asia and the Pacific will still be significant.* (ADB et al, 2006, p37)

Elements of this diagnosis can also be extended by analogy to other regions of the world. However, despite the increasing recognition that the main challenges for WSS are not technical or physical-natural, but rather, broadly speaking, social, there exist crucial systemic conditions in the social dimension that are largely neglected in the analysis. First, in the official literature (illustrated by the above quoted reports), the analysis systematically fails to address the structural processes and conditions underpinning such problems as the lack of political will showed by governments, the widespread institutional weakness, or the poverty that renders WSS unaffordable for a large proportion of the human population. The diagnosis is roughly correct, but it does not go deep enough: it only touches the surface of the actual causes of the crisis. Second, the prevailing approach to WSS illustrated by this literature remains silent about the failure of the mainstream policies implemented worldwide since the late 1980s, which have promoted the commercialization and even commodification of WSS[3] as the main solution. Not only have these policies failed to deliver the expected improvements, but they have also contributed to worsening the impact of negative systemic conditions, such as structural social inequalities, and have unleashed new systemic constraints, which are particularly harmful for the poorest countries.

In relation to structural inequalities, even in developed countries that long ago reached maturity in the organization and universalization of WSS, significant sections of the population frequently still lack access to a regular and reliable water supply and safe sanitation owing to persistent, and often worsening, social inequalities. In poorer countries, the picture is obviously much more severe, and a constellation of factors, including economic, gender, age and ethnic inequalities, explain the lack of access to WSS (Webb and Iskandarani, 1998; Laurie et al, 2002; Swyngedouw et al, 2002; Swyngedouw, 2004; Laurie, 2007). For instance, a report on the extreme asymmetries characterizing the access to WSS in La Paz-El Alto, Bolivia, argued that the local Aymara indigenous community continues to be precluded from access to reliable WSS because of ethnically grounded structural inequalities, a situation conceptualized as 'water racism' (Crespo Flores, 2007). Similarly, research carried out in Recife, Brazil, showed that a 20-year-old system of 'rationed delivery' of piped water that is nominally designed to ensure a fairer distribution of water is actually structurally unequal and is biased against the very poor. Officially, the city is divided into six delivery areas, where water is distributed in cycles of 20 hours of service followed by 28 hours without water; but the study found that the water utility has unofficially designated 'special delivery areas' where the cycle is often limited to 3 hours of service followed by up to 45 hours without water. These 'special' areas are located in the poorest neighbourhoods of the metropolis (around half of the city's houses are located in shanty towns), where in addition to the rationing the service is affected by low pressure and intermittence (Rocha Ferreira, 2006).

These structural social inequalities underpinning and shaping the organization of WSS in many settings, which are often embedded in the institutional frameworks and even in the physical infrastructure, and that can seldom be detected in the official statistics, are not the preserve of developing countries or poor regions. Thus, in Israel, a country that officially reports 100 per cent coverage for WSS (WHO and UNICEF, 2006, p33), it is estimated that the Arab citizens who compose about 18 per cent of the population only use around 2 per cent of the water delivered, but pay much higher prices than their Israeli counterparts (Libiszewsky, 1995; Isaac, 1997; see also Oxfam-UK and PWA, 2006, for the situation in the occupied Palestinian territories). Likewise, in the US, a country that has long ago achieved universal coverage for WSS (WHO and UNICEF, 2006, p38; see also Chapter 16), there is a well-documented pattern of structural, ethnic-, class- and gender-based inequalities that seriously affect the access to WSS by non-white minorities

(Bath et al, 1998; Berry, 1998; Whiteford and Cortez Lara, 2005; Wescoat et al, 2007). Similarly, in England and Wales, where WSS were universalized during the 1960s and later fully privatized in 1989, according to government reports, between 2 million and 4 million families live now in 'water poverty', which happens when families expend over 3 per cent of their income on WSS, while between 15 and 20 per cent of WSS customers are not paying their bills (Fitch and Price, 2002; Klein, 2003; UKP, 2003; Water UK, 2005; CCW, 2006; Fitch, 2006; Kearton, 2006).

These examples can be easily multiplied but are sufficient to illustrate our point: focusing merely on the techno-infrastructural and managerial aspects of WSS is not enough to understand the causes of the problems under consideration. However, the prevailing approaches to public policy and management in WSS not only tend to neglect the significance of the systemic social conditions, but also frequently reproduce and amplify their impact, or even contribute to the emergence of new constraints. This is the case, most notably, of the mainstream WSS policies that have been promoted and implemented worldwide since the 1980s as the best solution to these problems.

The inertial forces of mainstream WSS politics

As mentioned above, since the 1980s, the mainstream position adopted by the IFIs, international development agencies (IDAs), governments of developed countries, and other key actors has been centred on the promotion of commercialization, privatization, deregulation and other related policies as the best solution to solve the crisis of WSS. Unfortunately, the reports addressing the failure of the international community in relation to the MDGs remain silent about the intimate relationship between these policies and the lack of progress in the universalization of access. The prevailing approach can be illustrated with the following passage from the Keynote Address given by Willem-Alexander, the Prince of Orange from The Netherlands, at the Fourth World Water Forum in Mexico: 'I think it is safe to say that, at this point, we don't need more new policies. What we do need is swift action' (The Prince of Orange, 2006). Unfortunately, this dismissal of the need for policy change is part and parcel of the hegemonic techno-centric approach to water policy and management, which is compounded by a reluctance to recognize both the failure of the policies that have been promoted worldwide during the last two decades and the urgent need for radical change. However, the evidence suggests that mainstream WSS policies and the inertial forces that they have unleashed are posing new constraints for achieving the universalization of WSS, particularly where the needs are greater. There is certainly as much need for new water policies as there is for swift action.

The mainstream WSS policies, often labelled 'neoliberal' because of their insistence on the restructuring of WSS around market principles (detailed analyses of neoliberal WSS policies have been provided by, among others, Hall, 2002; Budds and McGranahan, 2003; Hukka and Katko, 2003; Bakker, 2004; Castro, 2004b; Hall et al, 2004; Swyngedouw, 2005; Laurie, 2007; Brown et al, 2008), did not emerge from the analysis of the particular problems and requirements of these services, but have rather been imposed on the WSS sector in the course of the process of economic globalization. Thus, recommendations by World Bank experts who argue that the 'complete privatization of water assets' and the creation of 'unregulated private monopolies' are the key to expand WSS coverage and improve infrastructure in the poorest countries (Brook Cowen and Cowen, 1998, pp22–23; see also World Bank, 1998) cannot be understood as being the result of a rigorous study of the specific problems facing public policy and management of WSS or a reflection of the historical experience of the countries where these services were successfully universalized. Rather, these policy reforms are part and parcel of the 'market-driven politics' (Leys, 2001) that have restructured the foundations for public policy globally since the 1980s. As stated by Joseph Stiglitz, former chief economist at the World Bank:

> *In setting the rules of the game, commercial and financial interests and mind-sets have seemingly prevailed within the international economic institutions. A particular view of the role of government and markets has come to prevail – a view which is not universally accepted within the developed countries, but which is being forced upon the developing countries and the economies in transition.* (Stiglitz, 2002, pp224–225)

The experience of WSS is particularly relevant to illustrate the pertinence of Stiglitz's statement: while in the majority of developed countries these services are mostly run as public-owned (mainly municipally owned) services (see Part II of this volume), developing countries and 'emerging economies' have been subject to enormous pressure to privatize their WSS. This is illustrated by the following statement from the US Agency for International Development (USAID):

> *[USAID] is defining future directions for privatization assistance based on the experience of the last 15 years – from a modest start in Latin America to the recent crescendo dominated by assistance to formerly Communist states. In the countries of Central and Eastern Europe and the former Soviet Union, the privatization of state-run enterprises and the dismantling of state monopolies are critical to these nations' transition to free markets... Poorer countries and governments ideologically resistant to opening their markets – largely in sub-Saharan Africa, the Middle East, and South Asia – will need continuing technical and policy support, as well as new, creative interventions.* (Farley, 1997, p10)

This global project, driven by what Stiglitz called 'commercial and financial interests', has informed water policies since the 1980s, setting the new rules of the game within which public services such as WSS are increasingly required to operate. These new rules of the game, and the forces that they have unleashed worldwide, are now part and parcel of the structuring systemic conditions that need to be incorporated within the analysis of public policy and management of WSS. I explore here two main aspects of these systemic conditions:

1 the drive to accelerate the process of commodification of WSS by cancelling their status of public good and social right; and
2 the impact of the inertial forces set in motion by these policy reforms.

The attack on the public good

As shown, for instance, in the chapters on the experience of the Nordic European countries, France, Canada or the US featured in this volume, the universalization of WSS was closely associated with the adoption of public policies centred on the principle that the provision of goods and services that are essential for collective well-being constitutes a social responsibility – a state duty – that cannot be subordinated to the logic of private market interests. In some countries, where the early development of WSS since the 18th century had been driven by private profit and market-oriented public policies, as was the case in England or the US, accepting the notion that these services must be universally available, provided as a public good and guaranteed by the state was the result of protracted social struggles over several decades (Laski et al, 1935; Mukhopadhyay, 1975; Schultz and McShane, 1978; Warner, 1987; Millward, 1991; Ward, 1997; Hassan, 1998; Ogle, 1999; Melosi, 2000). In the process, a new framework for the provision of essential services such as WSS was developed and consolidated in Western countries, characterized by what John Dryzek has called 'administrative rationalism' – a marriage between bureaucratic and scientific expertise that dominated much of 20th-century public policy (Dryzek, 1997).

Administrative rationalism provided the framework that allowed developed countries to expand their basic infrastructure, as well as the universalization of essential goods and services, including WSS. From another perspective, this outcome was neither intrinsic nor necessary to the inner development of the WSS sector, nor merely the consequence of more egalitarian approaches to the distribution of social wealth. Rather, the consolidation of the systemic conditions associated with the policy framework of administrative rationalism in developed countries was intertwined with the process of legitimatizing capitalist democracy driven by both internal and external forces, and particularly by the social and political struggles for substantive democratization and the requirement for collective (state-funded) provision of basic infrastructures and services to guarantee the reproduction of capital (see, among others, Castells, 1977; Decaillot et al, 1977; Cook and Kirkpatrick, 1988; Hobsbawm, 1994; McGowan, 1994; Taylor, 1999; Swyngedouw et al, 2002). Moreover, administrative rationalism also has significant drawbacks, notably the fact that it was developed as a mainly top-down and techno-centric approach – for instance, in the organization of WSS, which often adopted paternalistic, even authoritarian, practices that with rare exceptions have been largely hidden from citizen scrutiny and democratic social control.

Nevertheless, through a wide diversity of mostly country- or, at least, region-specific experiences and processes, in developed countries during the 20th century public policy and management of essential services such as WSS became structured around a number of key principles:

- the central role of the state in defending the 'public interest' (e.g. through the actions of the US Supreme Court in relation to the regulation of public services);

- the categorization of essential goods and services such as WSS as 'public goods', a 'market failure' requiring their removal from the sphere of market transactions in order to achieve socially efficient outcomes;
- the principle of social citizenship, where the access to essential goods and services such as WSS became enshrined as social goods that must be available to everyone independently of their market status; and
- the central role of the state in organizing and delivering these essential services to guarantee the conditions of universal provision.

Although these principles, and the policies derived from them, were always contested by free-market liberals, they came to provide a solid framework for public policy and management in the field of essential public services during the 20th century, and it was within this framework that the actual universalization of WSS took place in the developed world. As illustrated by several chapters in this volume, this framework also strongly influenced the organization of WSS in developing countries, although in practice, and with very few notable exceptions, the universalization of access was never achieved.

In this connection, mainstream WSS policies since the 1980s have been predicated on the need to reverse and cancel the core principles of administrative rationalism and social citizenship. Thus, as a justification for implementing neoliberal reforms in WSS, it has been argued that the principles of administrative rationalism and social citizenship would be a systemic constraint – an obstacle for the efficient running of these services that needs to be removed (see, for instance, Roth, 1988, pp240–242; Triche, 1990, p4). World Bank documents even suggest that people have been led to wrongly believe that WSS are 'a "public service" or even a "social good"', concepts that should be swiftly abandoned (WSP and PPIAF, 2002, pp8–10). These policies seek to transform the provision of essential public services such as WSS into commercial activities, which whenever possible should take the form of fully privatized profit-making ventures, with little or no regulation.

The inertial forces of neoliberal policy failure

The failure of these policies to achieve their stated objectives has been recognized, somewhat reluctantly, by many actors, including the World Bank: 'private-sector or NGO participation in health, education and infrastructure is not without problems – especially in reaching poor people' (World Bank, 2003, pp10–11). In addition, despite the significant efforts made on a global scale since the 1980s to dismantle the system of publicly provided WSS, the expected expansion of direct private provision of these services has not materialized and recent estimates show that private WSS companies serve less than 10 per cent of the world population (UN-Habitat, 2003, pp177–178; Hall et al, 2004, p25; Laurie, 2007; Brown et al, 2008). However, important aspects of the actual impact of these radical neoliberal reforms tend to be underplayed or left unexamined. I argue that despite their failure, these policies have set in motion significant transformations whose inertial forces came to constitute new systemic constraints with far-reaching consequences for the WSS sector. In this regard, I wish to highlight three main aspects:

1 the progressive weakening of the ethic of public service and social citizenship and its replacement by the individualistic values of the 'survival of the fittest' in the organization and provision of essential goods and services;
2 the long-term mortgage placed especially on the poorest countries as a result of failed neoliberal reforms;
3 the continuation of the same policies despite the evidence of failure, which is helped by the uncritical approach maintained by the leading international organizations, governments and influential sectors of the scientific community.

In relation to the first aspect, through changing the status of WSS from public or social goods into marketable commodities, these policies are driving far-reaching transformations in the organization and management structures of these services, whether they are in public or private hands. While certain changes are obviously welcome, such as the modernization of WSS by adopting the latest organizational technologies for user management (e.g. metering, billing, fee collection, etc.), the process often goes far beyond, and public utilities are being reorganized, giving priority to the principles of commercial and technical efficiency to the neglect of their main function as providers of an essential service. Thus, many public utilities are now almost indistinguishable from private entities as their main objectives have also become profit maximization, business expansion (e.g. by acquiring utilities in other countries where they effectively operate as multinational companies) and even private accumulation (e.g. through the flotation of equity shares in private capital markets). The citizen has been replaced by the customer as the formal subject of the services, and in the extreme these public utilities become more responsive to private investors than to their own users. These processes are relatively new and little is known about their actual scope and impact worldwide; but the trends already discussed in relation to the international failure in meeting the MDGs send a clear warning: the erosion of the ethics of public service and social citizenship and its replacement by an individualistic ethic that gives priority to market interests poses potentially intractable constraints to the sustainable universalization of WSS.

The second aspect that I wish to emphasize is the long-term mortgage especially facing the poorest countries as a result of failed neoliberal reforms. Little information has been provided – for instance, by institutions such as the World Bank – about the actual costs of policy failure that have to be paid by the countries where neoliberal WSS policies have collapsed or did not deliver the expected results. In addition to the costs incurred by the countries in the preparation of, for instance, privatization processes (consultancies for developing the bidding documentation; absorption of existing debts by the public sector to make the public utilities attractive to private investors; financial and other incentives to facilitate the privatization, such as undervaluing the assets, tax reductions or even public subsidies to the private investors; etc.) or as a result of the collapse of such projects, many countries are now facing the burden of paying compensation to private companies that have failed to deliver but are entitled by contract to receive such payments. These include some of the poorest countries in the world – for instance, Bolivia, which has been sued by multinational private companies, such as the French water monopoly Suez, seeking compensation for the termination of long-term concession contracts (most of these legal suits have been submitted to the World Bank's International Centre for the Settlement of Investment Disputes (ICSID)). Some countries such as Argentina face dozens of trials from private companies resulting from failed privatization contracts, most of which run into the hundreds of millions of dollars in compensation claims alone (e.g. Azpiazu and Bonofiglio, 2006). There is much silence about the actual impact of the burden of these debts on the countries' abilities to reorganize their public utilities and to achieve their MDG targets, as well as other long-term consequences of this policy failure.[4]

Finally, in relation to the third point, the IFIs and other key actors in the field of water policy continue to promote neoliberal reforms in WSS that disregard the lessons from recent experiences. A report from a British think tank linked to one of the main political parties, for instance, continues to argue that WSS privatization is needed 'both to meet the UN's Millennium Development Goals, [and] to actively contribute towards social justice the world over' (Balen, 2006, p4; see also Gilbert, 2007). More worryingly, the latest United Nations *World Water Report* classified domestic WSS as 'private commodities' (UNESCO–WWAP, 2006, p409), reproducing the mainstream understanding that these services should no longer be considered public or social goods. Moreover, the UNESCO report ignores the existence of a global debate where even other UN departments have declared WSS to be a human right and not a commodity (UN, 2002b; WHO, 2003). The insistence on promoting mainstream WSS policies not only ignores the evidence of failure emerging from the experience of the last two decades, but also neglects the overall systemic constraints created by the neoliberal project of economic globalization, of which neoliberal WSS policies are part and parcel. The evidence shows that neoliberal policies have significantly worsened already existing structural social inequalities (Stiglitz, 2002; Wade, 2004), thus eroding the very foundations required by a model that rhetorically promises to simultaneously achieve the sustainable universalization of access to essential services such as WSS and their full commodification (where users pay the full real cost of the services and the utilities comply with all of the requirements of sustainability, including ecological, economic-financial and infrastructural aspects, not forgetting the extraction of private profit, without resorting to public funding).

These are just a few examples to back up our argument that the inertial forces unleashed by neoliberal WSS policies have become an integral part of the systemic conditions that shape and even determine how WSS are organized and run, which needs to be incorporated within the analysis. In some respects, neoliberal WSS policies have just reproduced or reinforced existing constraints, such as the techno-centric approach to public policy and management or the top-down, often undemocratic and authoritarian, forms of organization that have historically prevailed in this field of activity. In others, they have unleashed significant inertial forces whose actual impact upon public policy and the management of WSS is still little understood and largely under-researched.

Conclusions

This chapter has focused on the need to incorporate into the analysis the systemic conditions that, to a large extent, structure the design and implementation of public policies and management in WSS, with particular emphasis on the social dimension. I have explored the example given by recent reports on the failure of the international community to live up to the commitments adopted in the WSS targets of the MDGs. Although these reports correctly point the finger at social and political factors such as weak institutional frameworks or undemocratic political practices in their diagnosis, they fail to address the actual causes underlying these issues, such as the structural social inequalities that preclude a large proportion of the human population from having access to adequate WSS even when these are readily available in their wider communities. Unfortunately, the predictions suggest that the MDGs will not be achieved in most of the poorest countries, which, in practice, means that over the next few decades millions of people will continue to suffer illness and death caused by preventable water-related diseases. Moreover, even in those countries that are expected to meet their MDG targets, large sections of the population will continue to be excluded from full access to these essential services, which is the direct result of protracted social inequalities that continue to be largely ignored in the analysis of public policy and management in the WSS sector.

In this regard, the prevalence of a techno-centric approach in the running of WSS, to the neglect of their multidimensional character, continues to be a major obstacle in achieving success. In particular, the prevailing techno-centric approaches embodied in mainstream WSS tend to promote one-size-fits-all solutions that are supposed to work in almost any situation, as if WSS utilities were somehow insulated from the effects of systemic conditions and constraints. Some of the clearest contemporary examples of the failure of this approach are the neoliberal policies that continue to promote the full commodification of WSS, with almost complete disregard for the fact that the social conditions required for a sustainable implementation of this model are largely fragmentary and unstable. On the one hand, it is only in relatively selected areas of the world that conditions may exist for organizing and delivering WSS in a fully commodified fashion. Probably, however, there are no pure examples of such a level of self-contained and fully sustainable commodification of WSS anywhere in the world, even today. On the other hand, even where these conditions partially exist, such as in the wealthy neighbourhoods of affluent urban centres or in some economically successful medium-sized cities, they are unstable and changing to the point that expectations of sustainably implementing such policies as full-cost recovery and full commodification, while simultaneously achieving the universalization of WSS, are likely to fail (and, again, pure examples of a sustainable implementation of these policies are probably non-existent). This is compounded by the fact that the mainstream WSS policies that continue to be promoted worldwide are an integral part of the process of neoliberal economic globalization, which has deepened structural social inequalities, thus eroding the very conditions required for the sustainable implementation of fully commodified WSS for all.

In this regard, and to end on a positive note, there is a need to give more centrality to the systemic conditions of the social dimension in public policy and management of WSS, not only for a better understanding of the processes at work but also to enhance the chances of success in practical matters. The starting point should be the promotion of a better understanding among policy-makers, managers and other relevant actors in the WSS sector of the conditions that help to explain the historical success of past WSS policies in those countries that have achieved and maintained the universal coverage of these essential services. This success was not only the product of sound techno-scientific development and managerial expertise in the running of WSS, but was, rather, the result of a particular configuration of systemic conditions in the social sphere, such as the development of a new policy framework grounded on principles that gave priority to the public good and the social rights of citizenship over individualistic market interests. These policies and the principles that inspired them were accepted and supported by a wide range of social and political forces, even by sectors that, in other respects, defended free-market liberalism but recognized that the universal provision of essential services such as WSS required different arrangements.

I argue that achieving success in the design and implementation of WSS policies and management models that give priority to the satisfaction of essential human needs over the egoistic pursuit of private profit can only happen through the amalgamation of a similarly broad and universalistic alliance of social forces. A number of different developments taking place worldwide indicate that such a process is already under way, through initiatives oriented at strengthening local capacities, fostering public–public cooperation and partnership, and consolidating the substantive democratization of the governance and management of water and life-sustaining services, such as water supply and sanitation. There is an urgent need for new policies and for swift action in the field of WSS; but it is also imperative to contribute towards the strengthening of the systemic conditions that may enable the successful defence of the common good.

Notes

1 This problem was examined in some detail in the PRINWASS project that studied the implementation of these policies during the 1990s in such different settings as the Thames River Basin in England (around 12.5 million inhabitants, including London), the Buenos Aires Metropolitan Area in Argentina (11.5 million people), Dar es Salaam in Tanzania (2.5 million people) or middle range cities such as Tucumán in Argentina (0.7 million people), Cochabamba in Bolivia (0.52 million people), Limeira in Brazil (0.25 million people), and Aguascalientes in Mexico (0.64 million people), among others (Castro 2004b). With the exception of the English case, where WSS are fully private (full divestiture), in all other cases the projects followed the model of long-term (20 to 30 years) concessions to private consortia led by multinational water monopolies. In all of these cases we found evidence that systemic conditions such as local water values and practices, political and institutional traditions and structures, socio-economic characteristics of the population, or ongoing socio-economic and political processes (such as increasing social and spatial polarization resulting from worsening socio-economic inequalities and rising unemployment and poverty during the 1990s) were neglected or even ignored in the design and implementation of these policies (Castro and Laurie, 2004; Crenzel and Forte, 2004). In fact, the neglect of systemic conditions in the design and implementation of these projects was often extended to those aspects that have historically been considered essential in the running of WSS, such as hydro-geological or technical-infrastructural constraints, as patently illustrated in the case of the failed concession granted to the private company Azurix in the province of Buenos Aires, Argentina (Amorebieta, 2005).

2 By 'techno-social', I mean the reduction of complex social processes involving cultural, socio-economic or political processes to legal or policy-institutional aspects. For instance, while considering the multidimensional character of the social and cultural values associated with water requires an approach that recognizes the essentially incommensurable character of ecological processes (e.g. their irreducibility to a market equivalent; on the problem of value incommensurability, see, for instance, Martinez Alier, 2002), much of the prevailing water policies being promoted worldwide depart from the opposite assumption that water's essential value is economic (in the narrow sense of being a market good). Authors promoting these policies, for instance, have argued 'that water is an economic good and [we should] not treat water as having "unique importance" but as one good among all others' and that, in consequence, it should be treated as a commodity and managed 'through the market' (Lee and Jouravlev, 1998, p7). This market reductionism of water values often leads to techno-social approaches in water policy that neglect the fundamental socio-economic, cultural and political aspects that are constitutive of the social dimension. For instance, a World Bank document suggests that the fact that WSS in most countries have been traditionally considered 'a 'public service' or even a 'social good'' constitutes a misperception created by governments that needs to be eradicated through making WSS a private good (WSP and PPIAF, 2002, pp8–10; see also, UNESCO–WWAP, 2006, p409).

3 I distinguish here between commercialization and commodification of WSS. The first refers to the introduction of commercial principles in the organization of WSS – for instance, in the running of public WSS utilities to enhance their overall performance and financial viability. Commodification of WSS refers to the specific case of converting these services into privately owned profit-making activities, where the delivery of essential water supply and sewerage services becomes the means for accumulating private capital. In both commercialization and commodification processes, but particularly in the latter, essential WSS are no longer considered to be a 'public good' that must be available to everyone independently of their capacity to pay and are transformed into a 'private good' or 'commodity'. In practice, the transformation from public to commercial or commodified WSS is far from straightforward, as is clearly illustrated in the case of England's privatized WSS utilities (e.g. Bakker, 2004).

4 There are other important aspects of the unwanted 'mortgage' imposed on many developing countries by failed privatization processes, such as the introduction of 'captive' WSS technologies during the privatization period. These are technologies controlled by the private operators (directly or through their subsidiary companies). In the event of a public takeover of the utility following the cancellation of the privatization contract, the public utility is often 'locked in' to that technology (and, thus, depends upon the technology provider) given the scale of the investments involved. There is emerging evidence of this type of situation from cases such as the failed privatization of Aguas Argentinas in Buenos Aires, which was cancelled in 2006.

References

ADB, WHO, ESCAP and UNDP (Asian Development Bank, World Health Organization, United Nations Economic and Social Commission for Asia and the Pacific and United Nations Development Programme) (2006) *Asia Water Watch 2015: Are Countries in Asia on Track to Meet Target 10 of the Millennium Development Goals?*, ADB, WHO, ESCAP and UNDP, Manila, Geneva, Bangkok and New York

Alexander, J. (2007) 'Environmental sustainability versus profit maximization: Overcoming systemic constraints on implementing normatively preferable alternatives', *Journal of Business Ethics*, vol 76, no 2, pp155–162

Amorebieta, G. (2005) 'Argentina: Worker's co-operative takes over post-Enron', in B. Balanyá, B. Brennan, O. Hoedeman, S. Kishimoto and P. Terhorst (eds) *Reclaiming Public Water: Achievements, Struggles and Visions from Around the World*, Transnational Institute and Corporate Europe Observatory, London, pp149–157

Azpiazu, D. and N. Bonofiglio (2006) 'New and old actors in public services: Capital transfers in the sectors of water and sanitation and electric energy distribution in post-convertibility Argentina' (in Spanish), *Working Documents Series*, no 16, Latin American Faculty of Social Sciences, Buenos Aires

Bakker, K. (2004) *An Uncooperative Commodity: Privatizing Water in England and Wales*, Oxford University Press, Oxford

Balen, M. (2006) *Water for Life: The Case for Private Investment and Management in Developing Country Water Systems*, The Globalization Institute, London

Bath, R. C., J. M. Tanski and R. Villarreal (1998) 'The failure to provide basic services to the colonias of El Paso County: A case of environmental racism?', in D. E. Camacho (ed) *Environmental Injustices, Political Struggles: Race, Class and the Environment*, Duke University Press, Durham and London, pp125–137

Berkeley, D. and J. Springett (2006) 'From rhetoric to reality: A systemic approach to understanding the constraints faced by Health For All initiatives in England', *Social Science and Medicine*, vol 63, pp2877–2889

Berry, K. A. (1998) 'Race for water? Native Americans, Eurocentrism, and Western water policy', in D. E. Camacho (ed) *Environmental Injustices, Political Struggles: Race, Class and the Environment*, Duke University Press, Durham and London, pp101–124

Brook Cowen, P. J. and T. Cowen (1998) 'Deregulated private water supply: A policy option for developing countries', *Cato Journal*, vol 18, no 1, pp21–41

Brown, E., J. Cloke and M. Ali (2008) 'Special Issue: GATS and development – the case of the water sector', *Progress in Development Studies*, vol 8, no 1

Budds, J. and G. McGranahan (2003) 'Are the debates on water privatization missing the point? Experiences from Africa, Asia and Latin America', *Environment and Urbanization*, vol 15, no 2, pp87–113

Castells, M. (1977) 'Equipements collectifs et consommation sociale', *International Journal of Urban and Regional Research*, vol 1, pp101–123

Castro, J. E. (2004a) 'Urban water and the politics of citizenship: The case of the Mexico City Metropolitan Area (1980s–1990s)', *Environment and Planning A*, vol 36, no 2, pp327–346

Castro, J. E. (2004b) 'Final report', in J. E. Castro (ed) *PRINWASS Project*, School of Geography and the Environment, University of Oxford, Oxford

Castro, J. E. and N. Laurie (2004) 'Cross-comparative report on the social, political and cultural dimension', in J. E. Castro (ed) *PRINWASS Project*, School of Geography and the Environment, University of Oxford, Oxford

CCW (Consumer Council for Water) (2006) *Rise in Consumer Debt Calls for Swift Action*, CCW, Birmingham

Cook, P. and C. Kirkpatrick (eds) (1988) *Privatization in Less Developed Countries*, Harvester Wheatsheaf, Hemel Hempstead

Crenzel, E. A. and G. Forte (2004) 'Cross-comparative report on the demo-geographic and socio-economic dimension', in J. E. Castro (ed) *PRINWASS Project*, School of Geography and the Environment, University of Oxford, Oxford

Crespo Flores, C. (2007) *Water Privatization and Environmental Racism*, British Academy Fellowship Report, School of Geography, Politics, and Sociology, Newcastle University, Newcastle-upon-Tyne

Decaillot, M., E. Preteceille, and J. P. Terrail (eds) (1977) *Besoins et Mode de Production*, Editions Sociales, Paris

Dryzek, J. S. (1997) *The Politics of the Earth. Environmental Discourses*, Oxford University Press, Oxford

Economic Perspectives (1997) 'Special issue on privatization', *Economic Perspectives: An International Journal of the US Information Agency*, vol 2, no 1

ESCAP (United Nations Economic and Social Commission for Asia and the Pacific) (2007) *Statistical Yearbook for Asia and the Pacific 2007*, ESCAP, Bangkok

Farley, P. (1997) 'USAID: Setting directions for the next decade of privatization', *Economic Perspectives: An International Journal of the US Information Agency*, vol 2, no 1, pp10–12

Fitch, M. (2006) *Fair and Affordable Water*, UNISON, London

Fitch, M. and H. Price (2002) *Water Poverty in England and Wales*, Centre for Utility Consumer Law and Chartered Institute of Environmental Health, London

Gilbert, A. (2007) 'Water for all: How to combine public management with commercial practice for the benefit of the poor?', *Urban Studies*, vol 44, no 8, pp1559–1579

GWP (Global Water Partnership) (2003) *Effective Water Governance: Learning from the Dialogues*, GWP, Stockholm

Hall, D. (2002) *The Water Multinationals 2002: Financial and Other Problems*, Public Services International Research Unit (PSIRU), University of Greenwich, Greenwich

Hall, D., K. Lanz, E. Lobina and R. de la Motte (2004) 'International context', in D. Hall (ed) *Watertime Project*, Public Services International Research Unit (PSIRU), University of Greenwich, Greenwich

Hassan, J. A. (1998) *A History of Water in Modern England and Wales*, Manchester University Press, Manchester

Heller, L. (2007) 'Basic sanitation in Brazil: Lessons from the past, opportunities from the present, challenges for the future', *Journal of Comparative Social Welfare*, vol 23, no 2, pp141–153

Hobsbawm, E. (1994) *Age of Extremes: The Short Twentieth Century 1914–1991*, Michael Joseph, London

Hukka, J. J. and T. S. Katko (2003) 'Water privatisation revisited: Panacea or pancake?', *Occasional Paper Series*, International Water and Sanitation Centre, Delft, The Netherlands

Isaac, J. (1997) *A Sober Approach to the Water Crisis in the Middle East*, Applied Research Institute, Jerusalem, Bethlehem, West Bank

Juuti, P. S. and T. S. Katko (eds) (2005) *Water, Time and European Cities: History Matters for the Futures*, University of Tampere and Tampere University of Technology, Tampere

Katko, T. S. (1997) *Water! Evolution of Water Supply and Sanitation in Finland from the Mid-1800s to 2000*, Finnish Water and Waste Water Works Association, Helsinki

Kearton, L. (2006) *Debt and Utility Bills*, Welsh Consumer Council, Cardiff

Klein, G. (2003) *Life Lines: The NCC's Agenda for Affordable Energy, Water, and Telephone Services*, National Consumer Council, London

Laski, H. J., W. I. Jennings and W. A. Robson (eds) (1935) *A Century of Municipal Progress 1835–1935*, George Allen and Unwin, London

Laurie, N. (2007) 'Special issue: The privatisation and global poverty debate', *Geoforum*, vol 38, no 5

Laurie, N., S. Radcliffe and R. Andolina (2002) 'The new excluded "indigenous"?: The implications of multi-ethnic policies for water reform in Bolivia', in R. Seider (ed) *Multiculturalism in Latin America: Indigenous Rights, Diversity and Democracy*, Palgrave-Macmillan, Houndmills, Basingstoke and New York, pp252–276

Lee, T. R. and A. Jouravlev (1998) 'Prices, property and markets in water allocation', *Environment and Development Series*, United Nations Economic Commission for Latin America and the Caribbean, Santiago de Chile

Leys, C. (2001) *Market-Driven Politics: Neoliberal Democracy and the Public Interest*, Verso, London

Libiszewsky, S. (1995) 'Water disputes in the Jordan Basin Region and their role in the resolution of the Arab–Israeli conflict', *Environment and Conflicts Project*, Swiss Peace Foundation, Zurich

Martinez Alier, J. (2002) *The Environmentalism of the Poor: A Study of Ecological Conflicts and Valuation*, Edward Elgar, Cheltenham, UK and Northampton, MA

McGowan, F. (1994) 'The internationalization of privatization', in T. Clarke (ed) *International Privatization: Strategies and Practices*, Walter de Gruyter, Berlin and New York, pp25–42

Melosi, M. V. (2000) *The Sanitary City: Urban Infrastructure in America from Colonial Times to the Present (Creating the North American Landscape)*, John Hopkins University Press, Baltimore

Millward, B. (1991) 'Emergence of gas and water monopolies in nineteenth century Britain: Contested markets and public control', in J. Foreman-Peck (ed) *New Perspectives in Late Victorian Economy: Essays in Quantitative Economic History 1860–1914*, Cambridge University Press, Cambridge

Moss, J., G. Wolff, G. Gladden and E. Gutierrez (2003) *Valuing Water for Better Governance: How to Promote Dialogue to Balance Social, Environmental, and Economic Values?*, Suez Business and Industry CEO Panel for Water, Paris

Mukhopadhyay, A. K. (1975) 'The politics of London water', *The London Journal*, vol 1, no 2, pp207–226

OECD (Organisation for Economic Co-operation and Development) (1994) DAC *Orientations for Development Co-operation in Support of Private Sector Development*, OECD, Paris

OECD (1995) *Private Sector Development: A Guide to Donor Support*, OECD, Paris

Ogle, M. (1999) 'Water supply, waste disposal, and the culture of privatism in the mid-nineteenth-century American city', *Journal of Urban History*, vol 25, no 3, pp321–347

Oxfam UK and PWA (Palestinian Water Authority) (2006) *Impact of the Current Crisis in the West Bank and Gaza Strip*, Palestinian Hydrology Group, Ramallah and Jerusalem

PPIAF (Public–Private Infrastructure Advisory Facility, and World Bank (2006) *Approaches to Private Participation in Water Services: A Toolkit*, PPIAF and World Bank, Washington, DC

Richard, B. and T. A. Triche (1994) 'Reducing regulatory barriers to private-sector participation in Latin America's water and sanitation services', *Policy Research Working Papers*, Transport, Water, and Urban Development Department, Water and Sanitation Division, World Bank, Washington, DC

Rocha Ferreira, H. (2006) *Water Rationing in the 'Special Service' Areas of Recife: The System of Rotation vs Egalitarian Access, Final Report: Specialization in Water Resources Management*, Federal University of Santa Catarina and Federal University of Alagoas, Florianópolis, Santa Catarina, Brazil

Roth, G. (1988) *The Private Provision of Public Services in Developing Countries*, World Bank and Oxford University Press, Washington, DC

Schultz, S. K. and C. McShane (1978) 'To engineer the metropolis: Sewers, sanitation, and city planning in late-nineteenth century America', *The Journal of American History*, vol 65, no 2, pp389–411

Stiglitz, J. E. (2002) *Globalization and its Discontents*, Penguin, London

Swyngedouw, E. (2004) *Social Power and the Urbanization of Water: Flows of Power*, Oxford University Press, Oxford

Swyngedouw, E. (2005) 'Dispossessing H2O: The contested terrain of water privatization', *Capitalism Nature Socialism*, vol 16, no 1, pp81–98

Swyngedouw, E., M. Kaïka and J. E. Castro (2002) 'Urban water: A political-ecology perspective', *Built Environment*, vol 28, no 2, pp124–137

Taylor, G. (1999) *State Regulation and the Politics of Public Service: The Case of the Water Industry*, Mansell, London and New York

The Prince of Orange, Willem-Alexander (2006) 'Keynote address', *Opening Ceremony, Fourth World Water Forum*, Mexico City, www.koninklijkhuis.nl/english/content.jsp?objectid=14597, accessed August 2008

Triche, T. A. (1990) 'Private participation in the delivery of Guinea's water supply services', *Working Papers*, Infrastructure and Urban Development Programme, World Bank, Washington, DC

UKP (United Kingdom Parliament) (2003) *Memorandum Submitted by the Public Utilities Access Forum*, UKP, London

UN (United Nations) (2000) *Millennium Declaration*, United Nations, New York

UN (2002a) 'Key commitments, targets and timetables from the Johannesburg Plan of Implementation', *World Summit on Sustainable Development*, UN, Johannesburg

UN (2002b) *The Right to Water: Articles 11 and 12 of the International Covenant on Economic, Social and Cultural Rights*, United Nations, New York

UN (2007) *The Millennium Development Goals Report 2007*, United Nations, New York

UN-Habitat (2003) *Water and Sanitation in the World's Cities: Local Action for Global Goals*, Earthscan, London

UN-Habitat (2006) *Meeting Development Goals in Small Urban Centres. Water and Sanitation in the World's Cities 2006*, Earthscan, London

UNESCO–WWAP (United Nations Educational, Scientific and Cultural Organization – World Water Assessment Programme) (2006) *Water, a Shared Responsibility: The United Nations World Water Report 2*, UNESCO and Berghahn Books, Paris and New York

UNICEF and WHO (United Nations Children's Fund and World Health Organization) (2008) *Progress on Drinking Water and Sanitation: Special Focus on Sanitation*, Joint Monitoring Programme for Water Supply and Sanitation (JMP), UNICEF and WHO, New York and Geneva

UNMP (United Nations Millennium Project) (2005) *Final Report on Health, Dignity and Development: What Will It Take?*, Task Force on Water and Sanitation, Swedish Water House, Stockholm International Water Institute, Stockholm

Wade, R. H. (2004) 'Is globalization reducing poverty and inequality?', *World Development*, vol 32, no 4, pp567–589

Ward, C. (1997) *Reflected in Water: A Crisis of Social Responsibility*, Cassell, London

Warner, B. (1987) *The Private City: Philadelphia in Three Periods of its Growth*, University of Pennsylvania Press, Philadelphia

Water UK (2005) *Household Debt in the Water Industry: A Water UK Briefing Paper*, Water UK, London

Webb, P. and M. Iskandarani (1998) 'Water insecurity and the poor', *Discussion Papers on Development Policy*, Centre for Development Research, Bonn

Wescoat Jr, J. L., L. Headington and R. Theobald (2007) 'Water and poverty in the United States', *Geoforum*, vol 38, no 5, pp801–814

Whiteford, S. and A. Cortez Lara (2005) 'Good to the last drop: The political ecology of water and health on the border', in L. Whiteford and S. Whiteford (eds) *Globalization, Water and Health: Resource Management in Times of Scarcity*, School of American Research Press and James Currey, Santa Fe, CA, and Oxford, pp231–254

WHO (World Health Organization) (2003) *The Right to Water*, WHO, Geneva

WHO and UNICEF (United Nations Children's Fund) (2006) *Meeting the MDG Drinking Water and Sanitation Target: The Urban and Rural Challenge of the Decade*, WHO and UNICEF, Geneva and New York

World Bank (1998) *Facilitating Private Involvement in Infrastructure: An Action Programme*, World Bank, Washington, DC

World Bank (2003) *World Development Report 2004: Making Services Work for Poor People*, World Bank, Washington, DC

WSP and PPIAF (Water and Sanitation Programme and Public–Private Infrastructure Advisory Facility) (2002) *New Designs for Water and Sanitation Transactions: Making Private Sector Participation Work for the Poor*, World Bank, Washington, DC

2

Troubled Waters: The Political Economy of Essential Public Services

Erik Swyngedouw

The urban water conundrum

To allow the market mechanism to be the sole director of the fate of human beings and their natural environment, indeed, even of the amount and use of purchasing power, would result in a demolition of society… Nature would be reduced to its elements, neighbourhoods and landscapes defiled, rivers polluted, military safety jeopardized, the power to produce food and raw materials destroyed. (Polanyi, 1954, p73)

I would like to have a brief look back at 2002 and 2003, whose disastrous results shook us to the core: the overly hasty expansion of water internationally ended in failures that were painful for all of us; ONDEO and SITA's acquisitions of companies that should have been sources of growth instead generated losses or were a cause of concern. We were forced to pull out of unprofitable projects (Puerto Rico, Atlanta, etc.) and to sell part or all of companies such as Northumbrian and Cespa, whose development we were no longer able to finance. This sorely tried our nevertheless proved business models and our certainties. (Chaussade, 2004)

With exactly 50 years between them and from radically different political positions, the above two quotes have an uncanny semblance. Jean-Paul Chaussade, executive vice-president of Suez, one of the world's leading water and waste companies, admitted that the privatization of public water and sanitation services had not yielded the expected high returns. Despite their 'proved business model', Suez did not easily succeed in turning water into profits. Chaussade repeated this claim at the World Water Forum in Mexico in March 2006, when World Bank representatives stubbornly insisted that the world's urban water problems had to be tackled through market-led strategies and private-sector investment. Indeed, as Odin Knudsen (2003) from the World Bank Environment Department insists:

> *Billions of dollars are needed [in order to provide clean water to the poor]. The public sector, civil society and multilateral financial institutions do not have the funds to meet the investment needs... The private sector needs to be involved.*

John F. Talbot, chair and chief executive officer (CEO) of SAUR, the world's fourth largest water company, also concluded in a speech to the World Bank in 2002 that the international water business is not an attractive business in light of 'a reduction in grants and subsidies, an often premature or unrealistic emphasis on concession contracts and full divestiture, and a belief that any business must be good business and that the private sector has unlimited funds' (Talbot, 2002). He further insisted that private water investment in the developing world had unrealistic expectations because of 'increased country risk, increased financial risk, increased contractual risk, unreasonable contractual constraints and unreasonable regulator power and involvement' (Talbot, 2002). Moreover, he argued that there is an emphasis on 'unrealistic service levels' that expected 'highly stringent water quality standards, attempts to apply European standards in developing countries, and demands "connections for all"'. All of this leads to 'overburdened private balance sheets, few new contracts, poor and diminishing returns for private investors, contract and even corporate failures, limited interest in the market, and investors turning to other, more lucrative, markets' (Talbot, 2002; see also Goldman, 2007).

This seems to be the world topsy-turvy. International and national governmental agencies insist on the market and the private sector as the main conduit to cure the world water's woes, while key private-sector representatives retort that, despite great willingness to invest if the profit prospects are right, they cannot and will not take charge; the profits are just not forthcoming, the risks too high, civil societies too demanding, political uncertainties too high (Vidal, 2006), contractual obligations too stringent, and subsidies have often been outlawed.

In addition to the rather limited success (as well as volume) of private-sector participation in the water sector, many of the publicly owned urban water service delivery organizations have been corporatized. While still under state ownership, many public service companies are increasingly forced to operate in a ring-fenced manner as autonomous, self-financing organizations that operate according to market logic. In operational terms, there is little, if any, difference between these two types of ownership. They share the same constraints and are subject to comparable conditions and dynamics. This has blurred the distinction between public or private ownership and management of water service companies. As Loftus (2004, 2006) has demonstrated for the case of Durban, public companies that are institutionally and legally forced to act as private corporations (i.e. operate in a competitive and profit-driven environment) behave, to a large extent, in ways that are comparable to private companies. In other words, the key problem with respect to urban water services is not the commodification of water (in fact, water has been sold as a commodity for a very long time) or their public versus private character, but rather the corporatization of water service delivery companies and the imposed requirement for profitability and 'full-cost' recovery. The latter has dramatic consequences for the ability of companies (public and private) to extend urban networks and improve connections to the poorer segments of the urban population. Karl Polanyi and successive generations of political economists had already shown a long time ago that a corporatized liberal market economy, left to its own devices, will ultimately self-destruct socially and environmentally, and cannot possibly deliver social services without proper embedding within a responsive public or social financial and regulatory infrastructure (see Castree, 2007a, 2007b; Heynen et al, 2007).

This observation leads to a paradoxical conclusion. While international organizations and national governments insist that market-led solutions to public-service organization and delivery will solve some of the most enduring and intractable socio-ecological problems of our times, the private-sector protagonists that are called upon to deliver on this promise have to admit defeat in the light of mounting evidence that delivering water to the poor or the unconnected remains excruciatingly slow (Pitman, 2002; Winpenny, 2003). Under a hegemonic frame that considers subsidizing socio-ecological processes (such as safe and reliable water supply) or transferring value from one place, social group, activity and/or sector to another that is economically ineffective and politically unacceptable, providing access to water to poor communities will remain a pipe dream. This chapter seeks to unravel some of the broader and deeply contradictory dynamics associated with the feeble attempts to improve urban water services worldwide – in particular, in the global South. Despite the promises of market-led solutions, the results have been less than encouraging, while no alternative to private-sector involvement is currently on offer. We shall insist that the public character of water services defies corporatization and requires a socialized response to deliver water effectively. We shall concentrate on the contradictions between corporatized provision (either by publicly or privately owned operators) of water and the collective or social nature of water delivery services.

URBAN WATER TROUBLES

While the Millennium Development Goals (MDGs) insist that the world's elites wish to see the number of people without improved access to water and sanitation halved by 2015 and maintain that the private sector is the preferred vehicle for achieving this, the urban water problem remains one of the most pressing socio-environmental issues of our time: 960 million urban dwellers do not have adequate access to potable water and over 1 billion lack proper sanitation services, resulting in water-related diseases being the number one cause of premature mortality in the developing world (UNDP, 2006). Hall and Lobina (2006, pp37–38) estimate that over the 1990 to 2005 period, the 'golden' age of water privatization, only 600,000 households received new water connections in cities under private water management, a number that falls to 250,000 if public subsidies and investment are excluded. Moreover, privatization has clearly failed in many cities around the world and is falling apart or faced with serious difficulties in others (Hall et al, 2004). While private-sector participation in the water sector was growing rapidly during the 1990s (from two projects in 1989 to a peak of 37 in 1999), it has dwindled to a mere trickle from 2001/2002 onwards (with the exception of China). Private investment in the water and sanitation sector almost halved in the first five years of the new millennium compared with the 1995 to 1999 period (World Bank, 2007). Only Chinese cities remain an attractive option for international investors – not surprisingly, one that is embedded in a strongly authoritarian and interventionist state that organizes all manner of direct and indirect subsidies and capital transfers from the public to the private sector (Loong Yu and Danqing, 2006; ICF International, 2007). By 2005, 109 of 383 World Bank-listed water and sewerage privatization projects were in China, while 44 of these 383 projects were cancelled or are considered to be in distress, accounting for 32.9 per cent of the total investment of US$50 billion between 1991 and 2005 (World Bank, 2007). There is, by now, a long and growing list of cities who saw their privatization experiments fail or their contracts in serious distress, including Buenos Aires, Atlanta, Belize City, Manila, Cochabamba, Jakarta, Nelspruit, Kelantan, Mozambique, Nkokebde, Conakry, Gambia, Panama, Paris, Wales, La Paz, Trinidad and Tobago, and Dar es Salaam (Castro, 2004; Prasad, 2006, pp682–683), while others face all manner of institutional, contractual or delivery difficulties (see Hall and Lobina, 2007).

Many cities in the developing world are now confronted with a double bind. Governmental neoliberal policies, often imposed or re-enforced by international organizations and donors, have turned to private capital for securing and upgrading urban social and physical infrastructures and a private sector that is faced with mounting problems. The market as a panacea for solving key problems seems to fail, while the public sector is still considered to be too incompetent, under-resourced and/or overburdened to be able or to be allowed to confront the situation. The untested assumption of state failure in service provisioning still haunts while the private sector strategically withdraws. There is an urgent need, therefore, to assess the contradictions of the current phase of neo-liberalizing water urbanization and the struggle over either replacing or restoring, via a series of chirurgical operations, the Washington Consensus, on the one hand (Mehrota and Delamonica, 2005), or radically re-politicizing urban socio-ecological processes, on the other. The inevitable conclusion of this experience has to be that clean water and adequate sanitation to the poor cannot be provided on market terms, but do require a public or socialized response, such as, for example, systematic and structural support or subsidies. In short, and this is what we argue throughout this chapter, sustained transfers of financial means from the better-off (socially and/or geographically) to the worse-off is a key precondition for alleviating the socio-ecological nightmare of the billion urban dwellers that lack adequate water and sanitation services.

THE TROUBLE WITH WATER FLOWS: FROM COLLECTIVE TO PRIVATE (AND BACK AGAIN?)

Water is a territorial 'flow' good that is embedded in a complex hydro-social arrangement (Swyngedouw et al, 2002; Swyngedouw, 2004). The fusion of human and non-human processes in the production, conduit system and consumption of water turns the terrestrial part of hydro-social circulation into a complex network (Kaika, 2005). Hydro-social circulation is a heterogeneously constituted networked good, which does not lend itself easily to homogenization, marketization, commoditization and corporatization. In other words, water, as Karen Bakker (2003) contends, is 'an uncooperative commodity'. It is bulky, non-substitutable, heavy, socially and economically contested, monopolistic and requires long-term fixed investment. While inevitably territorially organized and channelled through socio-technical conduits, its vital use resides not only in its particular 'locationality', but also in its flow character and constant socio-physical transformation and metabolism (Swyngedouw, 2006). It is this material territorial flow character of water that renders it particularly difficult to monopolize privately and turned into a commodity for sale with profit. While flowing, water acquires all manner of different, and often contradictory, meanings: it absorbs value as it is treated, purified and transported; it is subject to competing demands, varying from its use as production input in agriculture and industry to fulfilling basic physiological and metabolic needs; and ecological conditions and processes are affected by its flow. It may serve different purposes at once and be required to perform several functions throughout its hydro-social cycle (Bakker, 2003, p32). This indivisible and combined complex good character of the uses and functions of territorial water flows defy easy commercialization, while the functional multiplicity of its territorial-networked flow produce bundled (i.e. socialized) use values that are structured and made possible exactly because of its specific territorial organization. The 'collective' or 'social' (i.e. public) character of water resides, therefore, not solely in its vital necessity to sustain life, but in its complex and non-exclusive character, roles and purposes.

These territorial effects derive from the bundled, networked, multiple and coordinated arrangement of the flow of the resource are exactly what, in classical economics, is referred to as (positive or negative) externalities, but what I have called earlier 'territorial organization' effects (Swyngedouw, 1992). The latter concept permits understanding such effects not purely as 'external' to the economic process, but as an integral part of the bundled socio-economic territorial structure of the good; a structure that cannot be easily unbundled, the 'externalities' identified and isolated, and made subject to market forces (i.e. the market internalization of external effects). Urban water and sanitation networks are complex and bundled 'collective goods' that embed public and/or private fixed collective capital and constitute the collective equipment necessary for sustaining the economic production and circulation process (i.e. it is a collective means of production), as well as social reproduction (i.e. it functions as a collective means of reproduction) (Läpple, 1976; Perrat, 1987). Such an organizational pattern results in a highly creative and 'productive' ensemble, external to the individual agent or actor, but internal to the macro socio-spatial functioning in which the ensemble is embedded. This is a classic example of joint production and consumption (see Marshall, 1920) where the socialized or collective character of the network produces exactly the territorial effects referred to above. In value terms, the (positive or negative) value derived from the existence of the network is over and beyond the values that can be internalized through the price mechanism in each of its constituting points (Swyngedouw, 1992, pp421–422). Consider, for example, the quasi-impossibility to internalize the economic effect of improved sanitation (declining mortality, reduced health costs, improved productivity, changing gender roles and relations, etc.) within the water price mechanism. As Preteceille (1977, 1981) argued, bundled spatial or territorial organization produces:

- a set of use values that do not have a commodity character (or are difficult to commodify); and which
- constitute an indivisible and, hence, complex use value; and
- are essential preconditions for the 'development' of social relations, the socialization of individual and social groups, and the development of the productive forces; while
- the fixed capital investment required for the production of these complex territorial use values is often high and average turnover time very slow.

It may be useful in this context to retrace some of the arguments mobilized in early critical urban political economy that centred around the tensions between the need to build, maintain and expand collective means of (re)production, on the one hand, and the private appropriation of the positive value effects of such infrastructures, on the other (see Castells and Godard, 1974; Lojkine, 1974, 1977; Castells, 1975, 1976; Läpple, 1976; Lipietz, 1980). As Lipietz argued more than two decades ago, such socio-naturally produced territorial networks constitute the general rule of the urbanization process:

> *It should be emphasized from the outset that as soon as space intervenes, 'externalities' (of 'location' within the branch or of 'urbanization' external to the branch) become the general rule... This is where the contradiction between the social character of production and the private character of appropriation and economic ownership is expressed at its sharpest in its material form... As soon as fixed capital is installed, the field of distances and socio-economic space is irrevocably transformed: by materializing, private capital becomes collective fixed capital.* (Lipietz, 1980, pp71–72; see also Lipietz, 1983).

It is these contradictions that constituted the pivot of urban analysis around collective means of production and reproduction and the new social movements crystallizing around these issues (Castells, 1976). The argument rehearsed some of the key processes specific to Fordist/Keynesian urbanization and the tense relations between the provision of collective infrastructures, the city, the state and private capital. These contradictions, however, did not go away with the turn to neo liberal forms of urban service governing. On the contrary, they became more acute and accentuated in their present neoliberal form.

There is, indeed, a fundamental tension between the collective character of the circulation of socio-natural goods such as water and the private appropriation of the profits or surplus. It is captured centrally by the Marxist view of the contradiction between the increasing social character of the relations of production and reproduction, on the one hand, and the private organization of the accumulation process and appropriation of surplus, on the other. At the height of Fordist urbanization, this tension was mediated through the particular form of the state that took greater charge of ensuring the provision of the collective means of production and reproduction. Infrastructure of all kinds, social housing, health and recreation facilities, and all manner of social services were collectively organized, often debt financed, taken out of the private circulation of capital, and organized by public capital supervised and invested in by the state. It was exactly this collective provision of the means of production and reproduction that facilitated and supported the accumulation of capital and surplus production in other spheres. Although this state support could take the form of state ownership of collective means of production and reproduction, it also often took the form of particular public–private partnerships in which the state covered a substantial part of the investment or operational costs, while private-sector participants were guaranteed profitable returns (as is still the case of most of Europe's water and sanitation services). This proved to be an extraordinarily successful enterprise if one goes by the quality and comprehensiveness of all manner of collective infrastructures provided in the global North during the 20th century, based on private and profitable provision (and, in some countries, operation) of water infrastructure and public subsidies (see Graham and Marvin, 2001).

Of course, there were a series of often conflicting processes at work in the collective provision of these public (re)production services. First, the state invested in collective means of production, thereby reducing fixed capital investment requirements for private capital, lowering the organic composition of capital and, consequently, increasing surplus value production and general profitability. Castells and Godard's (1974) study of the steel complexes in Dunkirk became a classic example of such analysis. Second, the state as a collective investor could mobilize (primarily through tax transfers) sufficiently large volumes of capital (outside the profit-making logic) that private capital was unable or unwilling to provide (Lojkine, 1977). Third, the state marshalled judicial and legal powers to expropriate land and to control water resources and organize them in ways conducive to the provision of collective spatial infrastructures (Dunford, 1988). Fourth, investment in collective means of production and reproduction fuelled significant demand for capital goods from other sectors of the private capitalist economy and constituted a core part of state-led Keynesian demand management. Fifth, the provision of collective consumption goods at low or subsidized prices (such as housing, education, health, water, and the like) reduced pressure on wages and mitigated class struggle and conflict over working conditions and wages. The state ensured or guaranteed the basic collective reproduction of the workforce (Castells, 1976). And, finally, social struggle became increasingly mobilized around these collective means of production and reproduction, leading to new 'urban' social movements that cut across traditional class lines. The overall outcome was a corporatist public–private interaction that was characterized by the more or less generalized public provision of collective services and the private appropriation of the resulting positive socio-spatial externalities. In fact, this Fordist/Keynesian urban governance arrangement maintained social cohesion while ensuring relatively successful accumulation of private capital that was still largely organized on a national scale. This form of arranging the collective/private interplay and tension became a generalized model for urban governance, both in the capitalist global North as in the global South, although the latter often failed to achieve the same level of intervention compared with states in the global North.

With the crisis of this model – largely as a result of mounting fiscal problems for the local and/or national state, accelerating accumulation problems for private capital during the 1970s and 1980s, the 'failure' of the state in both the global North and South, and the rise of neoliberal hegemony – scholarly attention to this form of analysis waned as well. Of course, the contradiction between the social character of production and the private organization of the production process did not wither away. On the contrary, the neo-liberalization of urban governing in the last decades, combined with the rise of new collective social problems such as the ecological crisis, accentuated this contradiction as reproduction services became increasingly subject to market forces (Moulaert et al, 2002). Debt financing or cross-subsidization of services by the state became taboo as the market was considered to be a more effective and efficient service provider and allocating instrument. Partial or full privatizations taking place in the global North, together with state divestitures in the former Soviet Union and its satellite states, China and the global South, offered a vast terrain of activities for capital investment and the absorption of excess liquidity. During the 1990s, China became the hothouse of the world economy that attracted vast amounts of transnational capital for investment in urban infrastructures of a variety of kinds, among other investments, but with massive overcapacity and over-accumulation problems as a consequence (Duménil and Lévy, 2004; Bello, 2006).

The success of this 'accumulation by disposession' (Harvey, 2003) in the water sector was less than encouraging, both in terms of solving access problems for the poor and in terms of profitability for the water companies. These forms of organization generated new tensions. First, those dispossessed do not necessarily passively accept the theft of what they consider to be rightfully theirs. Second, once under the aegis of private capital accumulation, all manner of social tensions and conflicts arise. Predating competitors loom around the corner, recalcitrant workers raise the spectre of new forms of class struggle, disgruntled consumers mobilize the weapons of the weak when it becomes clear that the initial promises fail to materialize, and governments, for a variety of reasons, embed the private operators in a complex web of regulatory arrangements and contractual obligations. Third, the network character and the complex investment arrangements required to maintain, update and expand the network (particularly to poorer and more difficult to access areas, and with frequently unreliable policing of rate payment) and water production in a context of still problematic economic returns lowered profit expectations, particularly as subsidies or other forms of aid were ideologically considered unacceptable. This intensified the contradictions between the complex social and collective character of the territorial hydro-social flow and the private appropriation of surplus. In sum, by the early years of the 21st century, accumulation by dispossession of the kind pursued by global water companies was beginning to falter as the internal contradictions related to the provision of collective means of production and reproduction were exacerbated. Fourth, the financialization of the global economy unleashed a series of rampant but regionally specific regional crises, particularly in South-East Asia in the fall of 1998 and then spreading to Latin America, with Argentina as one of the worst hit economies. The problems with the Manila (see Lobina, 2005) and Jakarta (see Bakker, 2007) water contracts and the collapse of the Buenos Aires concession (Castro, 2007) were the direct result of the financial turmoil and associated currency devaluation (which ruined the US dollar-denominated spreadsheets of global corporations whose return was dependent upon domestic currency). Fifth, and to make matters worse, a socio-political onslaught against unbridled globalization and privatization had begun to spread around the world, from President Bush, who pursued a much more protectionist economic domestic stance than Clinton, and Uruguay, where the population voted to ban constitutionally the privatization of natural resources such as water, to water companies themselves becoming more reluctant to fulfil, without public financial support, the mission that international organizations had outlined for them. A dilemma of extraordinary proportions arises here, one that is replete with new and possibly explosive social, political and economic tensions. And this is what we turn to next.

RETOOLING THE WASHINGTON CONSENSUS: SUBSIDIZING CAPITAL?

According to Agenda 21, the report of the 1992 Rio Conference on Environment and Development, the estimated annual additional investment cost needed to achieve global water security was US$56 billion (see Cosgrove and Rijsberman, 2000). However, more recent estimates suggest that a much more significant effort is needed. The *World Water Vision* report (Cosgrove and Rijsberman, 2000) estimates that in the run-up to 2025, US$180 billion annually is required to achieve good water access for all (Cosgrove and Rijsberman, 2000). This includes an investment of US$550 billion in dams and irrigation schemes to feed the world's growing population, assuming a 40 per cent increase in world food production. The report also assumes a 1.5 billion increase in population, half of whom will live in cities. Combined with existing deficiencies (1.5 billion) and ongoing rural-to-urban migration (0.5 billion), this would bring the total of people who need to be serviced to 3 billion. In addition to that, industrial water use will expand, while urgent infrastructure replacement investment is required in the developed world and in the former socialist states. The summary of total annual investment needed up to 2025 (and compared to the actual situation in 1995) and their expected sources is provided in Table 2. 1.

In addition to the quite staggering magnitude of the investment required, 70 per cent of the total is expected to be raised by the private sector. The World Water Council asserts that private actors can thus provide the main source of infrastructure investment, and the World Bank endorsed this conclusion, which furthered their push to privatization as the main means by which to elicit private-sector participation (Pitman, 2002; Amann-Blake, 2004; Gleick, 2004; Gleick and Cooley, 2006). In light of the above argument, direct investment by the private sector in water and sanitation services is unlikely to be abundant or even forthcoming under the current configuration. Thus, the key challenge for the world's water elites becomes one of mobilizing public resources to shore up private profits. Therefore, retooling the Washington Consensus is very much on the agenda, and in such a way that the global neoliberal agenda is not fundamentally challenged.

Table 2.1 *Annual investment requirements for water resources and anticipated sources of investment*

Use	*Billions of US$* *1995*	*Vision 2025*	*Share (%)* *1995*	*Vision 2025*
Agriculture	30–35	30	43–50	17
Environment and Industry	10–15	75	13–21	41
Water Supply and Sanitation	30	75	38–43	41
Total	70–80	180	100	100
Source				
National				
Public Sector	45–50	30[a]	58–71	25
Private Firms (domestic)	12–15	90	15–21	45
International				
Private Investors	4	48	5–6	24
Donors	9	12	12–13	6
Total	70–80	180	100	100

Note: [a] This figure assumes an additional US$20 billion in direct subsidies to the poor.
Source: Cosgrove and Rijsberman (2000)

Two main strategies are actively pursued at this moment: first, a movement away from direct control through divestiture to the financialization of public services, keeping global investment flows at arm's length from the actual process of service delivery, but still ensuring its operation in conformity with market logic. The second strategy (and this seems to be the only option available to cities in the global South) is to complement private investment with all manner of subsidies and public support mechanisms or, in other words, the part restoration of an increasingly global Keynesian policy under an overall liberal agenda. We will briefly consider each of these strategies.

Option 1: From direct control to financialization

The world's capital markets are awash with liquidity. The International Monetary Fund (IMF) estimated that global capital assets mid-2007 amounted to circa US$76 trillion (Orłowski, 2008). There is no shortage of money in capital markets (although investors may be reluctant to lend it during periods of intense uncertainty, as exemplified by the financial crisis unleashed by the sub-prime mortgage market collapse and resulting in the US 'credit crunch' during the summer of 2007) and an increasing share of the world's liquidity is channelled through all manner of public and private infrastructures (such as real estate, roads, ports, railroads and public services infrastructures). Torrance (2006, p15) defines financialization in this context 'as the expansion of the nature and scope of financial markets and institutions to include the provision of urban infrastructures. It involves the continuous assessment of activities by financial markets' (see also Orléan, 1999, in Theurillat et al, 2006). The Organisation for Economic Co-operation and Development (OECD) considers this financialization of urban infrastructure to be a major potential investment outlet for all manner of investment funds:

> *Investments needed in infrastructure are extremely large, in part because a significant share of this infrastructure has suffered from benign neglect in the past (for example, water infrastructure), in part because of the large transformation expected in developing countries in the future (growth in population, growth in per capita income, rapid urbanization) and, in part, because of new demands that will be put on such infrastructure for security reasons and in response to growing concerns about the environment.* (OECD, 2006, p41).

Macquarie, the Australian investment company, for example, invested in a series of water and sewerage service activities (e.g. it acquired Thames Water plc, the water and sewerage company that services the London metropolitan region, from the German multi-utility provider RWE in 2006) and has become one of the leading infrastructure investment funds. As Torrance (2006) has documented comprehensively, pension and investment funds consider urban infrastructures as potentially attractive to secure long-term returns, diversify risk and generate new investment opportunities while maintaining a relatively flexible and balanced investment mix. The combination of lacklustre public investment in urban infrastructure, the rise of investment liquidity, and the stock market turbulence of the late 1990s that prompted a feverish search for alternative long-term investments made infrastructure an interesting option for investment fund strategists, a tendency that did not go unnoticed by the major international agencies, such as the OECD and World Bank, who quickly identified such strategies as potential sources to deal with all manner of public service delivery problems in the developing world.

However, private fund infrastructure investment depends crucially upon secure long-term return, a stable regulatory environment, strictly managed risk, efficient capital markets and reliable contractual arrangements (Torrance, 2006). Consequently, infrastructure asset investment has been almost exclusively restricted to advanced capitalist economies with solid and conducive regulatory environments. While divestiture and direct investment in water and sanitation services at least ensured a commitment from the investing company to manage in a market-effective manner, investment funds will only enter long-term commitments under conditions that the operational efficiency of the infrastructure can be ensured or the potential risks can be managed effectively and in a market-conforming manner. This all but rules out this option for enhancing service delivery for poor or unconnected parts of the urban population in the global South. All evidence suggests that this cannot easily be done in a profitable manner without subsidies or other forms of financial support or transfers, and depends vitally upon large initial state (or transnational) subsidies or other forms of redistribution. In other words, the financialization of urban water infrastructure is dependent upon the prior organization of water and sanitation delivery services in ways that guarantee relatively stable long-term profits. The financialization strategy as a solution for the water delivery crisis, therefore, requires opening up significant subsidized capital flows to enhance and expand existing infrastructure.

Option 2: Subsidizing private capital

A sustained private-sector involvement in water and sanitation services, therefore, requires opening up significant capital flows from the public to the private sector or, in other words, the subsidization of private-sector profits. This is, in fact, the tactic that the key elite water-sector leaders are pursuing with renewed energy. A recent report by the World Panel on Financing Water Infrastructure (the Camdessus report), a consortium of big water companies and leading international organizations, unashamedly called for greater public-sector involvement and direct or indirect subsidization. The report concluded that 'multilateral financial institutions will be the pillars of the new water financial architecture. They should do everything to reverse the recent decline in their water lending and make every effort to expand their use of guarantees and insurance' (Winpenny, 2003). The latter would include establishing a 'Devaluation Liquidity Backstop Facility'. This rather fancy name refers to the establishment of an international public body that would 'effectively guarantee the foreign loans and finance the additional debt service incurred from devaluation to be reimbursed by the authority responsible for setting the tariffs' (Winpenny, 2003). In straightforward language, as Amann-Blake (2004, 2006) argues, this would *de facto* mean that the international public sector would carry the risk of private investment and would recuperate potential losses from taxation in case devaluation occurs. In other words, the public would carry the brunt of unfavourable national and international political economic conditions; this will hardly improve the situation of the poor and disempowered. As Amann-Blake continues: 'through a review of the historical record [recent in case of the developing world; much longer in case of the developed world], we see that infrastructure … was not primarily financed by the private sector and therefore questions why this would be different in developing countries today as they face rapid urbanization, unstable economies and population growth' (Amann-Blake, 2006, p16).

The extraordinary expansion of capital liquidity in the market is faced with a condition that not all activities requiring substantial investments are profitable in the short to medium term, yet require urgent attention. As in the past, in such cases capital turns to the public sector, expecting sufficient financial support. However, the ideological stupor of the unreconstructed Washington Consensus does not permit an easy retooling of the public–private interface. Still, as exemplified above, water companies are actively positioning themselves with the key national and international agencies in an attempt to pave the way for a change of policy that would simultaneously improve profitability and ensure the companies' continuing expansion, but based on significant public subsidies through some form of global Keynesianism. The European Union's Water Initiative is a case in point (see www.corporateeurope.org/water/infobrief6.htm). In sum, the uncertain financing of urban water supply services opens up a new set of contradictions and a new front of social struggle. These will centre on the question of who will pay for the necessary investments, on the one hand, and who will receive these transfers. As the contradictions between the necessity of providing territorial networked complex commodities, on the one hand, and the inability of either the private sector to profitably do so or the state to effectively do so in the global South, on the other, intensify, the pressures to provide subsidized but privately managed public services will undoubtedly increase.

Irrespective of the outcome of the above struggle, the terms of the argument will remain decisively linked to questions of commodification, market efficiency, investment returns, and the articulation of the public–private interface. We conclude this contribution by arguing for the need to think 'out of the water box': to approach the water problematic in theoretically new and politically innovative manners.

Thinking out of the water box: Social power, water and the mythical debate over commodification

As discussed above, recent experiences with water privatization experiments have shown that turning water services into profitable and socially acceptable businesses is not an easy task. Moreover, demands for full-cost recovery of water-related activities reduce the possibilities for cross-financing and cross-subsidization. The very term 'full-cost recovery' is an oxymoron. It is self-evident that all investment project costs need to be recovered by someone somewhere. The key question is really a political one – that is, who will be responsible for the recovery of what kind of costs. When full-cost recovery is discussed in the context of water projects, it invariably refers to the view that water projects should be self-sufficient (i.e. that the cost of investment should be met fully through water rates – that cost recovery is organized via water consumers). This limits the possibilities of cross-subsidization to managing the tariff structure of water delivery in a redistributive manner. This, in turn, precludes the financing of projects from local, regional or national tax revenues or, through development aid, from tax revenues raised elsewhere. However, this has been the only way through which successful development of large-scale water works was achieved in the past in the global North, particularly in terms of solving the contradiction between the collective character of the system and the private organization of its management. There is no evidence that this will be any different in the developing world. Mobilizing tax revenues permits mobilizing resources obtained from elsewhere or from other activities into collectively more desirable ones. Therefore, the narrow definition of full-cost recovery needs to be replaced by a much wider social and political-economic understanding – one that permits systemic forms of redistribution of financial resources.

In sum, questions of investment in collective commodities such as water are never independent of the question of (re)distribution. To the extent that the water economy is publicly or privately organized (or a mixture of both), these modalities of redistribution will be organized differently. The pivotal social and political struggles for the years to come will exactly revolve around the modalities of subsidization. In other words, the struggle over mediating the tension between providing bundled territorialized and socialized hydro-social networks, on the one hand, and the private appropriation of surplus value, on the other, will be the pivot around which the social struggle over the construction of hydro-social infrastructure and access to water will be fought. While the private sector effectively claims that private-sector participation will be dependent upon public financial support, there is increasing pressure on public institutions to sustain private-sector investment by means of significant public financial support. While we have argued above that uneven access to water is primarily a question of economic or monetary power, achieving the Millennium Development Goals for water necessarily implies a major redistribution of capital resources. Guaranteeing access to clean and safe water requires the transfer of considerable amounts of investment capital whose return will have to be carried by the more wealthy sections of the world's population. This is independent of the question of whether the actual management of water supply and delivery should be publicly or privately organized. The latter question is one of efficient management. Around the world, both public and private (or mixed) companies have proven that they can be effective and efficient. However, the public–private debate should not overshadow (as it has done over the past decades) the question of the origin of the required investments to secure access to water. The private sector, because of the structural requirement for a normal return (profit) on investment, cannot guarantee access to water to social groups with insufficient effective buying power (or, in some cases, willingness to pay) or investment in projects of an uncertain return. The only strategy that can offer a mass solution is one based on subsidies and, thus, on redistribution of capital and income. Moreover, a public organization of investment and of distribution permits consideration of a much wider range of technological, organization and managerial options.

The key issue, therefore, is not about whether or not water is (or should be) a commodity or commoditized. Water is a commodity to the extent that delivering the right volume of water of the right quality to the right place requires major investments of capital and labour, and these have to be made available and paid for. The central concern is who will pay for what part of the hydro-social circulation process. Adequate and reliable access of water for those who lack access will require a major transfer of capital and systematic and sustained cross-subsidization. It is exactly the recognition of water as a commodity that permits effective cross-subsidization. However, the question of subsidization is necessarily a political one that needs to be addressed at local, national and transnational levels. Cross-subsidization of investment requires embedding issues of water access and distribution within appropriate institutional frameworks that discuss, democratically and openly, such questions of distribution. In fact, in the same way as a decision to privatize or ring-fence water services (and to insist on its full-cost recovery) is a political one, so are issues of cross-subsidization. Indeed, if the above argument is correct, then the question of who decides on both investment and distribution becomes an eminently political question, and one that relates directly to issues of democracy and of the distribution of political power.

In conclusion, private-sector participation in the water sector remains limited and the prospects for future private-sector investment rather dim. This leaves no other alternative than public financing to cover the bulk of the required investment. It would be a mirage to believe that the MDGs can be achieved on the basis of massively increased private-sector investment in the water sector. It has not happened in the recent past despite great pressure on all actors. The results of the existing experiments are mixed, to say the least, and the prospects for enhanced investment in the context of total privatization are not promising. Equally, the call from alternative 'people's non-governmental organizations (NGOs)' to improve water access by improving local-level stakeholder participation and citizen's involvement can easily prove to be a mirage, too, as a solution for solving the socio-hydraulic problems of the world's big cities. Without massively enhanced national and international public support, the MDGs will remain an empty promise.

Acknowledgements

I would like to thank José Esteban Castro and the PRINWASS project for providing much of the intellectual stimulus that led to this chapter. I am also grateful to Michael Ekers, Maria Kaika, Alex Loftus and Ame Ramos Castillo for their assistance, insights and creative discussions on this and related themes. I, of course, remain solely responsible for any remaining errors of reasoning or fact.

References

Amann-Blake, N. (2004) *Suez Lyonnaise des Eaux 1880–2004: French Water, Globalisation and the Discourse of Development*, MSc thesis, School of Geography and the Environment, Oxford University, Oxford

Amann-Blake, N. (2006) 'Turbid waters: Globalization, water, development ... and private capital?', Unpublished paper available from the author

Bakker, K. (2003) *An Uncooperative Commodity – Privatizing Water in England and Wales*, Oxford University Press, Oxford

Bakker, K. (2007) 'Trickle down? Private sector participation and the pro-poor water supply debate in Jakarta, Indonesia', *Geoforum*, vol 38, no 5, pp855–868

Bello, W. (2006) 'The capitalist conjuncture: Over-accumulation, financial crisis, and the retreat from globalisation', *Third World Quarterly*, vol 27, no 8, pp1345–1367

Castells, M. (1975) *Sociologie de l'Espace Industriel*, Anthropos, Paris

Castells, M. (1976) 'Krisis van de Staat, Kollectieve Consumptie en Tegenspraken op het Stedelijke Domein', *Te Elfder Ure*, no 23, pp759–790

Castells, M. and F. Godard (1974) *Monopolvile – Analyse des Rapports entre l'Entreprise, l'Etat et l'Urbain*, Mouton, Paris

Castree, N. (2007a) 'Neo-liberalising nature 1: The logics of deregulation and reregulation', *Environment and Planning A*, vol 40, no 1, pp131–152

Castree, N. (2007b) 'Neo-liberalising nature 2: Processes, outcomes and effects', *Environment and Planning A*, vol 40, no 1, pp153–173

Castro, J. E. (2004) 'Final report', in J. E. Castro (ed) *PRINWASS Project (European Commission, Framework V-INCO-DEV Project Contract: PL ICA4-2001-10041)*, University of Oxford, Oxford

Castro, J. E. (2007) 'Poverty and citizenship: Sociological perspectives on water services and public–private participation', in *Geoforum*, vol 38, no 5, pp756–771

Chaussade, J.-L. (2004) 'Address to the staff of Suez Environment', *W2 News: The Power of Waste and Water, Suez Environment Newsletter, Special Issue*, Suez, Paris
Cosgrove, W. and F. Rijsberman (2000) *World Water Vision: Making Water Everybody's Business*, Earthscan, London
Duménil, G. and D. Lévy (2004) *Capital Resurgent: Roots of the Neoliberal Revolution* (translated by D. Jeffers), Harvard University Press, Cambridge, MA
Dunford, M. (1988) *Capital, the State and Regional Development*, Pion, London
Gleick, P. (2004) *The World's Water 2004–2005: The Biennial Report on Freshwater Resources*, Island Press, Washington, DC
Gleick, P. H. and H. Cooley (2006) *The World's Water, 2006–2007: The Biennial Report on Freshwater Resources*, Island Press, Washington, DC
Graham, S. and S. Marvin (2001) *Splintering Urbanism Networked Infrastructures, Technological Mobilities and the Urban Condition*, Routledge, London
Goldman, M. (2007) 'How "water for all!" policy became hegemonic: The power of the World Bank and its transnational policy networks', *Geoforum*, vol 38, no 5, pp786–800
Hall, D. and E. Lobina (2006) *Pipe dreams: The Failure of the Private Sector to Invest in Water Services in Developing Countries*, Public Services International Research Unit, University of Greenwich/World Development Movement, Greenwich and London
Hall, D. and E. Lobina (2007) 'Profitability and the poor: Corporate strategies, innovation, and sustainabiltiy', *Geoforum*, vol 38, no5, pp772–785
Hall, D., V. Corral, E. Lobina and R. de la Motte (2004) *Water Privatisation and Restructuring in Asia-Pacific*, Public Services International, University of Greenwich, Greenwich
Harvey, D. (2003) *The New Imperialism*, Oxford University Press, Oxford
Heynen, N., J. McCarthy, S. Prudham and P. Robbins (eds) (2007) *Neoliberal Environments: False Promises and Unnatural Consequences*, Routledge, London and New York
ICF International (2007) *Privatisation and Regulation of China's Water Sector 2007*, ICF International, Fairfax, VA, www.icfi.com/Markets/Energy/doc_files/water-markets-china.pdf, accessed 18 June 2007
Kaika, M. (2005) *City of Flows*, Routledge, London
Knudsen, O. (2003) 'Letter: World Bank's role in water projects', *The Guardian*, 28 August 2003
Läpple, D. (1976) *Staat en Algemene Produktievoorwaarden: Grondslagen voor een Kritiek op de Infrastructuurtheorieën, Zone Special 1*, Stichting Zone, Amsterdam
Lipietz, A. (1980) 'The structuration of space, the problem of land, and spatial policy', in J. Carney, R. Hudson and J. Lewis (eds) *Regions in Crisis*, Croom Helm, London, pp60–75
Lipietz, A. (1983) *Le Capital et son Espace*, second edition with new preface, Maspéro, Paris
Lobina, E. (2005) 'Problems with private water concessions: A review of experiences and analysis of dynamics', *International Journal of Water Resources Development*, vol 21, no 1, pp55–87
Loftus, A. (2004) 'Free water as commodity: The paradoxes of Durban's water services transformation', in D. A. McDonald and G. Ruiters (eds) *The Age of Commodity: Water Privatization in Southern Africa*, Earthscan, London
Loftus, A. (2006) 'The metabolic processes of capital accumulation in Durban's waterscape', in N. Heynen, M. Kaïka and E. Swyngedouw (eds) *In the Nature of Cities: Urban Political Ecology and the Politics of Urban Metabolism*, Routledge, London
Lojkine, J. (1974) *La Politique Urbaine dans la Région Lyonnaise, 1945–1972, volume 6, La Recherche Urbaine*, Mouton, Paris
Lojkine, J. (1977) *Le Marxisme, l'Etat et la Question Urbaine*, Presses Universitaires Françaises, Paris

Loong Yu, A. and L. Danqing (2006) 'The privatization of water in China', in B. Balanyá, B. Brennan, O. Hoedeman, S. Kishimoto and P. Terhorst (eds) *Reclaiming Public Water (Chinese edition): Achievements, Struggles and Visions from Around the World*, Transnational Institute, Amsterdam, pp1–6

Marshall, A. (1920) *Principles of Economics*, Macmillan, London

Mehrota, S. and E. Delamonica (2005) 'The private sector and privatization in social services: Is the Washington Consensus dead?', *Global Social Policy*, vol 5, no 2, pp141–174

Moulaert, F., A. Rodriguez and E. Swyngedouw (eds) (2002) *The Globalized City: Economic Restructuring and Social Polarization in European Cities*, Oxford University Press, Oxford

OECD (Organisation for Economic Co-operation and Development) (2006) *Infrastructure to 2030: Telecom, Land Transport, Water and Electricity*, OECD Publishing, Paris

Orléan, A. (1999) *Le Pouvoir de la Finance*, Odile Jacob, Paris

Orłowski, L.T. (2008) 'Recent turmoil in financial markets – sources and systemic remedies', *CASE Network E-briefs* 4/2008 (May), www.case-research.eu/plik-20205120.pdf, accessed 12 June 2008

Perrat, J. (1987) *Technologies, Externalités et Nouveaux Rapports du Capital à l'Espace Régional*, PhD thesis on the economics of production, Université Lumière Lyon II, Lyon

Pitman, G. K. (2002) *Bridging Troubled Waters: Assessing the World Bank*, World Bank, Water Resources Strategy Operations Evaluation Department, Washington, DC

Polanyi, K. (1954) *The Great Transformation*, Beacon Press, Boston

Prasad, N. (2006) 'Privatisation results: Private sector participation in water services after 15 years', *Development Policy Review*, vol 24, no 6, pp669–692

Preteceille, E. (1977) 'Equipements collectifs et consommation sociale', *International Journal of Urban and Regional Research*, no 1, pp101–123

Preteceille, E. (1981) 'Collective consumption, the state and the crisis of capitalist society', in M. Harloe and E. Lebas (eds) *City, Class and Capital. New Developments in the Political Economy of Cities and Regions*, Edward Arnold, London, pp1–16

Swyngedouw, E. (1992) 'Territorial organization and the space/technology nexus', *Transactions of the Institute of British Geographers, New Series*, vol 17, no 4, pp417–433

Swyngedouw, E. (2004) *Social Power and the Urbanisation of Water: Flows of Power*, Oxford University Press, Oxford

Swyngedouw, E. (2006) 'Circulations and metabolisms: (Hybrid) natures and (cyborg) cities', *Science as Culture*, vol 15, no 2, pp105–122

Swyngedouw, E., M. Kaïka and J. E. Castro (2002) 'Urban water: A political-ecology perspective', *Built Environment*, vol 28, no 2, pp124–137

Talbot, J. (2002) *Is the International Water Business Really a Business?*, http://info.worldbank.org/etools/docs/voddocs/96/189/SAUR.pdf, accessed 27 March 2007

Theurillat, T., J. Corpataux and O. Crevoisier (2006) *Property Sector Financialisation: The Case of Swiss Pension Funds (1994–2005)*, School of Geography and the Environment, Oxford University, Oxford

Torrance, M. (2006) *The Financialisation of the Urban Infrastructure Landscape: Unravelling Financial Flows into Urban Geographies*, PhD thesis, School of Geography and the Environment, Oxford University, Oxford

UNDP (United Nations Development Programme) (2006) *Human Development Report 2006. Beyond Scarcity: Power, Poverty and the Global Water Crisis*, UNDP, New York

Vidal, J. (2006) 'Big water companies quit poor countries', *The Guardian*, 22 March, p14

Winpenny, J. (2003) *Financing Water for All: The Report on the World Panel on Financing Water Infrastructure* (chaired by M. Camdessus), World Water Council and Global Water Partnership, Marseille

World Bank (2007) *Private Participation in Infrastructure Database*, http://ppi.worldbank.org, accessed 15 June 2007

3

Public Policy Analysis in the Water and Sanitation Sector: Budgetary and Management Aspects

Andrés Sanz Mulas

Introduction

This chapter introduces the public policy approach to the analysis of water and sanitation services (WSS). Our public policy perspective is founded on two basic premises:

1 WSS services have a multidimensional character and their analysis and explanation require inter- and multidisciplinary approaches.
2 Universal access to these services, which are essential for life in a civilized society, constitutes a social right of citizenship and cannot be subjected to market criteria.

These principles form part of the development of what has been termed the New Water Culture (NWC), which requires significant transformations in the institutions and organizational processes involved in managing water and water services. Within this framework, the analysis focuses on two complementary aspects: budgetary and management policy. These include a consideration of systemic conditions, such as the budgetary constraints derived from the macro-economic goals adopted by the public sector, the constraints that the efficient allocation of resources and other requirements impose upon public-sector intervention, and the impact of political power reconfigurations upon budgetary policy. However, it must also be recognized that although public management is constricted by systemic conditions, it, in turn, restricts the scope of public policies in this field. More broadly, the chapter addresses the policy processes involved in the design and enforcement of political, legal and institutional frameworks for the organization and governance of WSS, with the emphasis on accountability and responsibility in public decision-making. This approach views budgeting as the expression of public-sector intervention in the economy. The aim of this chapter is to highlight those water policies in which the crucial component is public expenditure. This combined approach permits the incorporation of those aspects of the NWC that are linked to both budgetary and public economics, as well as public policy requirements.

WHY IS A PUBLIC POLICY APPROACH RELEVANT?

A significant proportion of citizens' concerns, needs and problems cannot be solved and, in fact, are not even addressed by the private sector – that is, through the market. Such issues as the minimum amount of water that must be supplied to meet people's essential needs or the regulation of immigration, among a very wide range of other social issues, cannot be properly understood unless a public policy approach is employed. In contemporary society, the public sector (and, as a result, public policy) tends to focus on a wide spectrum of policy problems from the macro level to everyday human activities. Thus, public policies may regulate behaviour, organize bureaucracies, distribute social benefits, collect taxes or do all of these things simultaneously. Given that not everything government does is reflected in public expenditure, the scope of the public policy approach is wider than its budgetary counterpart (Dye, 1998, p2). Although public policy is a very broad field, the following definition contains its most important elements: public policy consists of political decisions taken to implement programmes in order to achieve societal goals (Cochran and Malone, 1995, p1). These three key components must be taken into account when attempting to analyse any public policy, such as WSS policies.

The term political decision may be interpreted in many ways; it may be understood as being restricted to those decisions that produce legislation, or may refer to any decision in which the coercive powers of the state are exercised (Wood, 2000, p14). This chapter concentrates on political decisions as non-market decisions. Market decisions concern sales and purchases by private economic agents (companies and individuals) who attempt to maximize their profits or their individual utility. These economic agents make decisions according to their private interests, whose interaction results in the allocation of resources. The production and exchange of commodities is the central element: consumers demand commodities and supply labour services, while firms produce commodities on the basis of their technological knowledge. Commodities flow between agents by means of an exchange process performed via markets and prices. Political decisions involve mechanisms such as elections, public administration and budgets approved by the legislature.

The public policy approach is concerned with the implementation of programmes, using financial and material resources provided compulsorily by citizens within a framework of public objectives. This approach differs from the private approach, the aim of which is to produce goods and services to be sold in the market. The final aim of public policies is to achieve the societal goals defined by citizens through their elected representatives. We will take into account these three factors (political decisions, programme implementation and societal goals) throughout this chapter.

Public policy as a field of study began during the 1950s, motivated by a desire to understand human problems in a way that cuts across disciplinary boundaries. Public policy analysis draws on a variety of social sciences, including political science, public administration and management, economics, sociology and psychology; it must also take into consideration many other disciplines and knowledge areas linked to each specific policy. Thus, a public policy perspective of WSS services views them as multidimensional, where their analysis and explanation require inter- and multidisciplinary approaches. National water policy should be conceived and implemented within the framework of an interdisciplinary national economic, social and environmental development policy. As such, it should recognize that water development, including the development of water infrastructures, is an essential component in national development plans, and that land and water must be managed in an integrated manner. A national water policy also requires the definition of goals for the different water uses, including the supply of safe drinking water and wastewater treatment and disposal facilities, as well as water for agriculture, stock raising, industrial needs, transportation and hydroelectric power in a way that is compatible with the resources and characteristics of the area concerned (Salman and Bradlow, 2006, p5).

Current understandings of the public policy process usually separate it into several stages: problem identification; policy proposal; policy adoption; programme operation and evaluation (Cochran and Malone, 1995 p39); agenda-setting; policy formulation and implementation; policy assessment; policy change and termination (Lester and Stewart, 2000, p5). When we take these stages into account, it becomes clear that water-related public policy is difficult and challenging due to its complexity (Pahl-Wostl et al, 2008). Water systems must fulfil functions such as drinking water supply, irrigation and drainage, or the maintenance of vulnerable ecosystems. Moreover, the behaviour of water systems is difficult to understand as it is driven by complex hydrological and ecological processes, as well as by human activity, which may have various positive and negative impacts. Because of the complexity of the water cycle, systems-analysis methods are often used to support policy-making in the field of water resource management. Usually these methods are problem oriented, concerned with the causes of policy problems, the potential options for their resolution and the likely effects of these options on the initial problems and the wider environment (Hermans, 2003).

Presumably, a key stage in this process is when a problem is perceived and incorporated into the policy agenda that is part of the set of subjects, problems, needs or questions that must be addressed by public decision-makers. Some topics, such as the provision of essential water services to citizens, and water for agriculture and industry, can be easily included in the policy agenda; but others are more complicated and require more time and public discussion, such as the introduction of the full-cost recovery principle as a way of making water policy more efficient in economic and ecological terms. Multiple actors are responsible for the inclusion of items in the policy agenda; in the case of democratic societies it is expected that there will be a high degree of convergence between the goals of the citizens and the agendas set by public decision-making bodies (Pertschuk, 1987).

The public must be genuinely involved in the agenda-setting process rather than being left out as merely passive observers (Jones and Baumgartner, 2004).

Policy formulation tries to identify the actors, mechanisms, actions and resources available to respond to the challenges posed and to solve the particular social issues determined in the problem identification stage. Water policy formulation and decision structures involve a wide range of institutional actors in the legislative and executive branches of government; all such actors have relatively narrow responsibilities in the field of water policy (Hoornbeek, 2004).

Implementation is the putting into practice of the policies formulated. In this stage, multiple actors participate, with public administration playing a key role. There are at least three factors that should be highlighted. The first of these is that various levels of government work together and share competencies and responsibilities, which is required for the achievement of the policy goals that have been formulated. It is vital to consider which level of government is best equipped to implement the various elements of the policy (García Valiñas, 2007) and the coordination mechanism that must be established between governments and their public administrations (Woltjer and Niels, 2007). The second factor concerns the different ways in which water services may be provided. Usually, the supply of such services, for the reasons given below, is a public-sector responsibility; but this does not necessarily entail production by public bodies. The way in which public services are produced has extremely important effects upon their cost and the possibility of citizens' participation in decision-making; thus, a key policy decision is whether production should be public or private (Newig et al, 2005).

The final factor is the need for coordination among the various agents participating in such policies. At each level of government, multiple agencies, ministries and other public bodies are required to implement programmes that are complementary but, at the same time, are the result of many conflicting interests (Pierce, 1979). The evaluation and eventual introduction of changes to water policy are the final stages of the cycle. The true capacity to evaluate and change policy depends upon the quality of the available information systems, as well as upon the actual feasibility of citizen participation in the decision-making and monitoring processes.

Economics and the budgetary component

This chapter emphasizes that the budgetary component of water policy is the result of collective decision-making with a strong economic content. In this regard, the public finance approach helps us to understand four questions (Gruber, 2005, p2), which are usually not given sufficient consideration in the analysis of water policies:

1 When should governments intervene in the economy?
2 How can they intervene?
3 What is the effect of such intervention upon economic outcomes?
4 Why do governments choose to intervene in the way that they do?

The public finance approach allows us to understand the public policy options that are efficient and effective in terms of overall economic objectives such as stability, growth and development, and of the production of goods and services. In this connection, it is understood that governments must intervene in this sector for the following reasons, among others:

- to compensate for market failures;
- to encourage economic development and stability;
- to change income distribution.

Due to the specific characteristics of certain goods and services, it is impossible, in the real world, to find a competitive market capable of delivering them in sufficient quantity, with the required quality, at the right time and in conditions that are affordable for everyone. WSS are a natural monopoly, which is part of the explanation for the fact that they are usually provided by public entities. This market failure was already identified by John Stuart Mill when he discussed the case of London back in the 19th century:

> *It is obvious, for example, how great an economy of labour would be obtained if London were supplied by a single gas or water company instead of the existing plurality. While there are even as many as two, this implies double establishments of all sorts, when one only, with a small increase, could probably perform the whole operation equally well; double sets of machinery and works, when the whole of the gas or water required could generally be produced by one set only; even double sets of pipes, if the companies did not prevent this needless expense by agreeing upon a division of the territory.* (Stuart Mill, 2001, p155)

In this kind of market the expansion of a monopolist customer base allows the supplier company to spread fixed costs over a great number of users, and the average cost falls as a result. Thus, the most efficient system for the supply of such goods and services is a monopoly (Leach, 2004, p219).

From another angle, social and economic stability are linked to the provision of water (Boberg, 2005, p104), an input that is essential to guarantee economic development: 'effective water resource development and ecological management play a fundamental role in sustainable growth and poverty reduction, through different mechanisms' (World Bank, 2004, p93). Water has always played a central role in human societies as an engine for sustainable growth and poverty alleviation, as an overall input for productive activities (including agriculture, industry, energy and transport) and as the basic requirement for the reproduction of healthy people living in healthy ecosystems (Grey and Sadoff, 2006). Governments are responsible for guaranteeing the quantity and quality of water required by the economy since one of the most important objectives of political economy is to encourage economic stability and growth.

Water is also a human right. As Salman (2004, pix) states:

> *... there exists, within the legal framework of the International Covenant on Economic, Social and Cultural Rights, a human right to water because it is a right that inheres in several other rights, and a right without which key provisions of the Covenant would be rendered ineffectual.*

Water as a human right must be publicly provided if its supply has to be universally guaranteed regardless of individual income. In this sense, acknowledging that essential volumes of safe drinking water are a human right is one way of altering income distribution, especially in developing countries. To summarize, WSS are a service that must be provided by the public sector and require transfers in terms of public revenue and expenditure.

The budgetary process, as stated earlier, is restricted by the macro-economic objectives that all governments must take into account; public policies concerning water are also limited by the macro-economic framework. In the medium and long term, water resources are essential to national economic stability and growth. Empirical analysis suggests that in the vast majority of countries, current rates of freshwater consumption are not yet constraining economic growth. However, countries that are 'water stressed' (i.e. have limited freshwater supplies relative to current and future populations) may find it especially difficult to generate additional growth through increased water use. We cannot reject the hypothesis that the presence of moderate or extreme water scarcity adversely affects economic growth (Barbier, 2004). For example, current public-sector expenditure on water infrastructure will permit a higher level of growth in the medium term and, as a result, a greater capacity for funding future budgets. This can be illustrated by the fact that since the second half of the 19th century and through the first half of the 20th, the majority of the currently developed countries made significant investments in public water infrastructures to facilitate economic growth. A good example of this assessment is the case of Madrid, where the creation of the Canal de Isabel II (a quasi-autonomous public body responsible for the city's water supply) in 1851 made possible the public investment required to maintain the water supply to the capital and, as a result, fostered the economic development of the Madrid region (Ortega and Sanz, 2007).

Closely related to the public policy framework described above is the budget cycle, which is composed of three phases: elaboration, execution and control. The budget contains the revenue and expenditure that public managers can employ during the fiscal year. This chapter will not examine this subject in depth; but it is important to underline the fact that water-related public policy largely requires the use of financial resources that must be included in the budget. How these resources are obtained is also a key element in determining the possibilities of water policy. The budget regulates and defines how the funds required to implement water policy will be obtained. Two main methods can be employed to finance water and sanitation services – namely, cost recovery or funding via general taxation. It appears that in recent years the former has gained increasing support and has been chosen by the 2000 European Water Framework Directive (WFD) (EU, 2000). Cost-recovery policy applied to water involves attempting to transfer the cost to users; this transfer may be proportional to consumption or, instead, various tariffs may be established in order to encourage the responsible use of the resource or to achieve other political objectives. Traditionally, an important part of the cost of WSS has been funded from general revenue. In the case of Spain, for example, over the last century the bulk of the funding required for dam-building was provided by public resources, while, more recently, the European Union (EU) has also co-funded sanitation infrastructures jointly with the central and local governments. Each level of government must elaborate its budget, taking into account medium- and long-term financial constraints; due to the vast funds required by water infrastructures, the macro-economic setting can restrict the possibilities for effective management or production of this service.

In developed countries, continuous fiscal deficits have led to the setting of stricter limits upon the public administrations' spending and borrowing capacities. In developing countries, the fiscal restrictions due to external debt problems and the limited capacity for revenue collection in most countries often cause even greater financial constraints, which make the development of the necessary water infrastructure extremely difficult. Summing up, in this stage of the budget cycle the limited resources available for the coming financial year are allocated among various objectives and policies.

The conditions described above severely limit the room for manoeuvre in the elaboration phase of the budget and strongly influence the execution phase. The execution phase involves the deployment of the revenue and expenditure approved in the first phase. At least three criteria must be employed in execution: legality, effectiveness and efficiency. The first entails a series of requirements aimed at ensuring the correct use of public funds in legal terms and in accordance with the regulations approved by the body responsible for it. Furthermore, this criterion attempts to avoid the possibility of changing, during execution, decisions made in the first phase. For example, if the legislature has decided that a certain amount of money must be spent on water policy and related activities, the government must implement these decisions. The effectiveness criterion attempts to guarantee the effective achievement of the goals established by the legislature for the various programmes. Not only is it necessary to spend the funds allocated on the relevant policies, but the government must also achieve the agreed objectives. The third criterion aims to avoid the wasting of public funds. As stated above, the private provision of water services is inefficient; however, if the public sector is not able to provide them at a minimum cost the problem of efficient resource allocation cannot be solved.

The final phase concerns the monitoring of the results achieved. The executive must be accountable to the legislature and citizens. Usually, specialized institutions exist to defend the rights of citizens, while the legislature analyses the accounts and budgetary results. These control mechanisms must allow citizens to have enough information to evaluate the management performance of the government and to decide which representatives to vote for in elections. In this regard, there are three aspects of the budgetary cycle that are especially relevant to water services. The first of these is the way in which water services are financed. Those who pay are water consumers, taxpayers or a mixture of both. The sources of funds can be user tariffs, government subsidies or debt (whether public or private); the most common situation is a mixture of the three. It is essential to take into account that, as a result of the dual nature of such services (as both productive inputs and satisfiers of human and social needs), their method of financing strongly influences the redistributive effect of public policy. However, the resulting impact is sometimes far from obvious and must therefore be closely examined since equity should be one of the principal objectives of the use of public resources.

Financial questions can have important consequences in terms of the second aspect. They concern the type of management implemented to produce these services and to execute the budget for such programmes. One of the most common problems faced by public bodies responsible for the provision of WSS is the lack of funds available to meet their substantial investment needs. As a result, public managers must try to find mechanisms for the production and delivery of these services that allow access to finance without increasing their budget deficits or the limits placed upon their borrowing capacity.

The final question concerns citizen participation. As a result of the crisis of traditional democratic processes, a new system has emerged in the budgetary field, usually termed participatory budgeting. Water policy involves a set of factors that make accountability and responsibility in public decision-making a very complex process (Berger et al, 2007). Consequently, citizen participation has become a crucial element for effective, efficient and democratic water management. As described above, the multiplicity of agents, levels of government and budget programmes involved in WSS policy-making place severe restrictions upon the ability of common citizens to monitor the actual processes. Nevertheless, there are excellent examples of how citizen participation can help to improve the accountability and effectiveness of water policy, such as the 'yeast factory problem' episode in the Spanish city of Cordoba, which has been studied by the Watertime Project.[1]

The New Water Culture approach

The notion of a New Water Culture has its origins in the book *La Nueva Cultura del Agua en España* by Francisco Martínez Gil (1997). A useful summary of the basis and principles informing the NWC approach has been offered by Jean Margat (2001, p87):

> *There are three deeply rooted ideas about water: water is unlimited and hence inexhaustible, water has an essentially purifying function and finally, water is looked upon as a gift from heaven, in both the literal and the metaphorical sense, and is therefore necessarily free. The new water culture should break with these preconceptions. We must accept that resources are neither unlimited nor invulnerable.*

This chapter focuses on the socio-economic aspects of the NWC, and will be mainly based on the approach proposed by Arrojo (Arrojo and Martinez Gil, 1998; Arrojo, 1999, 2005).[2] The NWC proposes the development of a new mindset for managing water resources. In the past, one of the main obstacles for implementing sustainable water management policies has been the prevalence of a short-term utilitarian approach that does not take into account the specific characteristics of water. This is one of the key problems that continue to negatively affect water management practices and their significant impact upon development (UNDP, 2006). The NWC examines this problem by focusing on two different dimensions that are apparently contradictory but are, in fact, complementary. The first concerns the problem of water values: the NWC is an ecosystem-based approach that advocates new types of demand management based on sustainable strategies. As such, the NWC is critical of the liberal approach – still prevalent – which considers that water is a purely economic good that, like other commodities, has a value that can always be expressed in price terms. The traditional liberal conception of water management is incapable of perceiving water's non-market values. However, in practice, most values linked to water are non-monetary, as stated by Arrojo (2005):

> *Here are a few examples: the fundamental values of life, which are essential to dignified living conditions for people or communities; the values of preservation of the environment and aquatic ecosystems; the values of intra- and inter-generational equity or the values of social cohesion that water-distribution services bring. The value of these functions ought not to be administrated according to market rules since they cannot be measured in monetary terms.*

Similarly, Javier Martínez Gil has argued for the restoration of the sentimental value of rivers, which has been lost owing to their commercial exploitation (Martínez Gil, 1997).

The second dimension of the NWC is related to the uses or functions of water. The traditional approach to water management continues to be based on supply-side strategies that involve large-scale public investments in order to subsidize the building of water infrastructures but, among other issues, do not guarantee universal access to safe water and sanitation. Too often, the traditional approach has led to a very unequal order of priorities in relation to the functions of water that receive the bulk of public investment. The empirical evidence (e.g. in the Spanish case) reveals that a significant part of the public funds spent on water infrastructure have benefited those private agents (especially the agricultural sector) that use water essentially as a productive input in capital accumulation. Bearing in mind that agricultural water use represents the largest part of total consumption, a large share of the public financing destined for water infrastructure is transferred as a direct subsidy to this sector, which, in practice, represents a significant redistribution of income in favour a small group of citizens (the farmers). The predominance of agricultural uses in total water consumption is common to many Organisation for Economic Co-operation and Development (OECD) member countries (OECD, 2006, p24)

These trends in water use have led NWC economists to make a clear distinction between the different uses or functions of water in order to clarify the role of the public sector in water policy. To summarize, there are three principal uses or functions of water in terms of public policy interests:

1. water for life (to cover basic survival needs);
2. water for purposes of general public interest;
3. water for economic growth.

Rather than giving a detailed explanation, this chapter highlights the main features of each use or function. Human beings need a minimum volume of water to survive, and it is this first function that we associate with the notion of a human right to water. Taking only drinking water and essential sanitation needs into consideration, the amount of clean water required to maintain adequate human health ranges between 2 litres (l) and 80l per person per day, or up to approximately 30m^3 per person per year (Gleick, 1996). This variation is the result of a combination of factors, including the social characteristics of water use in different cultures and climatic conditions. However, this first function of water – water for life – also includes the water needed for the survival of non-human living beings and water ecosystems, which in the conception of the NWC is also a primary government responsibility.

Over and above these essential requirements for water for human survival in dignified conditions, there are other domestic uses that have become part of the accepted way of life, particularly in Western societies, such as water for gardening and other amenities that can be considered as social needs – part of the general public interest of communities. Although these uses cannot be considered part of the first category (water as a human right to cover survival needs), they have become part of the social needs accepted as legitimate in many societies. This is an important distinction in the conception of the NWC: while the government is responsible for guaranteeing the universal access to water for basic human needs (the first function), access to higher volumes as required for the second function is not considered a human right but as a right of citizenship, which must be subject to different criteria in terms of public policy. For instance, in the context of the European Union, the uses of water associated with this second function are said to remit to a broad concept that covers both market and non-market services, which the public authorities regard as being of general interest and which are subjected to specific public-service obligations. Within European Community law and practice, the concept of services of general interest refers to services of an economic nature that the member states or the European Community subject to specific public service obligations by virtue of a general interest criterion. It also extends to any other economic activity that is subjected to public service obligations. The term 'public service obligations' denotes specific requirements that are imposed by public authorities on the provider of the service in order to ensure that certain public interest objectives are met – for instance, in such matters as air, rail and road transport, and energy (Bovis, 2005).

It is clear that water has many functions beyond the very basic needs connected with daily survival that have consequences for society today and will be even more crucial in the future. According to these functions, the role of the public sector is not only to supply a set of water-based goods and services, but also to guarantee the social right to water with regard to the present and future sustainable provision of these goods and services. These include services that are of general economic interest, such as access by all citizens and collective actors, including business companies, to affordable high-quality water services that are essential to promote social and territorial cohesion, which includes the reduction of disadvantages caused by the lack of accessibility to water services in peripheral regions (EC, 2004).

The third use of water is for private profit-making purposes. As a matter of fact, most water abstracted is used as an economic input in productive profit-making processes that benefit property owners. In this perspective, it can be argued that this third function or use of water is similar to that of other commodities that intervene as inputs in the private economic process. Therefore, in the NWC approach, water as a commodity cannot be considered to be a human or a social right of citizenship, and, consequently, must be treated by the government in the same way as other commodities produced in a natural monopoly context. The efficient allocation of resources requires the transfer of the full cost to commercial consumers, without any kind of direct or indirect subsidies.

The functions of water mentioned above are those that have usually been defined in the NWC literature. However, for the analysis presented in this chapter, it is useful to further differentiate some domestic functions. In developed countries, various water uses that cannot be included in the preceding classification are of increasing importance. For example, individual water consumption for private swimming pools and other domestic sumptuary forms of water use should not be included in the category of social rights; consequently, their full cost must be transferred to the users.

CONCLUSIONS

The introduction to this chapter stressed the importance of a public policy and budgetary approach to WSS issues and their importance for understanding the constraints upon WSS decision-making and implementation processes. Behind the reality observable in any public policy, there can be found a correlation of power among the multiple agents involved, as Wildavsky (1964) states. In addition, as Wildavsky and Caiden (2001, pxv) suggest, a budget is the final result of compromises among numerous actors with varied interests:

> *... budgetary politics is incremental because it leaves untouched decisions about the great bulk of revenues and spending. The main budget story lies in the ongoing budgetary commitments, whose growth is largely determined by social and economic trends.*

The budgetary approach highlights three very important issues in WSS policy: how such services should be financed; the policy impact of WSS in terms of equity; and the importance of the mechanisms for the production and delivery of WSS.

Public expenditure in WSS can be financed through different budgetary resources and, depending upon the chosen options, more or less revenue can be collected. As stated earlier, budgets are constrained by the macro-economic framework that establishes expenditure limits; but some budgetary resources allow an increase of revenues while still respecting macro-economic constraints. For instance, if water fees are used (i.e. users are charged for the water they use), the cost can be transferred to users and the macro-economic restrictions will be circumvented. However, there are many problems inherent in this budgetary mechanism, which tends to reduce the complexity of water values due to ignorance of the many environmental externalities that are usually not included in the cost of water. In turn, different budgetary mechanisms also have different impacts in terms of income redistribution.

The second issue is equity, which involves two aspects: revenue and expenditure. The former has been addressed above, while the latter concerns income redistribution, as well as social equity. It implies fair access to resources and livelihood conditions. However, the concept of what is 'fair' reflects the ethical values prevalent in a society, as well as economic values associated with resource use. In the context of resource allocation, social equity refers to a set of rights and duties of governments, collectives and/or individuals; these are established in order to protect weak and vulnerable groups in society (Cai, 2008). Public water policy and its correlated budgetary programmes must be able to promote not only social equity, but also intergenerational and territorial equity, among other forms. Water resources infrastructure has important consequences for future generations; for instance, drainage works may improve the utility of land in the short run, but in the long run may result in land subsidence and an increase in the risk of floods (Mostert, 2008).

The final issue concerns water management. As explained above, due to the specific characteristics of WSS, the market is not able to allocate water efficiently in economic terms. If we employ the ecosystem-based approach promoted by the NWC, then it becomes clear that the new demand management and sustainability strategies required cannot be implemented through the market. It is also clear that the provision of WSS must be a public responsibility, although these services can be produced either publicly or privately. However, the way in which WSS are produced and delivered have a crucial impact upon the efficiency, effectiveness, equity and accountability of public decisions. Separating provision from production by methods such as the granting of private concessions results, in practice, in the creation of another type of monopoly since these policy instruments only allow the existence of one producer over 20, 30 or even 50 years. They also reduce the accountability of politicians as they are no longer responsible for the decision-making, management, production and delivery of WSS, which have been delegated for a considerable time (Sanz, 1998). The financial constraints that must be observed, due to macro-economic restrictions, stability agreements, requirements from multilateral financial institutions or other reasons, have led many public authorities to choose private production of these services as a way of obtaining revenue or avoiding an increase in fiscal deficits. However, the costs of such decisions in terms of finance and accountability are excessive since better options for maintaining public management and delivery of WSS exist.

With regard to the analytical tools offered by the NWC approach, it is essential to clearly distinguish between the different uses or functions of water when designing and implementing water policies and budgets. Business and 'sumptuary' uses require the strict application of the full-cost recovery principle (i.e. the service must be entirely funded by the users). However, other water uses linked to basic human survival or to the general public interest as reflected in the social rights of citizens must be universally provided by the public sector independently of the economic capacity of the citizen users. Thus, the full-cost recovery principle can be partially applied in conjunction with general taxes or public subsidies, the latter being applied to those uses where it is impossible to assign the benefits of use individually, especially when intergenerational effects are taken into consideration. The public policy and budgetary approach helps us to understand crucial aspects of WSS and, together with the principles of the NWC, contributes to promote a better, more ecological, more efficient and more equitable management of water resources.

Notes

1 See the Watertime Project report on the case of Cordoba (Watertime, 2005). A more detailed description of the Cordoba participatory framework can be found in Ortega and Sanz (2005).

2 A number of other authors whose work cannot be considered here for reasons of space have made extremely interesting contributions to this approach, including Castro (2007) and del Moral and Saurí (2001), among other signatories of the *European Declaration for a New Water Culture* that was made public in Madrid in 2005 (EUWATER, 2005).

References

Arrojo, P. (1999) 'El valor económico del agua', *Revista d'Affairs Internacionals*, nos 45–46, pp145–167

Arrojo, P. (2005) 'For a New Water Culture', *The Green Cross OPTIMIST*, winter 2005, pp9–13

Arrojo, P. and J. Martínez Gil (1998) *El agua a debate desde la Universidad: Hacia una nueva cultura del agua: 1er Congreso Ibérico sobre Gestión y Planificación de Aguas*, Prensas Universitarias de Zaragoza, Zaragoza, Spain

Barbier, E. (2004) 'Water and economic growth', *Economic Record*, March 2004, vol 80, no 248, pp1–16

Berger, T., B. Birner, N. Mccarthy, J. Díaz and H. Wittmer (2007) 'Capturing the complexity of water uses and water users within a multi-agent framework', *Water Resources Management*, no 21, pp129–148

Boberg, J. (2005) *Liquid Assets: How Demographic Changes and Water Management Policies Affect Freshwater Resources*, Rand Corporation, Santa Monica, CA

Bovis, C. (2005) 'Financing services of general interest in the EU: How do public procurement and state aids interact to demarcate between market forces and protection?', *European Law Journal*, January 2005, vol 11, no 1, pp79–109

Cai, X. (2008) 'Water stress, water transfer and social equity in Northern China – Implications for policy reforms', *Journal of Environmental Management*, vol 87, no 1, pp14–25

Castro J. (2007) 'Poverty and citizenship: Sociological perspectives on water services and public–private participation', *Geoforum*, vol 38, no 5, pp756–771

Cochran, C. and E. Malone (1995) *Public Policy: Perspectives and Choices*, McGraw-Hill, New York

del Moral, L. and D. Saurí (2001) 'Recent developments in Spanish water policy. Alternatives and conflicts at the end of the hydraulic age', *Geoforum*, vol 32, no 3, pp351–362

Dye, T. (1998) *Understanding Public Policy*, Prentice-Hall, New Jersey

EC (European Commission) (2004) *White Paper on Services of General Interest*, COM 374, European Commission, Brussels

EU (European Union) (2000) *Directive of the European Parliament and of the Council 2000/60/EC Establishing a Framework for Community Action in the Field of Water Policy*, Luxembourg: European Parliament and Council, http://europa.eu.int/comm/environment/water/water–framework/index_en.html, accessed May 2008

EUWATER (2005) *European Declaration for a New Water Culture*, Saragossa, Spain, New Water Culture Foundation, www.unizar.es/fnca/euwater/docu/europeandeclaration.pdf, accessed 1 June 2008

Garcia-Valiñas, M. (2007) 'What level of decentralization is better in an environmental context? An application to water policies', *Environmental and Resource Economics*, vol 38, no 2, pp213–229

Gleick, P. (1996) 'Basic water requirements for human activities: Meeting basic needs', *Water International*, no 21, pp83–92

Grey, D. and C. Sadoff (2006) 'Water for growth and development', in *Thematic Documents of the IV World Water Forum*, National Water Commission, Mexico City

Gruber, J. (2005) *Public Finance and Public Policy*, Worth Publishers, New York

Hermans, L. (2003) 'Agenda setting in policy analysis: Exploring conflict for a case of water resources management in the Philippines', in *Systems, Man, and Cybernetics Society 2003 Conference Proceedings*, Institute of Electrical and Electronics Engineers, Inc (IEEE), Washington, DC, pp3314–3321

Hoornbeek J. (2004) 'Policy-making institutions and water policy outputs in the European Union and the United States: A comparative analysis', *Journal of European Public Policy*, vol 11, no 3, pp461–496

Jones B. and F. Baumgartner (2004) 'Representation and agenda setting', *Policy Studies Journal*, vol 32, no 1, pp1–24

Leach, J. (2004) *Course in Public Economics*, Cambridge University Press, Cambridge

Lester, J. and J. Stewart (2000) *Public Policy: An Evolutionary Approach*, Wadsworth Publishing, Belmont

Margat, J. (2001) 'Towards a New Water Culture', in J. Binde (ed) *Keys to the 21st Century*, UNESCO Publishing/Berghahn Books, Oxford

Martínez Gil, F. (1997), *La Nueva Cultura del Agua en España*, Bakeaz, Bilbao

Mostert, E. (2008) 'Managing water resources infrastructure in the face of different values', *Physics and Chemistry of the Earth*, vol 33, nos 1–2, pp22–27

Newig, J., C. Pahl-Wostl and K. Sigel (2005) 'The role of public participation in managing uncertainty in the implementation of the Water Framework Directive', *European Environment*, vol 15, no 6, pp333–343

OECD (Organisation for Economic Co-operation and Development) (2006) *Environmental Performance Reviews – Water: The Experience in OECD Countries*, OECD, Paris

Ortega E. and A. Sanz (2005) 'La Gestión del Agua en el municipio de Córdoba: Un modelo de gestión pública participativa, eficaz y eficiente', in B. Balanyá, B. Brennan, O. Hoedeman, S. Kishimoto and P. Terhorst (eds) *Por un Modelo Público del Agua: Triunfos, Luchas y Sueños*, Ediciones de Intervención Cultural/El Viejo Topo, Barcelona, www.tni.org/books/waterspain.pdf, accessed June 2008

Ortega E. and A. Sanz (2007) 'A public sector multinational company: The case of Canal de Isabel II', *Utilities Policy*, vol 15, no 2, pp143–150

Pahl-Wostl, C., D. Tábara, R. Bouwen, M. Craps, A. Dewulf, E. Mostert, D. Ridder and T. Taillieu (2008) 'The importance of social learning and culture for sustainable water management', *Ecological Economics*, vol 64, no 3, pp484–495

Pertschuk, M. (1987) 'The role of public interest groups in setting the public agenda for the 90s', *Journal of Consumer Affairs*, vol 21, no 2, pp171–182

Pierce J. (1979) 'Conflict and consensus in water politics', *The Western Political Quarterly*, vol 32, no 3, pp307–319

Salman, M. (2004) *Human Right to Water: Legal and Policy Dimensions*, World Bank Publications, Washington, DC

Salman M. and D. Bradlow (2006) *Regulatory Frameworks for Water Resources Management: A Comparative Study*, World Bank, Washington, DC

Sanz, A. (1998) 'Las privatizaciones: Algunos aspectos generales', *Cuadernos de Relaciones Laborales*, no 13, pp19–52

Stuart Mill, J. (2001) *The Principles of Political Economy*, Batoche Books, Kitchener, Ontario

UNDP (United Nations Development Programme) (2006) *Human Development Report 2006: Beyond Scarcity: Power, Poverty and the Global Water Crisis*, Macmillan, New York

Watertime (2005) *Cordoba Case Study*, Project report, Greenwich, University of Greenwich, www.watertime.net/docs/WP2/D17_Cordoba.doc, accessed 1 June 2008

Wildawsky, A. (1964) *The Politics of the Budgetary Process*, Little, Brown & Company, Boston, MA

Wildawsky, A. and N. Caiden (2001) *The New Politics of the Budgetary Process*, Longman, New York

Woltjer, J. and A. Niels (2007) 'Integrating water management and spatial planning', *Journal of the American Planning Association*, spring, vol 73, no 2, pp211–222

Wood, P. (2000) *Biodiversity and Democracy: Rethinking Society and Nature*, University of British Columbia Press, Vancouver, BC

World Bank (2004) *Responsible Growth for the New Millennium: Integrating Society, Ecology and the Economy*, World Bank, Washington, DC

4

North–South Transfer of the Paradigm of Piped Water: The Role of the Public Sector in Water and Sanitation Services

Okke Braadbaart

THE PIPED WATER PARADIGM: EARLY INNOVATION IN THE NORTH

European foundries mastered the art of mass-producing cast iron pipes in the early 1800s (Cast Iron Soil Pipe Institute, 2006, p1). Slow sand filters, buried pipes with bell-and-spigot joints, steam-powered pumps and water towers make for a universally applicable technology for urban water supply. Piped water systems enjoy a long service life and modest operational costs, but require high initial outlays of capital. In addition, pipe-laying raises complex right-of-way issues (Meidinger, 1980). Around 1850 it was by no means certain that buried water infrastructure could hold its own against contending modes of urban water supply. Cost wise, it certainly could not compete against local water sources such as shallow wells and canals. By 1900, however, piped supply was establishing itself as the dominant source of domestic water in European and North American cities. Seventy years on, piped networks were the sole source of urban domestic water.[1] A parallel network of buried sewers carried the large volumes of piped water away from urban homes.

How did piped water and waste technologies come to dominate domestic water regimes in European and North American cities? Early literature attributes the rise of urban piped water and sewerage to an unfolding understanding of the nature of waterborne disease (Baker, 1949). However, subsequent investigations by urban environment historians show that public health concerns played only a minor role in early 19th-century decisions to embrace piped water infrastructure (Tarr et al, 1984; Troesken and Geddes, 2003; Millward, 2004; Juuti and Katko, 2005). The introduction of piped water supply, in enabling householders to install waterborne waste removal systems, may arguably even have aggravated the spread of waterborne diseases (Tarr et al, 1984).[2]

Urban environment historians explain the rise of piped water against a background of rapid industrial and population growth in European and North American towns. The environmental stress accompanying 19th-century industrialization triggered a crisis of urban water and waste that overwhelmed city administrations (Tarr, 1996). By the late 19th century, national and state governments responded by enacting laws and regulations that placed responsibility for solving the urban public health crisis on the shoulders of municipalities. City administrations responded by committing themselves to large-scale and centralized technologies for handling urban water and waste.

Much of the research published in English focuses on the US and the UK. This chapter uses a broader compass and describes developments on the European continent, putting the British experience into perspective as an early innovator, and the North American experience as a case of the late following of patterns established in the European context. We then look at how the paradigm of piped water is conveyed to the developing world in the first half of the 20th century.

The Industrial Revolution and urban water pollution in the North

The period of 1750 to 1825 saw a plethora of European initiatives aimed at solving the growing problem of urban water pollution. Engineers, chemists and medical doctors experimented with water filtration techniques, vigorously debated the nature of disease, carried out trials with piped water supply systems and argued about the best methods for transporting sewage out of cities (Ewbank, 1842; Baker, 1949; Rosen, 1958; Cast Iron Soil Pipe Institute, 2006). The first documented supply of water to an entire municipality using a purpose-built filtration plant reportedly occurred in Paisley, Scotland, around 1804 (Baker, 1949, p77). By this time, techniques for transporting water by means of steam-powered pumps were already well established.[3] Finding an affordable means of conveying water under pressure is another matter, however. Around 1800, bored wood and leaden pipes were the main means used for water conveyance (Ewbank, 1842, p283). But as pipe materials, both wood and lead have drawbacks. The history of the research and development that produced iron and, later, concrete pressure pipes for water supply remains to be written, but there are reasons to believe that this occurred in parallel with the development of city gas lighting (Falkus, 1982, p228). At any rate, by 1825 a viable set of technologies for piped water conveyance must have been in existence:

> *Cast iron pipe was first used in the United States at the beginning of the 19th century. It was imported from England and Scotland to be installed in the water-supply and gas-lighting systems of the larger cities, principally those in the north-eastern section of the country. One of the first cast iron pipe installations was at Bethlehem, Pennsylvania, where it was used to replace deteriorated wooden mains.* (Cast Iron Soil Pipe Institute, 2006, p2).

We do not know which municipality was first to combine all of the ingredients of 'modern' municipal water supply – that is, a purpose-built filtration plant, transport of purified water from plant to city through pipes using pumps powered by fossil fuels, and the feeding of water into urban homes through small-diameter service lines. What is certain, however, is that in Britain such systems were in use by the 1830s, spreading rapidly across the continent after 1850 (Juuti and Katko, 2005). In doing so, they imitated an example set by other municipal undertakings (Millward, 2004).

In some cities, drainage needs were more pressing than water supply. Sewer construction is a laborious process as long as trenching is done manually and sewers are constructed from brick and mortar. The advent of steam power and prefabricated concrete pipes boosted sewerage construction, and, as the 19th century drew to a close, these innovations were already transforming the sewerage industry (Clarke, 1885).

Piped water supply and piped sewerage are mutually reinforcing. Prior to the introduction of waterborne waste removal, substantial volumes of excrement and urine were disposed of in pit latrines or collected and sold as fertilizer. The advent of piped water supply disrupted this local metabolic cycle. Once connected to a piped water system, householders were quick to install water closets. The subsequent surge in wastewater created massive on-site problems in the shape of overflowing latrines and storm sewers. A 1909 assessment of trends in national water consumption in Britain observed that:

> *The increase in the domestic supply was at first due to the displacement of older methods of sewage disposal by the introduction of the water carriage system, the fitting of baths and WCs [water closets], not only in the better class houses, but also in the smaller dwellings, the provision of a separate tap within each cottage, instead of a single standpipe in a yard, supplying the block of dwellings enclosing it; in suburban districts, due to the custom of watering lawns and gardens.* (Baldman-Wiseman, 1909, pp259–260)

Municipal sewerage, in turn, increased wastewater loads on rivers. Wastewater discharges on rivers and streams created water pollution problems for downstream municipalities, forcing these to install purification works, in turn (Tarr et al, 1984). In this manner, piped water plus sewerage unwittingly propagated itself along watercourses.

MUNICIPALITIES AND PIPED WATER: DRIVERS OF EARLY CONCESSION EXPERIMENTS IN THE NORTH

Pollution of local water sources created vibrant competitive markets for freshwater in 19th-century cities. In The Netherlands, entrepreneurs hauled freshwater into cities with barges (Van den Noort and Blauw, 2000). In most cities, bulk water was delivered to households in barrels mounted on horse-drawn wagons. Piped technology made its entry as one more source of domestic water. The story of St Petersburg is not untypical:

> *Until 1825 the whole population of the city took water by pails straight from the Neva River and nearby watercourses. The first waterworks equipped with manual pumps were built in 1826. The water was obtained from the Neva River and then delivered to the citizens by water carts. It was expensive but rather pure and appreciated as well. In 1846 the first private water enterprise was established to construct a centralized water supply system in two parts of the city, but it turned out to be unprofitable and was soon closed down. The same fate was shared by another company which was established in 1853 but closed after a couple of years. The next attempt was more successful. The government approved the design in 1859, and on 30 November 1863 piped water was delivered for the first time.* (Krasnoborodko et al, 1999, p53)

Why did cities engage in such early experiments with piped water and sewerage technologies? Over the course of the 18th century, many inventors draw up designs or took out patents for components of piped water systems (Baker, 1949). But only from 1800 on were these plans put into practice. Municipal waterworks, as the combination of filtration plant and piped networks became known, were first constructed in Britain. British engineers and venture capital subsequently transferred this technology to the European continent (Juuti and Katko, 2005), as well as to North America (Cast Iron Soil Pipe Institute, 2006).

Public health concerns certainly do not tell the whole story of the early evolution of piped water. Piped water systems were initially introduced because they serve not just one, but a handful of uses: piped water is used for fire-fighting (Hassan, 1985, p543), for domestic water needs, for street cleaning, for the removal of human wastes, and for commercial and industrial purposes.

Table 4.1 *Drivers of investments in improved water and sewerage services in 23 European cities (1850–1900)*

Drivers	*Number*	*Share (percentage)*
Profit	13	57
Fire protection	9	39
Water scarcity (quantity and quality)	17	74
Environmental protection	2	9
Public health	9	39

Source: Juuti and Katko (2005, p223)

The analysis of European case histories illustrates this point. Table 4.1 offers a summary of the factors driving the introduction of piped water and sewerage in 23 towns and cities. Part of the reason why these municipalities committed themselves to piped supply is that local water sources were increasingly polluted. Moreover, piped supply offers protection against fire hazards. The significance of fire protection is overlooked in the water supply literature. In Finland, which featured fire-prone towns with many wooden structures:

> *The General Fire Assistance Company of the Grand Duchy of Finland was established in 1832. Later on, cities received funding from this company on good terms for establishing water works.* (Juuti and Katko, 2005, p39)

Profitability was also of paramount importance in early experiments. Waterworks require large initial investments, and municipal administrations must balance their budgets. Most administrations at the time were unwilling to commit municipal funds to risky experiments with novel water supply technology. This is why they left the task of developing early waterworks to the private sector. The predominance of the private sector in early waterworks development should be seen in terms of the riskiness of experiments with untried technology, doubts about the possibility of balancing the books over the longer run, and the exposure of public-sector administrators to public criticism. Consider also that urban water scarcity impinges upon commercial and industrial demand for piped water supply.

Table 4.2 illustrates these points: it presents the results of a factor analysis performed on data about the drivers of investment in water supply systems during the second half of the 19th century for 23 European cities. Fire protection and water scarcity go hand in hand as factors driving investment in piped water during this period. Furthermore, environmental protection and profit feature as separate factors. It is striking that the public health motive loads negatively on each of these drivers, emerging as a default factor – that is, public health concerns appear to drive investments only in the absence of other drivers.

Table 4.2 *Factor*[a] *matrix of investment drivers, improved water and sewerage services in 23 European cities (1850–1900)*

		Factor 1	*Factor 2*	*Factor 3*
1	Profit			0.94
2	Fire protection	0.75		
3	Water scarcity	0.91		
4	Environmental protection		0.87	
5	Public health	–0.41	–0.68	
	Eigenvalues	1.6	1.2	1.1
	Variance explained (percentage)	32.7	23.9	21.6

Note: [a] Extraction method is principal components analysis (PCA); rotation is varimax.
Source: Juuti and Katko (2005, p223)

Advent of the piped water paradigm in the North

During this early period, cities and towns on the European continent started to sign concession contracts with private waterworks operators. The Roman law concession instrument stems from the Middle Ages. A concession is a contract featuring a sovereign authority as the first party and a private entity, known as a concessionaire, as the second party. Pre-19th-century concessions typically involved activities such as mining and quarrying or the construction and exploitation of canals. Under concession contracts, concessionaires had the right to carry out a business undertaking that infringed on resources controlled by the state (Dankers-Hagenaars, 2000a, p66).

City administrations on the continent used the concession as legal vehicle for the new urban infrastructure industries (gas, water, electricity and the like) that emerged during the 19th century. As said, early waterworks concessions were granted in a competitive market for urban water. These waterworks concessions were akin to permits. Under the concession, the concessionaire had the right to construct purification plants and bury pipes in municipal soil for the supply of potable water. The concession, however, did not give the concessionaire a sole right to supply domestic water in the city. At most, early concessions contained clauses in which the municipal administration promised not to engage in rival concessions for waterworks in the concession area.

All of this changed during the second half of the 19th century. From the 1880s on, supra-local governments ratified laws that made city and town administrators responsible for solving the urban water and health crisis. In Germany, Schramm (2004, p11) describes how, at the end of the 19th century:

> *Legislators at both the State and Empire level required the municipalities to assume the role of guarantor, that is, to commit themselves to assuring the provision of drinking water (and water for industrial use), and to the maintenance of the quality of drinking water. Inside a few decades, then, what at first had been a voluntary assumption of responsibility on the part of the municipalities became the duty of a guarantor.*

In The Netherlands, the Housing Law of 1901 made municipal authorities responsible for the provision of adequate water and sanitation. A similar development occurred in Britain with the passing of the Public Health Act of 1875, in Sweden, which ratified a first Public Health Law in 1874, and elsewhere in Europe (Juuti and Katko, 2005). This legal pressure forced responsibility for urban public health upon municipal authorities in the North.

THE WATERWORKS CONCESSION: PERMIT, CONTRACT OR DECREE?

The public health legislation enacted at the supra-municipal level put a new spin on the system of waterworks concessions. From the perspective of the municipal administration, concession-granting now became not only a matter of permitting and contracting private operations, but also of regulating public health. Previously, concessionaires obtained permission from city governments to lay pipes in municipal soil and to sell piped water, and agreed with city governments to build and operate waterworks. In the new context, city administrations also insisted that the waterworks had to be operated in the interest of public health.

Municipal attempts to load generic responsibility for public health on existing concession arrangements created a crisis in the urban waterworks industry. Local governments tied in ongoing contracts with private concessionaires discovered that for-profit concessionaires were unwilling to accept responsibility over and beyond contract terms. Municipal pressure on concessionaires led to a plethora of lawsuits on the continent (e.g. Wiarda 1939).[4] In this context, continental Europe discovered that adapting the concession vehicle to the new circumstances was not easy. Legal experts had a difficult time coming to grips with the novel public health dimension of concession-granting activities. In The Netherlands, this caused the concession to drift away from private law to find a new mooring somewhere in between private and public law, a process that unfolded over the space of a century (Dankers-Hagenaars, 2000a). Writing in 1898, Nap observed that, even though concessions for public enterprises are contractual agreements, they are no part of private law proper because 'in nearly all cases they are granted in the public interest' (Nap, 1898, 16). Nevertheless, well into the 1930s there was still no consensus among Dutch experts whether water concessions were part of public or private law (Wiarda, 1939).

It is noteworthy that France followed a different path. Unlike in other European countries, in the French case concessions were anchored in public law at the outset. Early French waterworks concessions, along with other urban infrastructure concessions, were classified as public works concessions (*concession de travaux public*). These arrangements were slowly transformed into public service concessions (*concession de service public*). A public service concession was a concession that satisfied a number of criteria. First, the first party to the concession contract was a sovereign authority; second, the concession concerned a task in the public interest that would otherwise be executed by the government; third, the concession activity was intended for public consumption; fourth, the concessionaire derived an income from payments made by users/customers; fifth, the concession existed by decree (Dankers-Hagenaars, 2000a, p272). Dankers-Hagenaars, referring to Llorens (1992), claims that the public service concession vehicle received official recognition in France early in the 20th century (2000a, p218). This created considerable confusion both in the European context (Commission of the European Communities, 2000), as well as in the context of transferring European legal codes to developing countries (Braadbaart, 2005).

MUNICIPALIZATION

The rise of the public service concession was accompanied by a gradual shift in utility ownership from private entities to municipal administrations. This phenomenon is known as municipalization in the US and as municipal socialism in the UK. Municipalization was also in evidence on the European continent (see Table 4.3). The takeover of private waterworks by city administrations during the early 20th century has generated a voluminous American literature. One issue was why city administrations ousted private operators (Jacobson, 2000; Troesken and Geddes, 2003), another how this takeover affected the efficiency of service delivery (Demsetz, 1968). A large number of empirical tests comparing the relative efficiency of public and private water utilities was carried out during the 1970s and 1980s. These tests show, by and large, that whether a utility is publicly or privately owned does not matter for the price or productivity of water services (Braadbaart, 2002).

Municipalization occurred for various reasons. First, city administrations discovered that transferring generic responsibilities to private concessionaires was not without problems. Private concessionaires, as for-profit entities, will only accept limited responsibility and are prone to cherry-picking behaviour. This behaviour was incompatible with the generic responsibility for safeguarding public health that now rested on the shoulders of municipal administrations (Hassan, 1985). This friction created bad blood between administrations and concessionaires and resulted in much litigation. Second, municipal administrations had acquired sufficient experience with waterworks concessions to know that the piped water business was profitable and had a low-risk profile. Third, the extension of water networks throughout the city required a planning time horizon that was too long for private entities (Behling, 1938; Jacobson, 2000).

Table 4.3 *Municipalization of waterworks in the US, Britain and The Netherlands, circa 1800–1930*

Country	*Year*	*Number of waterworks surveyed*	*Share of public owned works (percentage)*
US	1800	16	6
	1927	9800	70
Britain	1801	12	58
	1851	60	27
	1901	81	90
The Netherlands	1890	37	30
	1900	65	40
	1910	111	54
	1920	144	72
	1930	195	81

Source: Eutsler (1939, p89): US; Hassan (1985, p536): Britain; Blokland et al (1999, p38): The Netherlands

THE PUBLIC SERVICE CONCESSION AND EURO-AMERICAN PERCEPTIONS OF DOMESTIC WATER

The transfer of generic responsibility for urban public health to local administrations had a number of important consequences. Responsibility for public service water concession gave waterworks operators a virtual monopoly on domestic water supply in North American and European cities. Competing supplies were gradually phased out and forgotten. By the middle of the 20th century, the piped water paradigm was firmly entrenched in the Euro-American mindset. Thus, the European Standards for Drinking-Water of 1970 state that:

> *This report is concerned with the minimal chemical and bacterial quality that might reasonably be expected of piped supplies of water for domestic use. By a piped supply is meant a drinking water which is supplied through a distribution system and which is under the control of, or subject to regulation made by, communal or local authorities.* (WHO, 1970, p11)

The World Health Organization (WHO)'s global standards for water quality, first appearing in 1958, likewise equate tap water with domestic water. Compare the following observation made in a study of domestic water commissioned by the WHO:

> *In its Guidelines for Drinking-Water Quality, WHO defines domestic water as being 'water used for all usual domestic purposes, including consumption, bathing and food preparation'... This implies that the requirements with regard to the adequacy of water apply across all these uses and not solely in relation to consumption of water.* (Howard and Bartram, 2003, p2)

Last, but not least, the term drinking water (German: *trinkwasser*; Dutch: *drinkwater*; Spanish: *agua potable*; French: *eau potable*) becomes synonymous with tap or piped water (German: *kran/leitungswasser*; Dutch: *kraan/leidingwater*; Spanish: *agua del grifo/corriente*; French: *eau du robinet/de distribution*). Table 4.4 summarizes these developments.

Table 4.4 *Evolution of urban water concessions in Europe and North America*

Stage	*Concession scope*	*Market power of concessionaire*	*Perception of piped water*
Concession contract	Piped water for public use	Monopoly, sometimes contested, on supply of piped water for public use	Piped water is one source of drinking water; piped water ≠ domestic water
Early public service concession	Drinking water	Monopoly on supply of drinking water	Piped water ≈ drinking water; piped water ≠ domestic water
Mature public service concession	Domestic water	Monopoly on supply of domestic water	Piped water = drinking water; piped water ≈ domestic water

Source: author

TRANSFER OF THE PARADIGM OF PIPED WATER TO THE SOUTH

'The rapidity of urbanization in most of the pre-industrial areas is surprising... The general picture is ... one of fast urbanization comparable to that experienced at earlier periods in the now industrialized nations' is the observation of Davis and Golden (1954, p20) during the early 1950s. Sovani (1964, p113) finds the phenomenon worrying enough to coin the term 'over-urbanization' and thinks that this has happened 'because rural migrants have been "pushed" rather than "pulled" into the urban areas in these countries, as a result of great and mounting population pressure in the rural areas'.

Even though the recent evidence clearly points out health gains from access to piped water and sanitation (Esrey, 1996; Esrey et al, 1985, 1991), this evidence is not borne out directly by tallying access to piped water and population health. The reasons for this are, first, that how households use water in their homes has an effect on health (Sobsey, 2002) and, second, that water availability varies from one location to another (UNESCO–IHP, 2006). The result of these two factors yields the mixed pattern of Table 4.5, which shows data for selected Latin American countries.

Table 4.5 *Access to piped water and population health, Latin America*

Country	*Percentage of total population supplied by piped water*[a]		*Health of population*	
	Urban	*Rural*	*Urban*	*Rural*
			Crude death rate[b]	
El Salvador	41.1	4.7	15.4	10.1
Dominican Republic	100.0	0.0	12.7	9.2
Mexico	60.0	6.3	11.1	
			11.9	
			Deaths of children under one year of age[c]	
Argentina	71.5	0.0	32	61
Chile	73.9	0.0	83-88	120
Venezuela	50.7	3.5	53	55

Notes: a = 1956.
b = 1959–1960.
c = El Salvador, 1962; Dominican Republic, 1949–1951; Mexico, 1959–1961.
Source: Arriaga and Davis (1967, p101); Johnson (1964, pp296, 306)

Nevertheless the overall effect of the flushing of the city with piped water on urban death rates is momentous:

> *The evidence from the Americas suggests that the 'epidemiological transition' (the shift from a predominance of communicable disease to a predominance of non-communicable disease) is taking place fastest in countries with highest levels of urbanization, and that the transition generally occurs first in urban areas... This pattern is supported by studies in other developing countries that demonstrate higher rates of malaria, malnutrition, maternal mortality and respiratory diseases in rural compared with urban populations.* (Harpham and Molyneux, 2001, p115)

Table 4.6 provides meta-evidence on the relationship between a variety of hygiene measures and diarrhoea in support of this point.

Table 4.6 *Meta-analysis of the effect of hygiene measures on diarrhoea*

Area	*Number of studies*	*Relative risk*
Hygiene	11	0.63
Sanitation	2	0.68
Water supply	6	0.75
Water quality	15	0.69
Multiple	5	0.69

Source: Fewtrell et al (2005)

What we see happening is the application of principles of urban health planning experimented in Western Europe and North America to developing country cities, with a great onset of urban population as an immediate effect. Note how this phenomenon initially results from the endeavours of Western-bred and trained urban health specialists residing in the capital cities of protectorates and other colonial centres of power. Subsequently, local administrators take over the urban health effort, partly for selfish reasons and partly for nation-building purposes. It is they who take up residence in the fully plumbed mansions abandoned by the departing colonial elite. It is they who come to equate piped water supply and wastewater removal with modern living that should become available to all.

The result of this effort is to make people live longer in cities than they do in the countryside, with city populations growing spectacularly as a result, a phenomenon that we struggle with to this day (McMichael et al, 1999). Furthermore, the piped water paradigm has settled itself at the heart of local political debates and has become a main recipient of international aid (Braadbaart, 2005).

CONCLUSIONS

19th-century innovation in piped urban supply in the North. Piped water and sewerage systems made their first appearance in European and North American cities during the 19th century. After a phase of early innovation driven by private entrepreneurship, piped water became the sole source of domestic water in cities and towns. Household multi-sourcing of water and multiple waste disposal strategies gave way to mono-sourcing and disposal. Households and businesses increasingly solely sourced water from the town's waterworks. And, increasingly, human wastes were disposed of through water-carriage systems. This monopolization of the urban market for domestic water was completed by the middle of the 20th century.

The historical record traces the emergence of piped water and sewerage as a paradigm for domestic water to the 19th-century urban water and waste crisis. Evidence marshalled from the European continent shows that, from the early 1800s, continental municipalities encouraged private entrepreneurs to pioneer the development of waterworks – that is, piped transport and distribution systems preceded, in most cases, by purification plant. The large volumes of water available through piped networks occasion investment in a parallel network of sewer lines. Investments in piped water supply were made for a variety of reasons, including fire protection, the profit motive, the increasing scarcity of urban water resources, household demand, and commercial and industrial needs. Public health concerns were a minor driver of this innovation.

From the 1880s on, municipalities became more deeply involved in the fast-growing piped water business. They did so in response to legislation that placed responsibility for solving the urban public health crisis on the shoulders of municipalities. Under pressure, city administrations committed themselves fully to large-scale and centralized piped technologies for handling urban water and waste.

20th-century transfer of the piped water paradigm from North to South. The North–South transfer of the piped water paradigm early in the 20th century had a number of important consequences for cities in the South. In the first place, Southern cities became hydrocephalous. Thanks to liquid removal systems and the rapid removal of solid wastes, cities became successful in terms of guaranteeing high rates of survival in their domain (Esrey, 1996). This experience was very different from that experienced by Northern cities in the late 19th century (Haines, 2002). The high urban survival rates achieved in the South are made even more marked by the circumstance that medical expertise tended to concentrate on urban centres (Stolnitz 1958, 1965). The consequences of this high rate of urban concentration became visible at the end of the 20th century when developing world cities teeming with human life came to be seen as a premier global problem (McMichael et al, 1999).

Second, the piped water paradigm was pioneered in the North, which has the water resources for copious flushing required to make the system work. Cities in the South do not always possess the necessary water resources, however. In cases where water resources lack in abundance, severe crises ensue (UNESCO–IHP, 2006).

Third, as was the case in the North, local governments in the South are key to putting the piped water paradigm into place. It is local governments who solve the numerous title and ownership issues that arise from diverting large amounts of water into man-made water and sanitation systems. It is local governments who contract with private-sector firms for a variety of technical services, such as the roll-out of piped networks and the construction of treatment plants.

Whether local governments succeed in these tasks is largely a matter of their autonomy from the indigenous power structure (Braadbaart et al, 2007). Local and national politics tend to throw a spanner in the works of local government trying to establish a viable piped water paradigm (Braadbaart, 2005, 2007; Braadbaart et al, 2008).

It does not help, fourth and finally, that rolling out the piped water paradigm in premier cities is a costly undertaking. Much of the cost involved is borne by foreign parties in the form of grants or long-term loans (Braadbaart, 2005). This creates ample opportunities for foreign interests to work themselves in the skein, adding to a sense of permanent crisis surrounding the local politics of essential water services.

Notes

1 I do not consider bottled water a product that competes with tap water as far as domestic water consumption is concerned.

2 We now know that improvements in household hygiene and waste removal must have contributed more to falling mortality rates than improvements in the quality of water supplied to the home (Esrey, 1996).

3 Indeed, this technology as developed for the draining of mine shafts in Britain figures as a catalyst of the Industrial Revolution (Landes, 1969, pp99–102).

4 As it does in the US (Troesken and Geddes, 2003).

References

Arriaga, E. R. and K. Davis (1967) 'Rural–urban mortality in developing countries: An index for detecting rural under-registration', *Demography*, vol 4, no 1, pp98–107

Baker, M. N. (1949) *The Quest for Pure Water*, American Water Works Association, New York

Baldman-Wiseman, W. R. (1909) 'The increase in the national consumption of water', *Journal of the Royal Statistical Society*, vol 72, pp248–303

Behling, B. R. (1938) *Competition and Monopoly in Public Utility Industries*, University of Illinois, Urbana, IL

Blokland, M., O. Braadbaart and K. Schwartz (1999) *Private Business, Public Owners: Government Shareholdings in Water Enterprises*, Netherlands Ministry of Housing, Spatial Planning, and the Environment (VROM), The Hague

Braadbaart, O. (2002) 'Private versus public provision of water services: Does ownership matter for utility efficiency?', *Journal of Water Supply: Research and Technology – AQUA*, vol 51, pp375–388

Braadbaart, O. (2005) 'Privatizing water and wastewater in developing countries: Assessing the 1990s' experiments', *Water Policy*, vol 7, no 4, pp329–344

Braadbaart, O. (2007) 'Privatizing water: The Jakarta concession and the limits of contract', in P. Boomgaard (ed) *A World of Water: Rain, River and Seas in Southeast Asian Histories*, Royal Netherlands Institute of Southeast Asian and Caribbean Studies (KITLV), Leiden, pp297–320

Braadbaart, O., N. van Eybergen and J. Hoffer (2007) 'Managerial autonomy: Does it matter for the performance of water utilities?', *Public Administration and Development*, vol 27, pp111–121

Braadbaart, O., M. Zhang and Y. Wang (2008) 'Managing urban wastewater in China: A survey of build–operate–transfer contracts', *Water and Environment Journal*, doi:10.1111/j.1747-6593.2008.00108.x

Cast Iron Soil Pipe Institute (2006) *Cast Iron Soil Pipe and Fittings Handbook*, Cast Iron Soil Pipe Institute, Chattanooga, TN

Clarke, E. C. (1885) *Main Drainage Works of the City of Boston*, Rockwell and Churchill, Boston

Commission of the European Communities (2000) *Commission Interpretative Communication on Concessions under Community Law*, Commission of the European Communities, Brussels

Dankers-Hagenaars, D. L. M. T. (2000a) *Op het spoor van de concessie: Een onderzoek naar het rechtskarakter van de concessie in Nederland en in Frankrijk*, Boom, The Hague

Dankers-Hagenaars, D. L. M. T. (2000b) 'Beknopte schets van de concessie in het verleden en van de concessie in Frankrijk', in *De concessie, 'revival' van een oud instrument?*, Ministerie van Verkeer en Waterstaat, The Hague, pp35–49

Davis, K. and H. Hertz Golden (1954) 'Urbanisation and the development of pre-industrial areas,' *Economic Development and Cultural Change*, vol 3, no 1, pp6–26

Demsetz, H. (1968) 'Why regulate utilities?', *Journal of Law and Economics*, vol 11, pp55–66

Esrey, S.A. (1996) 'Water, waste and well-being: A multi-country study', *American Journal of Epidemiology*, vol 143, no 6, pp606–623

Esrey, S. A., R. G. Feacham and J. M. Hughes (1985) 'Interventions for the control of diarrhoeal diseases among young children: Improving water supplies and excreta disposal facilities', *Bulletin of the World Health Organization*, vol 63, no 4, pp757–772

Esrey, S. A., J. B. Potash, L. Roberts and C. Shiff (1991) 'Effects of improved water supply and sanitation on ascariasis, diarrhoea, dracunculiasis, hookworm infection, schistosomiasis, and trachoma', *Bulletin of the World Health Organization*, vol 59, no 5, pp609–621

Eutsler, R. B. (1939) 'Public and private ownership of water supply utilities', *Annals of the American Academy of Political and Social Science*, vol 201, pp89–95

Ewbank, T. (1842) *A Descriptive and Historical Account of Hydraulic and Other Machines for Raising Water, Ancient and Modern*, D. Appleton and Company, New York

Falkus, M. E. (1982) 'The early development of the British gas industry, 1790–1815', *The Economic History Review*, vol 35, pp217–234

Fewtrell, L., R. B. Kaufmann, D. Kay, W. Enanoria, L. Haller and J. McColford Jr (2005) 'Water, sanitation, and hygiene interventions to reduce diarrhoea in less developed countries: A systematic review and meta-analysis', *The Lancet*, vol 5, pp42–52

Haines, M. R. (2002) *The Great Modern Mortality Transition*, Social Science History Association, Fort Worth, TX

Harpham T. and C. Molyneaux (2001) 'Urban health in developing countries: A review'. *Progress in Development Studies*, vol 1, no 2, pp113–137

Hassan, J. A. (1985) 'The growth and impact of the British water industry in the nineteenth century', *The Economic History Review*, vol 38, pp531–547

Howard, G. and J. Bartram (2003) *Domestic Water Quantity, Service Level and Health*, World Health Organization, Geneva

Jacobson, C. D. (2000) *Ties that Bind: Economic and Political Dilemmas of Urban Utility Networks, 1800–1990*, University of Pittsburgh Press, Pittsburgh, PA

Johnson, G. (1964) 'Health conditions in rural and urban areas of developing countries', *Population Studies*, vol 17, pp293–309

Juuti, P. S. and T. S. Katko (2005) *Water, Time and European Cities: History Matters for the Future*, Tampere University of Technology, Tampere, Finland

Krasnoborodko, K. I., A. M. Alexeev, L. I. Tsvetkova and L. I. Zhukova (1999) 'The development of water supply and sewerage systems in St Petersburg', *European Water Management*, vol 2, pp51–61

Landes, D. (1969) *The Unbound Prometheus: Technological Change and Industrial Development in Western Europe from 1750 to the Present*, Cambridge University Press, Cambridge

McMichael, A. J., B. Bolin, R. Costanza, G. C. Daily, C. Folke, K. Lindahl-Kiessling, E. Lindgren and B. Niklasson (1999) 'Globalization and the sustainability of human health', *BioScience*, vol 49, no 3, pp205–210

Meidinger, E. E. (1980) 'The "public uses" of eminent domain: History and policy', *Environmental Law*, vol 11, pp1–66

Millward, R. (2004) 'European governments and the infrastructure industries, c.1840–1914', *European Review of European History*, vol 8, pp3–28

Nap, J. N. (1898) *Concessies voor publieke ondernemingen*, Universiteit Groningen, Groningen, The Netherlands

Rosen, G. (1958) *A History of Public Health*, MD Monographs on Medical History, MD Publications, New York

Schramm, E. (2004) 'Privatization of German Urban Water Infrastructure in the 19th and 21st centuries', *International Summer Academy on Technological Studies*. Institut für sozial–ökologische Forschung, Deutschlandsberg, Austria

Sobsey, D. (2002) *Managing Water in the Home: Accelerated Health Gains from Improved Water Supply*, Water, Sanitation and Health, Department of Protection of the Human Environment, World Health Organization, Geneva

Sovani, N. V. (1964) 'The analysis of "over urbanization"', *Economic Developmental and Cultural Change*, vol 7, pp113–122

Stolnitz, G. J. (1958) 'The revolution in death control in non-industrial countries', *Annals of the American Academy of Political and Social Science*, vol 316, pp94–101

Stolnitz, G. J. (1965) 'Recent mortality trends in Latin America, Asia and Africa: Review and re-interpretation', *Population Studies*, vol 19, no 2, pp117–138

Tarr, J. A. (1996) *The Search for the Ultimate Sink: Urban Pollution in Historical Perspective*, University of Akron Press, Akron, OH

Tarr, J. A., J. McCurley III, F. C. McMichael and T. Yosie (1984) 'Water and wastes: A retrospective assessment of wastewater technology in the United States, 1800–1932', *Technology and Culture*, vol 25, pp226–263

Troesken, W. and R. Geddes (2003) 'Municipalizing American Waterworks, 1897–1915', *The Journal of Law, Economics and Organization*, vol 19, pp373–400

UNESCO–IHP (United Nations Educational, Scientific and Cultural Organization–International Hydrological Programme) (2006) *Urban Water Conflicts: An Analysis of the Origins and Nature of Water-Related Unrest and Conflicts in the Urban Context*, UNESCO–IHP, Paris

Van den Noort, J. and M. Blauw (2000) *Water naar de zee: Geschiedenis van Waterbedrijf Europoort 1874–1999*, Jan van den Noort, Rotterdam

WHO (World Health Organization) (1970) *European Standards for Drinking Water*, second edition, WHO, Geneva

Wiarda, G. J. (1939) *Overeenkomsten met overheidslichamen*, Tjeenk Willink, Zwolle, The Netherlands

5

Management and Organization of Water and Sanitation Services: European Experiences

Osmo T. Seppälä and Tapio S. Katko

Water services within the framework of integrated water resources management (IWRM)

When discussing water management, we should distinguish between water resources, on the one hand, and water and wastewater services, on the other. Yet, they are closely interlinked. After water is used for community or other purposes, it is discharged via sewerage and wastewater treatment systems to downstream watercourses, which, again, are sources of other users. Therefore, the quantity and quality of water resources are of utmost importance to suppliers and users.

During the last decade, integrated water resources management (IWRM) has become one of the basic water service-related policies. IWRM calls for the planning of water resources use on a larger geographical scale than typical water services areas. While water resource planning has a wide geographic perspective, the subsidiarity principle calls for water services management at the lowest appropriate level, as pointed out in Figure 5.1 (Pietilä, 2006, p29).

Although IWRM has been emphasized as a systemic approach to water sector development, it may not, as such, provide solutions to water and sanitation services (WSS) sector paradigms (Ghosh, 1999). Obviously, an adequate balance should be achieved between the WSS sub-sector and the water resources sector (Seppälä, 2002, p380).

Global Water Partnership (GWP) defines IWRM as 'a process, which promotes the coordinated development and management of water, land and related resources in order

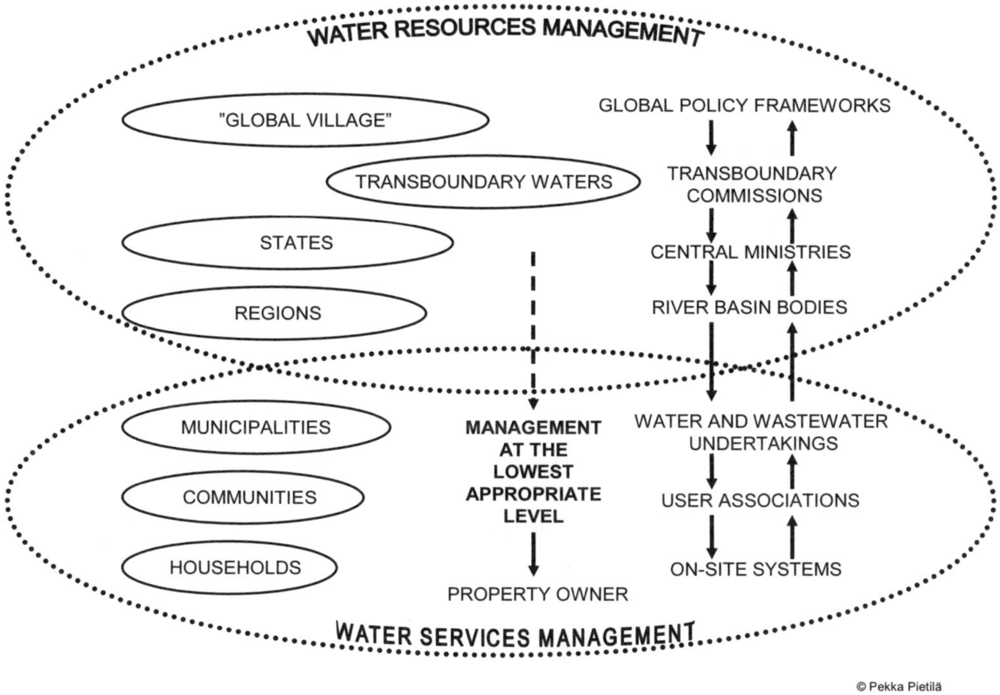

Figure 5.1 *Managing water resources and services at various levels*

Source: Pietilä (2006, p30)

to maximize the resultant economic and social welfare in an equitable manner without compromising the sustainability of vital eco-systems' (GWP, 2003, p1). This principle, as such, is widely accepted. Yet, surprisingly little attention has been paid to the priorities of water-use purposes and the potential conflicts between them.

A study by Katko and Rajala (2005) explored the priorities of ten different water-use purposes in ten countries/regions on five continents with different water resources and use, socio-economic and cultural conditions. On average, community water supply was ranked as the first priority – even in a case where about 80 per cent of the water was used for irrigation, while nature conservation was ranked second and hydropower third. Thus, in addition to water quantity, sustainable water resources management has to consider water quality, priorities of water-use purposes and their potential conflicts, as well as local conditions and needs. In addition, the legislation in force in several countries clearly defines that the first priority should be community water supply. Moreover, as the Dublin Principles adopted in preparation for the 1992 United Nations Conference on Environment and Development (UNCED) in Rio de Janeiro rightly emphasize, water resources planning should be done at the regional level and management should be implemented at the lowest possible level (UN, 1992).

One drawback of IWRM is that it tends to focus solely on surface waters. In many countries, groundwater is used especially for community supply. Although groundwater is over-mined in some megacities and regions, it still has far more potential for use. This need is currently noted, for instance, by the UNESCO-coordinated project on International Shared Aquifer Resource Management (UNESCO, 2008). Similarly, regarding the integration between water and wastewater services, there is some degree of diversity in the European context: while in Nordic urban systems they are normally managed by one joint organization, in many other cases they operate under separate entities (Juuti and Katko, 2005, pp236–237). In the latter case, it is important that the idea of integrated management is taken seriously through other forms of cooperation.

After discussing the connections between water resources management and water services, we will explore water services and their management from the following perspectives: ontology (the special features of water services); institutional approaches; time frames of decision-making and management; path dependence; scopes and elements of management; management and leadership and their interconnections; and diversity of options for water services management, followed by concluding remarks. This chapter is based on research carried out by the authors and colleagues, who are members of the capacity development in water and environmental services (CADWES) research team (www.cadwes.org) at Tampere University of Technology, Finland, including involvement in various research projects such as PRINWASS (www.prinwass.org) and Watertime (www.watertime.net).

Ontology and the unique nature of water and sanitation services

Undoubtedly, water and sanitation services are a basic requirement for all human activity. One of the special features of water and wastewater services is that water is clearly a local resource to be mined, processed and distributed locally. Occasionally, water services are mistakenly considered similar to energy services. Instead of making such simplified assumptions, we should, rather, have a more analytical approach and explore the special features of any services and only then consider the options for their management.

Due to the local nature of water and wastewater services, the subsidiarity principle – decision-making at the lowest appropriate level – is highly applicable to them (Pietilä, 2006, p102). Water is a satisfier of a basic need and cannot be replaced with any other product; operational water services are a prerequisite for all economic activity; water supply is and will remain a natural monopoly; water services are limited and highly susceptible to local natural conditions; and there are competing uses of water that may be mutually exclusive (Pietilä and Katko, 2006).

Furthermore, water and sewerage services have certain special features not typical of other infrastructure. First, they are exceptionally capital intensive: up to between 65 and 75 per cent of annual costs can be capital costs. Second, fixed costs – those that do not vary with the volume of production – often account for about 80 per cent of operating expenses. If water consumption per capita or total consumption declines, tariff structures based merely on water consumption have to be revised. Third, and perhaps most importantly, the WSS infrastructure is a natural monopoly – a concept first introduced by John Stuart Mill in 1848 (Sharkey, 1982, p14). Certainly, it is not feasible to construct several parallel networks and facilities for a single service area.

The role of water as a basic need, a merit good, and a social, economic, financial and environmental resource makes its pricing exceptionally controversial. Another economics-based classification of water as a good is based on its single or joint use versus its exclusion or non-exclusion. In single use, water can be used in on-site systems as a private good and through a water cooperative as a toll or club good where it is delivered to members only (some authors have also argued that WSS in Western Europe currently resemble a club good; see Chapter 14 in this volume). Yet, in the majority of cases, water is supplied through public utilities as a common pool resource, a public and social good that may often be subsidized to different degrees and that, in some cases, like public standpoints or fountains, could be even delivered free of charge.

Unfortunately, the need to consider water as an economic good as stated in the 1992 Dublin Principles (UN, 1992) has been often overemphasized, forgetting that water has also other important social, public policy and environmental values and requirements. Or at least the principles have been often interpreted in this way. This is a common problem that we have identified in our research on international water policies, which shows that over the years some policies become fashionable and overemphasized for a certain period, only to be sidelined or altogether abandoned later on, which runs counter to the rhetorically accepted ideal of a long-term, holistic and balanced approach to WSS policy (Seppälä, 2004, p83).

Nevertheless, the costs incurred by the provision of water supply and sanitation services must be covered in some way, preferably at least partly by users and beneficiaries. As for their positive direct and indirect impacts, investments in water supply and sewerage/sanitation, including adequate treatment and disposal, are highly justified. It is estimated that such investments pay themselves back tenfold through improved health benefits alone, in addition to providing environmental and social benefits. Different cost and benefit assumptions changed cost–benefit ratios considerably; but even in pessimistic scenarios, potential economic benefits generally outweigh the costs (Hutton and Haller, 2004).

From another angle, since sustainable water and sanitation services are subject to a variety of societal demands, they must meet several criteria. In this regard, Silfverberg (2007) suggests that in order to be considered functioning and sustainable, water services should be:

- *Socially sustainable*: fair and equitable, fulfilling the customers' varying needs and promoting sustainable development of communities, while being sensitive to changing needs.
- *Secure and operationally reliable*: of high techno-operational reliability and up to increasingly stringent health and safety requirements. Services should also be reliable in special circumstances.
- *Environmentally sustainable*: raw water supply, water treatment, wastewater sewerage, treatment and disposal should be environmentally sustainable and developed in

accordance with the increasingly stringent environmental requirements being adopted internationally. Impacts of climate change should also be considered.
- *Economically viable and efficient*: financing of water utilities should be secured, enabling long-term operation, management and development, while providing reasonably priced and equitable services.
- *Flexible*: good-quality water services should be also secured in a changing operational environment.
- *Well-managed and resourced*: customer orientation and owner policy of water utilities must be transparent and efficient, and the internal management of the professional and human resources of WSS utilities must be adequate to ensure reliable and sustainable services.

This complexity involved in the management and operation of WSS can be also captured by applying the PESTEL framework, also used by futures researchers. PESTEL stands for the 'political and policy, economic, social, technological, environmental and legislative' influences on water services, which we have adopted in several studies. Among other outcomes of this research, we have found that some of the prevailing principles informing current WSS policy reforms worldwide, such as the assumption of neoclassical economists about the centrality of rational choice in explaining human behaviour, seem not to hold true in real life, nor are they able to explain the institutional and organizational diversity that can be identified in water resources and services management (Hukka et al, 2005, p64). It is this institutional and organization diversity that we place at the centre of our interest, which we consider next.

INSTITUTIONAL APPROACH TO WATER SERVICES MANAGEMENT

Our approach to water and sanitation services management and organization is institutional. From this perspective, an enabling institutional framework for operational and sustainable water and sanitation services requires a balanced policy, a legal and administrative framework, and favourable informal institutions (see Figure 5.2).

The institutions related to water and sanitation services can broadly be categorized as:

- policies;
- legislation and regulations;
- administrative structure; and
- informal institutions (Seppälä, 2004, p45).

North (1990) recognizes institutions – both formal and informal – and institutional incentives as the underlying determinants of the long-run performance of economies. Informal institutions may account for a substantial share of the total costs of any organization.

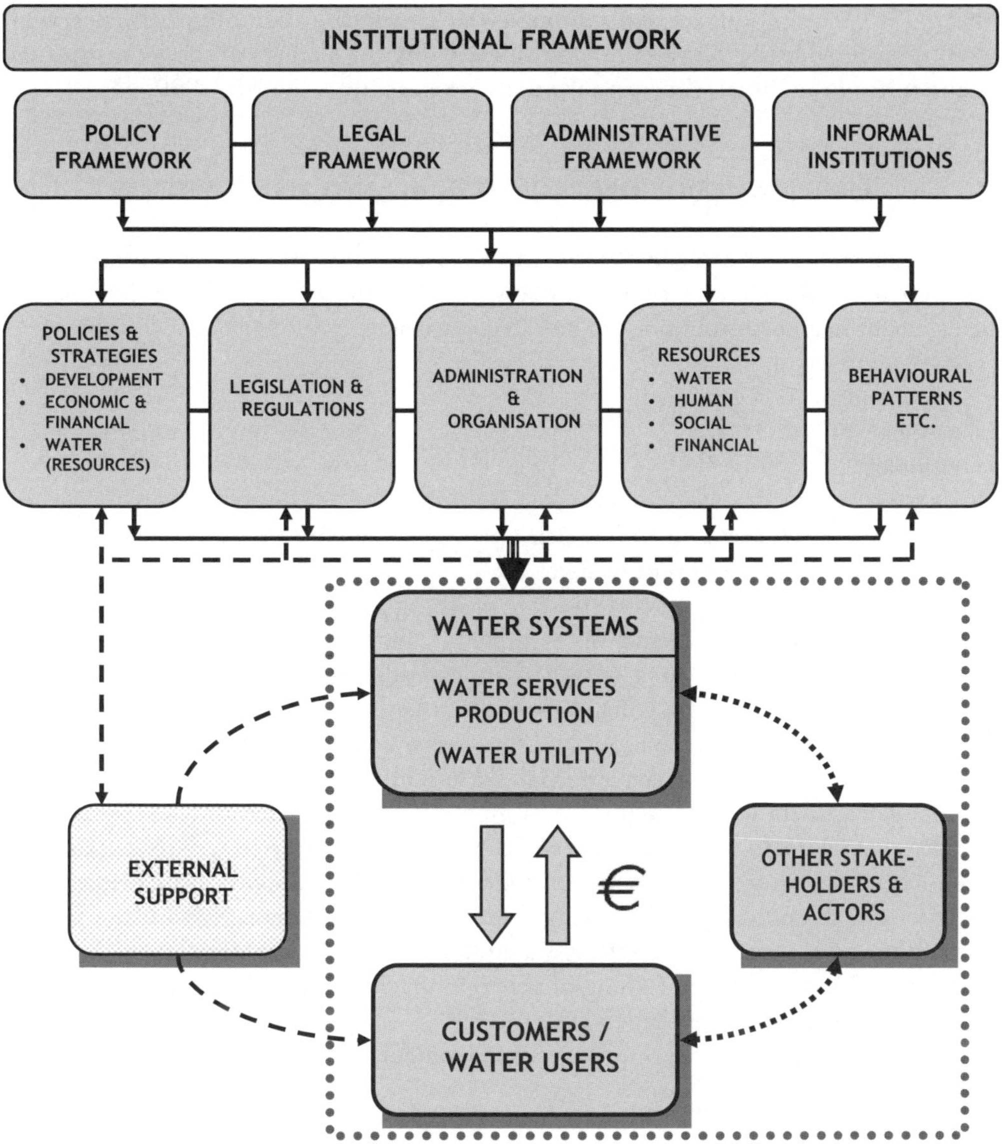

Figure 5.2 *Institutional framework for water services*

Source: Seppälä (2004, p32)

The institutional environment means the background constraints that guide individual and organizational behaviour, which is based on both formal and informal institutions. Institutional arrangements are the structured set of linked or interdependent institutions, which comprise the social system of the economic, social and political domains, and guidelines or governance structures through which decisions are implemented (Seppälä,

2004, p45). Organizations are actors through which institutions come into existence, while institutions provide the 'playground and rules'. Using the analogy of soccer, institutions are the rules of the game, while organizations are the players (North, 1990, p3).

TIME FRAMES OF DECISION-MAKING AND MANAGEMENT

In decision-making and management, three distinct time frames can be recognized. On their basis, the characteristics of decisions, and the models of decision-making management in an organization, can be categorized as:

1 opportunistic or operational;
2 strategic; or
3 visionary.

The first category focuses on short-term operational activity of a maximum duration of one to two years. The second refers to the fact that preparing for future changes and challenges requires more long-term strategic planning and management, normally over a period of three to ten years (Kaivo-oja et al, 2004). In turn, the third category, visionary management, constitutes a long-term activity extending over 10 to 30 years or more. The main challenge of visionary management is to enable the organization to utilize new opportunities in its future operations and to manage uncertainty.

Instead of operative management, Malaska and Holstius (1999) use the term opportunistic management for referring to day-to-day decisions that do not take into account considerations about futures. In the case of water and sanitation services, the requirement of a long-term perspective on operation and management renders opportunistic management inadequate as the only decision-making option. In strategic management, the focus is on medium-term strategic interests and adaptation. In visionary decision-making, the aim is long-term survival and finding of new options through the reframing of business, envisioning and creating new capabilities.

Long-term visionary development, management and thinking are necessary, for instance, for the following reasons:

- safeguarding the availability of water resources in the long term;
- long-term investments (long lifetime of assets);
- need to consider regional cooperation (e.g. supra-municipal);
- customer or citizen service demands;
- irreversible and path-dependent nature of decisions.

PATH DEPENDENCE

In relation to longer time frames, the path-dependence theory suggests that decisions made in the past are likely to have long-term impacts. Liebowitz and Margolis (1995) recognize

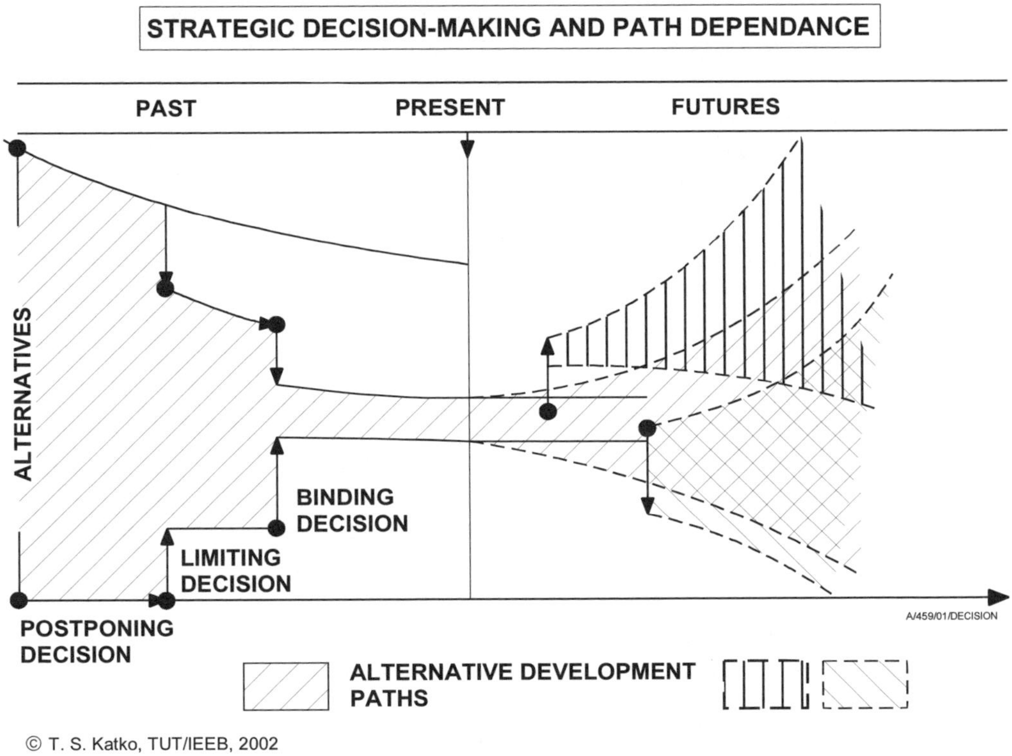

Figure 5.3 *Path dependence and related decision-making*

Source: Kaivo-oja et al (2004, p536)

three degrees of path dependence. The first degree implies no inefficiency in outcomes; the second degree leads to outcomes that are suboptimal and costly to change; and the strongest, third degree, leads to inefficient outcomes.

Path dependence is very obvious in infrastructure sectors such as water and sanitation services (Melosi, 2000). Strategic decisions are linked to path dependence and inevitably affect the available future development options or paths. By their nature, strategic decisions can be classified as binding, limiting or postponing (see Figure 5.3).

A good example of a binding decision is the selection of raw water source between ground and surface water. A far-sighted example is the decision in favour of water metering and related pricing systems. Another fortunate one is the abandoning of lead pipes in the Nordic countries – which are currently a big problem in many cities of the countries that first developed piped water supply.

As early as 1877 to 1878, Helsinki Water Works conducted experiments and found out that excess lead was diluted to their raw water from lead pipes. Therefore, by 1890, the use of lead pipes was practically abandoned, followed soon by other cities and water works in the country (Lillja, 1938, pp301–302; Katko et al, 2006), which in historical perspective has proved to be a truly visionary approach. Thus, while much of the literature often presents path dependence as binding and limiting – a sort of negative technological lock-in, as shown above – path dependencies can also be positive if good strategies have been forecasted and selected.

Naturally, the selection of a given path may also have limiting impacts. The decision-making process may cause delays or postponement when decisions cannot be made. In the context of water systems, the planning period is often about 30 years, sometimes even 50, while the actual life time of the infrastructure may exceed 100 years. These time frames leave ample scope for both positive and negative outcomes, enabling and constraining path dependencies affecting the management and operation of WSS.

Various scopes and elements of management

According to our research findings, sound water management principles are reflected in policies, roles of organizations, legislation and strategies (Seppälä, 2004, p86). These principles include, *inter alia*, the following:

- an integrated approach to water resource management;
- adequate emphasis on the environment and integration of water management within environmental management;
- publicity and appropriate participation of interested stakeholders and parties;
- water resources management at the lowest appropriate level;
- clear delineation and separation of responsibilities between policy-making and service provision, and participation of public and private sectors;
- appropriate implementation of cost-recovery policies, adequate financial autonomy, and the adoption of the user/polluter pays principle, whenever feasible;
- recognition of the limits imposed by user affordability on the range of policy and management instruments that can be deployed.

When we look at the challenges and most appropriate methods of managing water and sanitation services, there is a need to recognize the various scopes, elements and needs of management. The system of WSS consists of both the physical water and sanitation infrastructure and the management system, with all its human activity components, including the customer–citizen interface.

Asset management

In many countries, the water infrastructure is ageing, and one of the biggest challenges is how to convince the decision-makers that sufficient investments are needed for replacement and repairs. In this context, we also have to consider the vulnerability of infrastructure, including the protection of critical infrastructure or security of supply. For instance, in Finland the legislation in force allows the public water and sewerage utilities to make a 'reasonable rate of return'. Yet, legislation does not specify what exactly this means. As a consequence of having to share part of their profit with the municipal owner, WSS utilities are often left with insufficient funds for meeting the increasing investment needs of infrastructure rehabilitation. In this regard, the current system of economic regulation is rather passive and narrow in scope, and tends to focus only on customer charges. A recent study by Vinnari (2006) showed that the majority of the WSS managers interviewed were in favour of broadening regulation to also include a clearer treatment of the allowed rate of return, which they believe may contribute to alleviating the current limitations. Infrastructure renewal, though, continues to be a major challenge in most countries, and notably so in relation to ageing WSS systems in Europe, the US and, increasingly, in many developing countries too. Moreover, the evidence shows that the financial bottlenecks affecting funding for infrastructure renewal are not only the preserve of publicly managed WSS systems, but have been also identified in privately run water utilities (see, for instance, Chapter 6).

Knowledge management

In addition to their capital intensiveness, water and sewerage systems are also very knowledge intensive. Operating and managing water and sanitation services requires skilled utilization of data and information related to water resources, water and wastewater treatment processes, distribution networks, water quality control, and asset management, as well as operational and strategic management and thinking (Seppälä, 2004, p64).

In practice, most water utilities do not, however, manage their knowledge and information processes – communication, core competencies and change management – in systematic ways. Our research shows that WSS utilities often lack clear direction and purpose in the acquisition of strategic information and suffer from inefficient management of quantitative information. Knowledge management offers a long-term approach that allows utilities to leverage their most valuable assets – collective know-how, talent and expertise. Only by focusing on these resources can WSS utilities be innovative and meet new threats and opportunities (Seppälä, 2004, p64).

Water and sewerage utilities typically have substantial *tacit knowledge* accumulated in experienced employees. It has been estimated that this may represent as much as 95 per cent of all available knowledge. The great challenge is elucidating how this valuable tacit expertise can be adequately utilized and transferred. Tacit knowledge is also related to organizational memory, which is at risk in many WSS utilities due to the retirement of employees in the absence of adequate policies for the transfer of their valuable knowledge to the new generation (Seppälä, 2004, p65). Globally, it is estimated that about 40 per cent of WSS utility personnel will retire by the year 2015, while in some countries such as Finland the retirement boom will arrive even sooner. Thus, WSS utilities should pay more attention to knowledge management, especially for capturing tacit knowledge and sharing it with new employees (Sandelin, 2006, p2). Knowledge management should be seen as a necessary and practical function in every organization, not a passing fad. It should be an integral element of an organization's management system.

Both management and leadership are needed

Water and sanitation utilities have two different roles and tasks – management and leadership, which should not, however, be confused. The basic processes and functions of management include planning, organizing, directing and controlling (Grigg, 1986). Leadership, in turn, is 'the process of influencing others to understand and agree about what needs to be done, and how it can be done effectively, as well as facilitating individual and collective efforts to accomplish shared objectives' (Yukl, 2002, p7). Management consists basically of activities and processes, while leadership is about influencing and directing people and organizations, and getting things done.

It should also be remembered that water and sanitation utilities are, essentially, service organizations and, thus, leadership is equally important to them as management. Indeed, more often than not, WSS utilities have been both over-managed and under-led. However, while the technical operations of a water and sanitation utility require a high degree of organization and management skills, managing human resources (people) requires competence that goes beyond traditional management expertise – it requires leadership.

In this regard, the water and sanitation utilities of most countries have traditionally been fairly conservative in their management and leadership approaches, and have largely focused their efforts on managing technical systems. Yet, they are increasingly adopting more innovative, integrated and systemic approaches and styles of management and leadership. It is our argument that this is partly the result of the example and inspiration provided by private-sector organizations, which public WSS utilities have started to imitate and follow, along with the introduction of commercial principles in the water-service sector (Seppälä, 2004). This development, though, may also have negative impacts if the practices of the private sector are adopted uncritically without first assessing their potential applicability to WSS utilities (Windischhofer, 2007).

There is a vast diversity of management and leadership styles and methods that can and have been applied in the WSS sector, and the selection of the style best suited to a particular organization depends upon a number of issues. The importance of the various time frames of management should be kept in mind, as well as the need for a holistic and balanced approach to various elements of management. We consider this in the next section.

Diversity of options to organize and manage water and sanitation services

Considering the divergence of traditions in urban planning, water resources management, legislation and administration, cultures and organizations, and the variation in the roles of local governments in the European context, there exists a wide diversity of options for WSS management. A study by Juuti and Katko (2005) suggested that there is probably even more variety in water services management in Europe than, for instance, in the traditions of urban planning. Thus, if water services are managed at the local level, then it is likely that we can explain their form of management by studying the role of local governments in the relevant society. This would apply, for instance, to the examples of the Stadtwerke in Germany or the development of autonomous public companies operating at local and regional levels and involving various forms of private–public arrangements in many countries, which is normally associated with the central role played by local governments as it happens, for instance, in Nordic Europe or France. Contrastingly, the exceptional case of the private WSS companies of England and Wales has taken place in the context of a significant curtailment of local government authority in those countries since the late 1970s.

Some of the main options commonly applied include the following:

- 'purely public management' (e.g. direct municipal management or, in some cases, management by state organizations);
- commercialized public management (e.g. public limited companies (PLCs) or municipal enterprises);
- various forms of public–private cooperation and partnership (e.g. outsourcing of services and mixed public–private management companies);
- private management (e.g. through concessions or lease contracts);
- 'fully private options', including both private ownership and operation (see Figure 5.4).

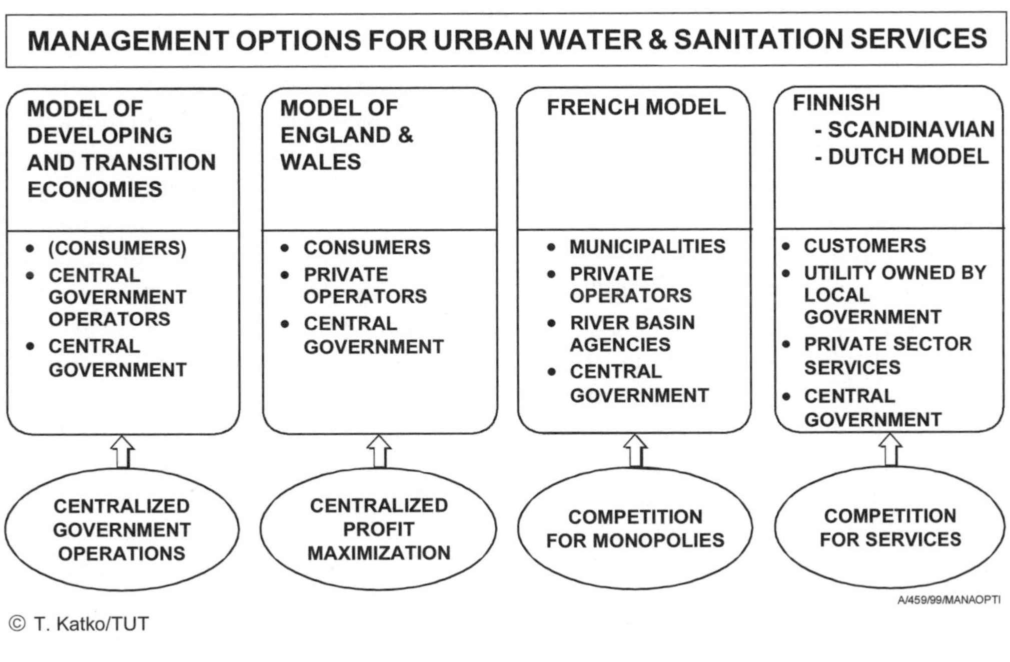

Figure 5.4 *The four main models for urban water and sewerage services: Finnish–Scandinavian–Dutch, French, English–Welsh, and developing and transition economies*

Source: Hukka and Katko (2003, p115)

Until recently, the international discussion on private-sector participation has centred far too much on two models: various forms of maintenance and operation contracts (the 'French model') and total privatization (the 'English–Welsh model'). In the French model, oligopolistic competition occurs between monopolies, and in the English–Welsh model, profit maximization is the actual driving force and premise. Indeed, in ongoing debates on the need for public–private arrangements in WSS, too often the relationship between these two sectors has been narrowly understood as covering either the French or the English–Welsh models. These debates, often under the banner of 'public–private partnerships', continue to disregard the existence of a third option: the longstanding traditional forms of public–private cooperation where public utilities buy goods and services from the private sector, which some authors wrongly consider to be inefficient and old-fashioned. However, instead of granting very long-term operational or management contracts to private companies, which effectively cancels the possibility of competitive practices, the traditional forms of public–private cooperation found, for instance, in Nordic Europe are based on continuous competitive bidding. This eliminates the possibility of the monopolistic practices that can be identified in long-term concessions or in the full divestiture model of WSS ownership and management, where the private companies tend to concentrate the sourcing of goods and services on a very small number of providers and even develop vertically integrated monopolies where most outsourced goods and services involve suppliers controlled either by the WSS utilities themselves or by their mother companies (Hukka and Katko, 2004).

Due to the limitations of a natural monopoly such as WSS, most competition tends to occur through the third option: traditional forms of public–private cooperation where, apart from the core operations, all or most of the significant production-related or non-core activities are outsourced to the private sector based on regular competitive bidding. The core operations consist particularly of strategic planning and management, outsourcing and contracting out, and overall administration (Hukka and Katko, 2003, p119).

Our research on these matters highlights the importance of examining the evolution of public–private interactions in WSS according to a historical perspective. In almost all Western countries, except France and, to some extent, England and Wales, most privately owned water systems were taken over by local governments in the mid- and late 19th century. There were serious reasons why this process took place; but to put it in a nutshell, it was mainly because the private companies were not able to meet the growing needs of society in a time of rapid urban expansion and rising living standards. Interestingly, about a century later, since the 1980s, neoliberal policy-makers have reintroduced the privatization of water and sewage systems as an 'innovation' to solve the problems of WSS worldwide, which has been promoted particularly by international financial bodies and multinational companies (Hukka and Katko, 2003). The introduction of these policies awakened much enthusiasm in many countries and led to sweet promises about investment flows, efficiency improvements and expansion of coverage to the un-served. However, at the time of writing, almost two decades later, even representatives of the institutions that have been the main promoters of privatization, such as the World Bank, have publicly admitted that the expected outcomes, like the pouring of private financing for WSS, were never realized

(Annez, 2006). From the viewpoint of citizens' and societies' expectations and seen from a professional ethics perspective, the promises were unjustified from the start.

To conclude this section on organizational and management options for WSS, it should be emphasized that regional cooperation between WSS utilities has become a widely applied option for improving services and it is being promoted in many countries. Such cooperation can take various forms, from merging smaller utilities into larger units in order to improve service delivery, economy and efficiency, to the more informal sharing of know-how. Twinning between utilities and other forms of public–public cooperation have also become common and have, increasingly, been seen by many as the way forward to improve the situation of WSS.

NEED FOR REGULATION

Whatever model is applied, the need for appropriate regulation and ground rules is a crucial requirement. For instance, Rees concluded that unregulated water and sanitation companies are simply not an option, and that regulation is needed to ensure customer protection, especially in privatized water service arrangements (Rees, 1998, pp22, 29). The social, developmental and environmental importance of the water sector means that continued public regulation is necessary, a responsibility that governments cannot escape. As several chapters in this volume demonstrate, regulation of WSS activities is being tightened in most countries, although closing the gap between the formal legal and regulatory frameworks and their actual enforcement remains an elusive target everywhere and one of the most significant challenges facing the WSS sector.

Both the private and public sectors have played their own important roles in the development of water and sanitation services, and will continue to do so in the future. In public–private partnership – or, rather, cooperation – the role of the public sector includes maintaining a strong political commitment to the universal provision of the services, as well as integrating environmental and other value considerations within the planning and decision-making process. The public sector has fundamental social and economic tasks and liabilities, and its assumption of a fundamental role in safeguarding and providing for the economic and social welfare of the citizens is a precondition of democratic government (Martin, 1994).

In conclusion, it seems inevitable that independently from the particular institutional option adopted – public, privatized or mixed – regulation will continue to be a crucial requirement. In particular, the introduction of competition (to the extent that it is possible within such a natural monopoly as WSS) requires appropriate sector-specific regulation. More importantly, perhaps, and especially in the vast majority of developing countries, reaching the un-served poor and sustainably maintaining adequate service standards, including affordable pricing of WSS, would not be possible without strict and well-designed regulation arrangements. The interests of the ultimate beneficiaries and citizens are the paramount concern to be taken into account.

CONCLUSIONS

In this chapter we have discussed the multidisciplinary and interdisciplinary nature of water management, and highlighted some of the key aspects that should be considered in the management and organization of water and sanitation services. We wish to make the following concluding remarks:

- Water resource management and water and sanitation services are closely interlinked. While the former should be viewed in a regional, national and even international perspective, the latter should essentially be managed at the local level.
- Water services have several special features compared to other infrastructure services. The role of water as a basic need, a merit good, and a social, economic, financial and environmental resource makes its pricing exceptionally controversial. Water and sewerage services are exceptionally capital intensive, the share of fixed costs is very high, and they constitute a natural monopoly.
- A distinction is to be made between institutions and organizations. The former provide the 'playground and rules' for the responsible organizations to manage services, while the latter are the actors – 'players'. Besides, informal institutions (rules, norms, behavioural codes, attitudes and other ethical standards of conduct) can have a significant effect on the total costs of an organization.
- There are distinct time frames for decision-making and management within the water and sanitation sector referred to as opportunistic or operational; strategic; and visionary. A long-term view is especially important due to the irreversible and path-dependent nature of infrastructure investment decisions and other strategic choices.
- At the utility level, the various targets, scopes and elements of management are diverse and include, for instance, technical systems, material and intellectual assets, human resources and the organizational sphere, and financial aspects. More attention should be paid to asset management (to prevent infrastructure deterioration), appropriate knowledge management tools, and use of the accumulated tacit knowledge.
- Water services organizations need both management and leadership. The former deals more with activities and processes, while the latter has to do with leading people and organizations and getting things done. The fact is that water utilities have often been over-managed but under-led.
- There is a vast diversity of institutional and management options to organize water and sanitation services, whose applicability should be considered on the basis of local circumstances. Due to the limitations of a natural monopoly, most competition occurs in the model, where, apart from the core operations, all or most of the significant production-related non-core activities are outsourced to the private sector based on competition.
- Whatever type of institutional, management and ownership arrangement is used, at least some type of sector-specific regulation is needed.

This chapter has focused on urban areas. Yet, we should remember that Europe has many countries with large rural areas where even in the future, permanent or free-time housing will, at least to some extent, use on-site systems. In addition, although some of the experiences and conclusions presented here may not be directly relevant to the cases of developing countries, the lessons learned and the principles discussed may contribute to ongoing debates in these countries and to the development of autonomous thinking about WSS organization and management that is grounded on past experience.

As a final comment, in the selection of the mode(s) of WSS implementation, one must consider the long-term experiences and nature of these services. Appropriate management options for water and sanitation services depend upon local circumstances. There is no 'one-type-fits-all' solution and the diversity of institutional arrangements and management options – '*insdiversity*' – should be cherished.

ACKNOWLEDGEMENTS

The authors wish to thank the reviewers for their comments, partners of the PRINWASS and Watertime projects for their cooperation, and the Academy of Finland (nos 115397, 210816) for its financial support.

REFERENCES

Annez, P. C. (2006) *Urban Infrastructure Finance from Private Operators: What Have We Learnt from Recent Experience?*, Policy Research Working Paper 4045, World Bank, Washington, DC

Ghosh, G. (1999) 'Some observations on the status of water policy reforms and hydropolitics', Presentation to the Vision 21 Conference, Water Supply and Sanitation Collaborative Council, Wageningen, The Netherlands, April 1999

Grigg, N. S. (1986) *Urban Water Infrastructure: Planning, Management and Operations*, John Wiley and Sons, New York

GWP (Global Water Partnership) (2003) 'Sharing knowledge for equitable, efficient and sustainable water resources management', in *Integrated Water Resources Management Tool Box*, second edition, Press Works, UK

Hukka, J. and T. Katko (2003) 'Water privatisation revisited – panacea or pancake?', *Occasional Paper Series*, no 33, International Water and Sanitation Centre (IRC), Delft, The Netherlands, www.irc.nl/page/6003, accessed May 2008

Hukka, J. and T. Katko (2004) 'Liberalisation of water sector – a way to market economy or to monopoly market?', *Water and Wastewater International*, vol 19, no 9, pp23–25

Hukka, J. J., P. S. Juuti, T. S. Katko, A. Mohamed, E.-N. Nyangeri and O. T. Seppälä (2005) 'Science for water services and related institutions', in O. Hietanen (ed) *The University Partnership Network for International Development (UniPID)*, Finnish Development Knowledge FFRC Publications, Jyväskylän, Finland, no 6, pp61–91

Hutton, G. and L. Haller (2004) *Evaluation of the Costs and Benefits of Water and Sanitation Improvements at the Global Level*, World Health Organization, Geneva

Juuti, P. and T. Katko (eds) (2005) *Water, Time and European Cities: History Matters for the Futures*, www.watertime.net/Docs/WP3/WTEC.pdf, accessed May 2008

Kaivo-oja, J. Y., T. S. Katko and O. T. Seppälä (2004) 'Seeking for convergence between history and futures research', *Futures, Journal of Policy, Planning and Futures Studies*, vol 36, pp527–547

Katko, T. S. and R. Rajala (2005) 'Priorities for fresh water use purposes in selected countries with policy implications', *International Journal of Water Resources Development*, vol 21, no 2, pp311–323

Katko, T. S, P. S. Juuti and P. E. Pietilä (2006) 'Key long-term decisions and principles in water services management in Finland, 1860–2003', *Boreal Environment Research*, vol 11, no 5, pp389–400

Liebowitz, S. J. and S. Margolis (1995) 'Path dependence, lock-in and history', *The Journal of Law Economics and Organization*, vol 11, pp205–226

Lillja, J. L. V. (1938) *Helsinki Water Works 1876–1936* (original in Finnish: *Helsingin kaupungin vesijohtolaitos 1876–1936*), Otava, Helsinki

Malaska, P. and K. Holstius (1999) 'Visionary management', *Foresight*, vol 1, no 4, pp353–361

Martin, B. (1994) *In the Public Interest? Privatisation and the Public Sector Reform*, Zed Books Ltd, London

Melosi, M. V. (2000) *The Sanitary City: Urban Infrastructure in America from Colonial Times to the Present,* Johns Hopkins University Press, Baltimore, MD

North, D. C. (1990) *Institutions, Institutional Change and Economic Performance*, Cambridge University Press, Cambridge

Pietilä, P. (2006) 'Role of municipalities in water services', PhD thesis, Tampere University of Technology Publications no 617, Tampere, Finland

Pietilä, P. E. and T. S. Katko (2006) 'Ontology of water and sanitation services', in P. Pietilä (ed) *Role of Municipalities in Water Services*, Tampere University of Technology Publications no 617, Tampere, Finland, pp215–226

Rees, J. A. (1998) *Regulation and Private Participation in the Water and Sanitation Sector*, TAC Background Papers no 1, Global Water Partnership (GWP), Stockholm, Sweden

Sandelin, S. (2006) *Knowledge Management and Networking in a Water Utility: The Case of Pori Water*, Licentiate thesis, Tampere University of Technology, Tampere, Finland

Seppälä, O. T. (2002) 'Effective water and sanitation policy reform implementation: Need for systemic approach and stakeholder participation', *Water Policy*, vol 4, p367–388

Seppälä, O. T. (2004) *Visionary Management in Water Services: Reform and Development of Institutional Frameworks*, PhD thesis, Tampere University of Technology Publications, no 457, Tampere, Finland

Sharkey, W. W. (1982) *The Theory of Natural Monopoly*, Cambridge University Press, Cambridge

Silfverberg, P. (2007) 'Evaluation and benchmarking as a performance tool for small and medium size water utilities' (original in Finnish: 'Arviointi pienten ja keskisuurten vesihuoltolaitosten ohjauskeinona'), Draft paper for the Ministry of Agriculture and Forestry, Helsinki, Finland, 12 March 2007

UN (United Nations) (1992) *The Dublin Statement on Water and Sustainable Development*, International Conference on Water and the Environment (ICWE), UN, Dublin, www.gdrc.org/uem/water/dublin-statement.html, accessed March 2008

UNESCO (United Nations Educational, Scientific and Cultural Organization) International Hydrological Programme (ed) (2008) *International Shared Aquifer Resource Management (ISARM) Project*, UNESCO, Paris, www.iah.org/isarm/links_isarm.htm, accessed May 2008

Vinnari, E. M. (2006) 'The economic regulation of publicly owned water utilities: The case of Finland', *Utilities Policy*, vol 14, pp158–165

Windischhofer, R. (2007) Municipal Entrepreneurialism and the Commercialization of the Finnish Water Sector, PhD thesis, University of Tampere, Tampere, Finland, http://acta.uta.fi/english/teos.phtml?10996, accessed May 2008

Yukl, G. (2002) *Leadership in Organizations*, fifth edition, Prentice Hall, Upper Saddle River, NJ

6

Public Policy Options for Financing Sewerage Systems

David Hall and Emanuele Lobina

Introduction: Sewers and sanitation

Much current discussion of sanitation marginalizes sewerage, usually on the basis that it is too expensive for most developing countries. Yet, sewerage systems have a massive impact upon public health, especially child health, and are a classic example of a public good. This chapter examines how these systems have developed in the North and how they are being developed in the South. The public sector, and public finance, is central to this development as it solves the problems of insufficient demand from even richer consumers, unaffordability to the poor, and the impossibility of making investment in universal systems profitable. The taxation capacity of the central and/or local state is thus more important for sewerage than pricing issues, and cannot be substituted by either private investment or by community organization, which lacks both the necessary tax base and the necessary spending power. Avoidance of tax finance requires retreating to sanitation without sewerage at the cost of much higher levels of infant mortality – especially in cities.

In January 2007, over 11,000 readers of the *British Medical Journal* chose 'the sanitary revolution' (i.e. connecting people's homes both to clean piped water and to sewers to dispose of their waste) as the most important medical milestone since 1840. They thought it was more important than antibiotics, vaccination or the discovery of the structure of DNA (Ferriman, 2007).

Urban sewerage was first developed in the ancient cities of the Indus Valley around 4000 BC and is thus a South Asian invention. The first modern system was introduced in London during the 19th century and included four key features:

1 the technology of a network of sewers throughout the city, flushed by water;
2 public administrative structures to finance, build and manage these 'expensive works';
3 a public environmental measure, rather than an attempt to alter individual behaviour;
4 a universal public measure applied to everyone, not selectively targeted (Mackenbach, 2007).

These same principles have been applied in every high-income country in the world. Cities, towns and even villages in these countries have piped water and sewerage connections to all houses (EU, 1991). It was very expensive to develop the system – as it is today in developing countries – and it was financed through taxation or massive cross-subsidies (Barraqué, 2007). But in developing countries, many cities and towns still lack the sewerage systems that protect Northern cities, and the rate of death and disease continues to reflect this.

The need for new urban sewerage is highly concentrated in relatively few countries. Table 6.1 shows that half of all the new sewerage connections needed to meet a target of halving the urban population without a household sewerage connection are in just four countries: India, China, Indonesia and Brazil. Three quarters of all the connections required are in just 20 countries.

The need for sewers in cities is thus as compelling now as it was for cities in the North a century ago. The next question is how this need can be met – in particular, how it should be financed. The next section discusses how this has happened in practice, and contrasts the evidence of successful use of public finance with the failures of full-cost recovery from users and encouragement of the private sector: the policies often recommended by donors and development banks.

THE NEED FOR PUBLIC FINANCE

Donor policy obsessions: Cost recovery and the private sector

The main policy advice of donors and development banks emphasizes three key policy positions:

1 the insistence on the need to finance developments through cost recovery from users;
2 the preference for a central role for the private sector;
3 the assumption that sewer systems are too expensive and therefore unaffordable.

The United Nations *World Water Development Report* (WWDR), for example, states:

> *Population growth and burgeoning water demand have convinced most policy-makers that the cost of water system development will increasingly have to*

Table 6.1 *The 20 countries needing most urban sewerage connections*

	Millions	*%*	*%*
	Numbers of people needing connection to achieve target (millions)	*2004 % urban household sewerage connection*	*2015 % urban household sewerage connection after achieving target*
China	251	50	75
India	184	25	63
Indonesia	73	2	51
Brazil	60	53	77
Nigeria	43	23	62
Philippines	34	7	54
Pakistan	32	40	70
Bangladesh	27	7	54
Iran	25	19	60
Congo DR	15	4	52
Viet Nam	14	14	57
Argentina	13	48	74
Thailand	12	0	50
Sudan	11	1	51
Egypt	11	68	84
Venezuela	10	61	81
Ethiopia	10	2	51
Malaysia	10	41	71
Myanmar	9	10	55
Korea Rep	9	65	83
Total of top 4 (China, India, Indonesia, Brazil)	568		
Total of 20 countries with greatest needs	851		
Total connections needed in all developing countries (from table 5)	1141		

Source: PSIRU calculations from WHO–UNICEF (2006) and UNESA (2008)

> *be met by users, especially if the Millennium Development Goals are to be achieved … with private-sector participation – ranging from small water vendors to large private utilities – projected to increase in the next decades, the issue of pricing is critical.* (UNESCO, 2006, p419)

The United Nations Development Programme (UNDP) taskforce report even suggests that choosing to use the private sector is the most fundamental choice, which then determines whether sewers are appropriate:

> *... the use of self-provision, informal provision, public provision, or private-sector provision determines, in part, the scale of service. This, in turn, determines what type of infrastructure or technological option would be appropriate. Thus, sewerage is not a technology of choice for private provision.* (UNDP, 2005, p96)

It also treats public policy as a matter of influencing individual consumer choices, urging public authorities to 'broaden their focus toward an emphasis on influencing citizen/consumer behaviour, as well as toward engaging community-level institutions in planning appropriate interventions' (UNDP, 2005, p83).

Angel Gurría, the Organisation for Economic Co-operation and Development (OECD)'s general secretary, told an OECD forum in 2007 that, although there may be a temporary need for some use of public finance:

> *The first requirement is to make better use of market mechanisms, both to ensure the most efficient use of water supplies and to help finance water infrastructure to encourage greater supply. This means pricing water so that there is full-cost recovery and capacity for reinvestment... True pricing of water also provides an incentive for the development of new technologies and for greater participation by private investors in helping to build and operate water supply systems.* (Gurría, 2007)

The problem extends beyond the operation of water supply and sewerage. Donors have encouraged the development of public toilets by local groups on a private enterprise basis. But these commercial ventures create similar problems of affordability and service delivery. The Wateraid representative in India has stated that:

> *Public toilets are being viewed as profitable sub-contracted works and are being increasingly contracted out, resulting in both unaffordable and badly maintained infrastructure for the urban poor... The trend to privatize and contract out public infrastructure in slums and even in mixed areas (commercial and slums), needs to be curtailed.* (Kapur, 2007)

In Ghana, the policy of contracting out management of public toilets to 'community businesses' led to local political elites creating 'front' businesses to run these lucrative contracts as a prime source of funding, which they would not easily give up – thus creating an obstacle to developing better services (Crook and Ayee, 2006).

But this emphasis on the private sector and cost recovery is contrary to the experience of high-income countries, all of which have developed sewerage systems using public finance; ignores the failure of privatization in the South to deliver any significant investment; and ignores the fact that successful extensions of sewerage systems in the South also use public finance. Such policy advice is in itself an obstacle to the development of sewerage and sanitation in the South because it directs aid and government efforts into developing schemes that will not deliver the investment needed for sanitation.

The problems of cost recovery

Insisting on full-cost recovery means that sanitation programmes are only worth doing if people are willing and able to pay the full cost themselves. There are three great flaws in this approach.

First, the benefits of sewerage connections are public. Improvements in health and the environment are benefits to the community as a whole, not just the households connected, and therefore should be financed collectively. The United Nations taskforce report notes that many households are reluctant to pay for the health benefits of sanitation: 'among the reasons that people invest in improved [sanitation] services, health does not figure particularly prominently' (UNDP, 2005, p83). But, as the taskforce also states, this was exactly the same in the countries of the North in the 19th century: the citizens of Boston, Massachusetts, in 1850 were also: 'unable or unwilling to take on personal responsibility to conduct their lives in accord with recommended sanitary principles' (UNDP, 2005, p83).

The second problem with user charges is affordability. The poorest, who usually are in greatest need of connections, will be least likely to afford them. An insistence on cost recovery from users of the system thus becomes an obstacle to achieving improvements in sanitation: 'If international donors wish to pursue a policy of universal access, they should acknowledge that the costs of improved services are far beyond the reach of many households' (Whittington, 2006, p141). Even low levels of connection charges act as deterrents for poor households, who suffer most from the disease consequences of poor sanitation because of unequal resources. The death rate within the same city varies according to the income of residents: a study found that infant mortality rates in seven different areas of Karachi varied from 33 to 209 per 1000 live births (Bartlett, 2005).

The third problem is fairness. The rich are the greatest beneficiaries of full-cost recovery from users: they contribute nothing to the extension of services to the poor.

The solution of public finance and compulsory connections

These were the core reasons why, in all developed countries, the idea of financing sanitation through cost recovery from users was abandoned. The sewerage systems in Europe, the US and Japan were not developed through full-cost recovery from users – they were paid for by distributing the costs amongst the public, using taxation and cross-subsidy: 'public financing of sanitation infrastructure was seen as the only option for ensuring investment adequate to protect public health' (UNDP, 2005, p83).

The use of public finance deals with the problems of affordability and reluctance to pay by public spending based on redistribution through taxation. This redistribution is key; but it means that the wealthy have to contribute more. The same principle of cross-subsidy continues to be applied in Europe at a transnational level. The European Union raises taxes across all the countries of Europe to support the cost of water and sanitation improvements in the poorer countries, the equivalent of 20 Euros per person per year (Hall and Lobina, 2007).

An important step was to move away from private consumer choice to collective public decisions to connect all households. Connection was not allowed to remain a matter of individual choice, but was required as a matter of public policy. In France, for example, from the 19th century, sewerage became a matter of public health policy, and so:

> *Connection to a main sewer was compulsory for households, and therefore it was covered by local taxes (as in Germany)... It required an intense effort, supported by government subsidies, to catch up with the rapid urbanization and industrialization process that took place after 1945 ... investment in sewage treatment led to the creation of the Agences de l'Eau ... levying water pollution and abstraction charges from water bills at river basin level, through a mutualization of investment needs.* (Barraqué, 2007, p124)

The development of sewers was financed through local taxation, subsidies from central government taxation and cross-subsidies through special taxes at regional river basin level.

Japan expanded sewerage coverage from 8 per cent in 1965 to 69 per cent in 2006 (and projected to reach 72 per cent in 2007), using public finance, public operations and domestic public–public partnerships (PUPs), mainly technical and financial assistance provided by a central governmental agency to local authorities (ADB, 2007a).

The city of Toronto in Canada provides a clear example of how the process worked. Before the 1870s the city was mainly dependent upon private contractors for water supply. Sewers were not automatically laid in new streets: householders had to petition the local council for connections, and then pay the cost themselves. Neither water nor sewerage connections grew fast enough. Deaths from waterborne diseases, especially typhoid, were common (Benidickson, 2001). During the 1870s, despite an economic recession, the city council not only municipalized the water service, it authorized the city engineer to install new sewers for public health reasons, whether householders asked for it or not, financed by the municipality. The benefits were immense:

> *This unprecedented power ... led to tremendous sewer development in the 1880s... The effects of the typhoid fever epidemic were greatly reduced by the presence of a complete clean sewage system. At the beginning of the 20th century, most of the streets in the city had been serviced and the operational costs were met through direct taxation.* (Pharasi and Kennedy, 2002)

The same approach was taken up across the province of Ontario, where public water systems grew rapidly by the turn of the century. The 1912 Public Health Act enshrined the Toronto principles of public finance and compulsory connection by giving the provincial board of health the right not only to decide when a water or sewerage system was necessary 'in the interest of public health', but also to require local councils to finance it (Benidickson, 2001).

The illusion of private-sector investment

The idea that the private sector can or will invest significant money in developing sanitation or sewerage systems is misleading. In developed countries, as described above, the private sector played almost no role in financing the sanitation systems. The constant donor advice to involve the private sector is thus contrary to the experience of successful development of sanitation and sewerage.

In developing countries, despite the encouragement and support from donors and development banks, the private sector has contributed only a trivial amount to investment in urban infrastructure during the last 20 years. A key problem has been that the private sector has to make profits that cover the cost of its capital and the associated risks. It is therefore selective about the countries it chooses to operate in – only one third of developing countries have received any kind of private investment in water and sanitation; it has frequently relied on governments to revise contracts and provide guarantees, and therefore has created fiscal risks for governments in the shape of unexpected liabilities; it has been unable to get the necessary rate of return, due at least, in part, to public resistance to paying the prices required to deliver this rate of return; and it has retreated even from those areas where it has invested. Even in middle-income countries where the presence of the private sector has been greater, private-sector investment is minimal in comparison to investment by the state. In South Africa, for example, total private investment in urban infrastructure from 1983 to 2004 'has been quite insignificant ... much less than 1 per cent of one year's local government spending' (Clarke Annez, 2006, pp4, 9; Estache, 2006).

A World Bank research paper in 2006, reviewing actual private investment during a 22-year period from 1983 to 2004, concluded bluntly that:

> *PPI [private participation in infrastructure] has disappointed – playing a far less significant role in financing infrastructure in cities than was hoped for, and which might be expected given the attention it has received and continues to receive in strategies to mobilize financing for infrastructure... PPI is inherently limited in scope for financing urban infrastructure for the wide array of non-commercial infrastructure services cities need.* (Clarke Annez, 2006)

This failure is confirmed by reviewing the actual cases of private-sector involvement in sanitation in developing countries. Asia needs about 1 billion sewerage connections to achieve 80 per cent urban coverage; but the great majority of cities have received no sewerage extensions from the international or local private sector. There have been private water concessions in the capital cities of Indonesia and the Philippines – but in Jakarta, where only 1 per cent of the population is connected to a sewer, the concessions do not cover sanitation at all; and in Manila, sewerage coverage has barely changed from pre-privatization levels of 7 per cent to 10 per cent in one half of the city and 3 per cent in the other (AIM, 2008; MWSS, 2008). Malaysia set up a private concession for developing sewerage, which was re-nationalized after it failed to deliver investment, partly due to consumer unwillingness to pay charges (CAG, 2004; Clarke Annez, 2006; *New Straits Times*, 2007). A project in Tamil Nadu, India, used a private company but only as a construction contractor in building a sewerage system. In China, as of 2002, there have been a number of private build–operate–transfer (BOT) concessions for wastewater treatment plants, but no private concessions for extending sewerage connections (CAG, 2004).

In Africa, the only case of private investment in sewerage is the concession at Nelspruit, South Africa, where the company laid 35 kilometres (km) of sewer mains and most residents gained access to waterborne sanitation (Smith et al, 2003). In Senegal, responsibility for urban sanitation and sewerage was excluded from the privatization of the water services because 'including the sanitation sector in the responsibilities of the private operator would be too burdensome, given its poor state' (Brocklehurst and Janssens, 2004, p27): between 1998 and 2007, the World Bank helped to finance new sewerage connections for 212,250 additional people – an example of public finance for investment in sewers. In Côte d'Ivoire, the nationwide water privatization, which covers sewerage as well as water, is also a lease contract, not a concession, and therefore there is no private responsibility for investment in sewerage. Abidjan, the capital, has a higher level of sewerage connections than most other African cities; but these were financed by World Bank loans and other public finance (Obrist et al, 2006). In Latin America, water privatization has been most common, but has contributed very little to extension of the sewerage system. In some cities the levels of sewerage connection are high; but this is due to public investment, not to the private concessions. In Chile, the high levels of coverage were achieved under public ownership before privatization took place. In Argentina, the flagship concession in Buenos Aires managed to increase a pre-existing connection level of 58 per cent to only 63 per cent in nine years – 1 million connections short of the original target. In Brazil, sewerage connection in Manaus was at just 3 per cent when the private concession started: by 2005, the company had increased this to just 12 per cent, compared with a target of 31 per cent. This compares badly with the progress in the comparable city of Salvador, state of Bahia, which in almost the same time period, under the public sector, increased the connection level to 80 per cent. In Colombia, significant sewerage extensions have occurred in Cartagena, reaching an official level of 95 per cent; but these were overwhelmingly financed by public finance from the World Bank and the government, with the private company contributing little. The concessions in Bolivia, Ecuador and Peru also fell short of targets by varying amounts (Hall and Lobina, 2008, pp22–24).

Development in the South: Public sector and public finance

The continuing importance of public finance for the development of sanitation can be seen in the actual policies being pursued by the four countries that are of the greatest importance for connecting urban populations: Brazil, China, India and Indonesia.

Three of these – Brazil, China and India – are investing in sanitation, including sewerage connections, using public finance. As a result, Brazil and China are investing enough to achieve the Millennium Development Goals (MDGs) in full and 80 per cent urban sewerage connections by 2015; India may need further investment, but is actively increasing its tax revenues, which will permit this. The fourth country, Indonesia, has no national programme of investment in sewers using public finance, despite having very healthy government finances, with growing tax revenues. It is being advised by the World

Bank to focus on increasing user charges. If it does so, Indonesia will fail to improve its urban sewerage connections to anywhere near the proposed target level.

Brazil

In January 2007, Brazil announced a new four-year programme for economic growth, the Programme for Accelerating Growth (PAC), based on investment of US$236 billion (504 billion Brazilian reais) in infrastructure, especially in roads and electricity, but also water, sanitation and housing (Hall and Lobina, 2008, p26). The water and sanitation investment budget of US$18.7 billion represents an annual rate of 4.7 billion investments, which is 0.53 per cent of Brazil's gross national income (GNI) – which is sufficient for Brazil to meet its MDG targets in full and the urban sewerage connections target (*Business News Americas*, 2007).

The development of sewerage connections in the city of Salvador, state of Bahia, provides an outstanding example of the expansion of urban sewerage connections in a short space of time using public finance, and demonstrates the same lessons evident from the introduction of the sewerage system in London during the mid-19th century. In 1996, only 26 per cent of the city's households were connected to a safe sewerage system, mainly the upper and middle classes in the oldest part of the city. The primary objective of the new programme was to extend the sewerage system to 80 per cent of households. This was completed in eight years at a cost of about US$220 million (Hall and Lobina, 2008, pp26–27). The overall reduction in childhood diarrhoea was 22 per cent, and 43 per cent in the highest risk areas inhabited by the poorest. A recent study on this case concluded that:

> *Because sewerage is mainly external to houses and the fact that it prevents disease transmission in the public domain, public responsibility is to ensure that sewerage is installed. At a typical cost per person of $160, investment in sewerage is too large to be left to cash-strapped municipalities, and needs the involvement of international organizations, and central government and its agencies.* (Barreto et al, 2007)

China

The urban sewerage connection rate in China rose from 30 per cent in 1990 to 50 per cent in 2002 (WHO–UNICEF, 2006). Public spending on infrastructure has not only kept pace with the growth of the Chinese economy, it has increased twice as fast: 'Since 1995, China's GNI has almost tripled while overall annual municipal infrastructure spending, including roads, has increased sixfold' (Browder et al, 2007). The total length of urban sewerage networks increased by nearly 225 per cent between 1991 and 1998, but less than 4 per cent of all the investment in water and sanitation was financed through the private sector (Clarke Annez, 2006). China is now investing over US$10 billion per year (0.4 per cent of GNI) and spending another 0.6 per cent of GNI in operating costs in water and sanitation. This combined total of 1 per cent of GNI (US$25 billion) is sufficient not only to achieve the MDGs, but also the urban sewerage target (see below). The contribution from development banks and aid has been large in absolute terms, but small as a proportion of total cost. Between 1992 and 2013, the World Bank has loaned about US$7 billion, an average of US$0.3 billion per year, which is just over 1 per cent of the current level of China's spending on water and sanitation (Browder et al, 2007).

China's approach has been favourably compared to that of the World Bank by economist Jeffrey Sachs:

> *Unlike the Chinese, the [World] Bank has too often forgotten the most basic lessons of development... The Bank can regain its relevance only if it becomes practical once again by returning its focus to financing public investments in priority sectors, just as the Chinese leadership is prepared to do.* (Sachs, 2007)

India

India is developing new plans for investment in water and sanitation as part of the current five-year plan for the economy. Water and sanitation has been given priority in its urban infrastructure programme, and the new plans propose to nearly double the previous finance from central and state governments. The new plans amount to US$31.75 billion, about US$6.4 billion per year, which is the equivalent of 0.7 per cent of GNI. This could be sufficient to achieve the MDGs and the urban sewerage target. Over 90 per cent of this is to be financed by central and state governments and national financial institutions, with only 8 per cent funded by aid and only 1.5 per cent from the private sector.

Table 6.2 *Financing India's water and sanitation plans*

Financed by	*Rupiah (crore)*	*US$ (billions)*	*Percentage*
Central government	70,000	17.50	55
State governments	35,000	8.75	28
National banks	10,000	2.50	8
Aid	10,000	2.50	8
Foreign direct investment (FDI)/private sector	2025	0.50	1.5
Total	127,025	31.75	100

Note: Currency converted at 40 Indian rupiah = US$1.
Source: Government of India (2007, pp55–56)

Indonesia

Indonesia has no comparable national programme for water and sanitation. The national finances, however, allow it to do so. A World Bank analysis of Indonesia's public finances in 2007 estimated that Indonesia could spend an extra US$15 billion per year, and that there is a particular need to do so in infrastructure spending because it has fallen to low levels, partly because of the refusal of the World Bank to lend further money to public authorities already in arrears – which includes most of the country's water and sewerage authorities (World Bank, 2007). The report's main suggestion is that Indonesia should charge higher prices to users of water and sanitation. But the lack of a public investment programme is not being compensated for by private-sector investment: the World Bank itself shows clearly that the private sector is not investing in infrastructure in Indonesia.

Japan: Donor funding and training to support sewerage development

The experience with sanitation in Asia also illustrates the potential for a supportive role by donors.

The Japan Bank for International Cooperation (JBIC) has acquired considerable experience in assisting developing countries to develop sewerage systems (ADB, 2007b). On the basis of this experience, the JBIC highlights the importance of raising public finance through central governments, and for donors to provide capacity-building and training, including the use of public–public partnerships. The JBIC recommends the following:

- establishing a national financial support system for sewerage development;
- developing capacity.

Establish national financial support system for sewerage development

> *Since sewerage systems are very expensive and are sometimes not affordable for the majority of residents, financial support of the central government is indispensable. Economic externality of sewerage – the necessity of preserving the water quality of public water bodies – would justify the financial support by the central government.* (ADB, 2007b, p1)

Capacity development

> *In order for sewerage systems to work effectively, capacity development in various areas – that is, training of engineers in designing; operation and maintenance of sewage treatment plants; strengthening of the administrative capacity of local government in order to keep book-keeping of the basic data and records on sewer networks and house connections; strengthening of the administrative capacity of the central government in creating regulatory frameworks for house connections, industrial water, aquatic water quality control, and in creating financial support systems for sewerage development; and the environment and sanitation education.* (ADB, 2007b)

Public–public partnerships (PUPs) have been used to enhance local capacity-building in the design and operation of sanitation systems. A twinning arrangement between Tokyo Metropolitan Sewerage Bureau and Beijing Municipal Design and Research Institute was instrumental to the design of the Gaobei Dian wastewater treatment plant, but was then extended to include a sewerage component:

> *The first phase of construction work had started in 1990, and Beijing City itself executed the entire work under its direct management. In March 1993, when the work was almost 80 per cent completed, Beijing City requested Tokyo Metropolitan Sewerage Bureau to provide them with training for sewerage operation and management.* (ADB, 2007c)

The training was funded by the JBIC.

Taxation needed

As emphasized by the JBIC and the experiences detailed above, raising taxes is central to finance public spending on sewerage connections, or other investment in infrastructure and public services: 'Small government and low taxes are not the answer for reaching the MDGs' (McKinley, 2005).

Higher levels of taxation are associated with higher levels of economic performance: tax revenue as a share of GNI is about 14 per cent in low-income countries, 19 per cent in lower middle-income countries, 23 per cent in upper middle-income countries and 38 per cent in high-income countries (World Bank, 2004). The most unequal societies have tended to resist proposals for higher taxation because the rich would have to pay most, and is one reason why developing countries have not made enough public investment in education, health or water and sanitation (Engerman and Sokoloff, 2006). Some of the countries with greatest need for urban sewerage connections have very low taxation levels. In 2002, India only raised 9.9 per cent of GNI in taxes, Bangladesh only 7 per cent of GNI (World Bank, 2003, Table 5.6).

A combination of economic growth and more active public investment policies can raise taxes enough to make a difference. In China, economic growth is producing a growth in personal incomes, which means that income tax can start to grow; China may be able to collect 4.5 per cent of GNI in income tax by 2010, with total taxes worth over 18 per cent of GNI. This would reflect a similar process in Northern countries during the first half of the 20th century, the same period when much of the investment in sewerage systems was made. In those countries: 'moving from an elite income tax raising less than 1 per cent of GNI to a mass income tax raising around 4 to 5 per cent of GNI is exactly the kind of process through which Western countries went during the 1914–1950 period' (Piketty and Qian, 2006).

Similar growth in taxable personal incomes is expected in India (Piketty and Qian, 2006). The current Indian government is already increasing tax revenue: between 2006 and 2008, tax revenues have increased by around 50 per cent, with special attention to increased taxation of multinational company profits and capital gains (*Financial Times*, 2008). Indonesia is also experiencing an increase in tax revenues: the World Bank forecasts that non-oil tax revenues will rise from 13 per cent of GNI in 2005 to 14.5 per cent in 2010 (World Bank, 2007).

Affordability: Economic capacity

The explicit or implicit position of most of the official donor publications is that, despite the clear balance of benefits, household sewerage connections cannot be afforded. The WWDR can be taken as typical of this position. It argues that the option of full household connections to sewers and water supply cannot, and will not, be financed:

> *In many nations, at least in the next five to ten years, it will not be possible for the provision deficiencies in most urban areas to be addressed by the conventional model of a (public or private) water utility extending piped water supplies and sewers to individual households.* (Hall and Lobina, 2008, p33)

National affordability is the most important issue. In practice, the great majority of the resources for extending water and sanitation come from national resources. There is a political reason for this too. The countries concerned are all sovereign states, and so decisions are – or should be – taken by governments of those states. The key decisions are taken in Beijing, Delhi, Brasilia, Jakarta and other capital cities, not in Washington, London or Paris. Table 6.3 estimates the costs facing countries. It sets out estimates of the annual costs of the MDGs, with household connections, urban and rural – plus the extra needed for the proposed urban sewerage target – for the 20 countries requiring the greatest number of urban sewerage extensions, as estimated in Table 6.3 (these countries between them cover nearly 90 per cent of the need for urban sewerage connection).

For many of the middle-income countries, the cost is less than 0.5 per cent of GNI per annum. China, Brazil and India are already planning to spend as much on development of water and sanitation as these estimates suggest is needed for the MDGs, including full household connections for water and sewerage (see above). Even including the running and depreciation costs of existing services, which effectively includes the running costs of all water and sanitation services, only half of these countries would need to spend more than 1 per cent of GNI. These levels of spending are affordable elements of public investment in relation to the size of economies, especially in light of recent growth rates. Between 2001 and 2006, China averaged growth of 9.7 per cent per annum and India 7.6 per cent; but other much poorer countries have also grown, including Mozambique (average annual growth rate of 8.6 per cent between 2001 and 2006), Vietnam (7.6 per cent), Tanzania (6.4 per cent), Bangladesh, Iran and Nigeria (5.6 per cent), Ghana and Pakistan (5.2 per cent), Indonesia (4.9 per cent) the Philippines (4.6 per cent), Democratic Republic of the Congo (4.2 per cent), and Brazil (2.9 per cent) (World Bank, 2008). Spending an extra 1 per cent of GNI on investment in water and sanitation is thus allocating part of this growth, not a claim on other uses of existing income.

This level of spending makes greater demands on the taxation systems of countries. The taxation collected by some countries is adequate for these levels of public spending; but others need to increase the tax collected. India collected only 12.5 per cent of GNI as tax in 2004, Bangladesh 10 per cent, Democratic Republic of the Congo 8 per cent, Pakistan 13 per cent and the Philippines 15 per cent; other low-income countries collect more – for example, Ghana collected 24 per cent (World Bank, 2008). Establishing

Table 6.3 *National affordability: Costs as a percentage of national income*

	National income group	*GNI 2006 $billion*	*Urban sewer target m*	*Annual cost of MDG HC + urban sewer target $m*	*%GNI*
China	ML	2641.6	251	7878	0.30
India	L	906.5	184	5764	0.64
Indonesia	ML	315.8	73	2291	0.73
Brazil	ML	892.8	60	1881	0.21
Nigeria	L	92.4	43	1364	1.48
Philippines	ML	120.2	34	1069	0.89
Pakistan	L	122.3	32	1000	0.82
Bangladesh	L	69.9	27	855	1.22
Iran	ML	207.6	25	790	0.38
Congo DR	L	7.7	15	485	6.29
Viet Nam	L	58.1	14	450	0.77
Argentina	MU	201.4	13	403	0.20
Thailand	ML	193.7	12	379	0.20
Sudan	L	29.9	11	352	1.18
Egypt	ML	101.7	11	340	0.33
Venezuela	MU	164.0	10	310	0.19
Ethiopia	L	12.9	10	306	2.37
Malaysia	MU	141.4	10	299	0.21
Myanmar	L		9	288	
Korea Rep	H	856.6	9	269	0.03
Total of above				26,773	
Total for all developing countries				34,900	

Note: Income groups: L = low income; ML = lower middle; MU = upper middle; H = high.
Source: PSIRU calculations based on WHO–UNICEF (2006), UNESA (2008) and World Bank (2008)

sustainable public revenues and building the capacity of public authorities are important elements in development. Water and sanitation investments can drive these developments as they did in European and North American countries a century ago (Hall and Lobina, 2007).

Conclusions

The costs of the programmes mentioned in this chapter are affordable for national economies, with the majority of people needing connections. These programmes should be treated as an investment that is justified based on the health and economic benefits achievable. Developing countries should continue to plan for the development of household water and sewerage connections. The important financial issue is to ensure that sufficient taxes are raised to finance programmes. Attempts to finance programmes through user-charge recovering costs, or attempts to involve the private sector in investment, are likely to be expensive irrelevances that will slow down achievements. Countries such as Indonesia and the Philippines need to develop major public spending programmes to develop urban sewer systems.

Donors should stop encouraging countries to try to finance the development of sewerage systems through cost recovery from users, and should stop encouraging countries to believe that the private sector will make any significant contribution to investment in sanitation. They should, instead, encourage countries to build the taxation capacity needed to finance this investment, and provide capacity-building support and training through public–public partnerships, following the model of Japan.

Donors should focus aid on the countries in greatest need of assistance in order to meet the costs of urban sanitation, particularly low-income African countries, led by Nigeria and the Democratic Republic of Congo. The need for aid should be assessed in relation to national needs and affordability, not by reference to the total global cost of developing water and sanitation systems, most of which will be met by national resources.

Finally, community organization of sanitation cannot provide an adequate substitute for the tax revenues of the central and local states. Even the best-known example of urban community-led sanitation, the Orangi project, has campaigned – successfully – for its approach to be adopted as a city-wide plan for sewerage throughout Karachi, financed by the provincial and central government (OPP–RTI, 2008). The people of modern Pakistan are playing as important a role in developing modern urban sewerage policy as their ancestors did in developing the first urban sewerage systems in the Indus Valley.

References

ADB (Asian Development Bank) (2007a) 'Summary of the legal framework, finance, and institutional arrangement of Japan's sewerage system', in *Asian Water Development Outlook*, www.adb.org/Documents/Books/AWDO/2007/AWDO.pdf, accessed June 2008

ADB (2007b) 'Lessons learnt from JBIC's experience in assisting sewerage development in Asian countries', in *Asian Water Development Outlook*, www.adb.org/Documents/Books/AWDO/2007/AWDO.pdf, accessed June 2008

ADB (2007c) 'Development of sewage treatment system in Beijing, China', in *Asian Water Development Outlook*, www.adb.org/Documents/Books/AWDO/2007/AWDO.pdf, accessed June 2008

AIM (Asian Institute of Management) (2008) *Lessons from a Water Utility Privatization*, Makati City, the Philippines, www.aim.edu/home/announcementc.asp?id=622, accessed June 2008

Barraqué, B. (2007) 'Small communes, centralisation, and delegation to private companies: The French experience', *Journal of Comparative Social Welfare*, vol 23, no 2, pp121–130

Barreto, M. L., B. Genser, A. Strina, M. G. Teixeira, A. M. O. Assis, R. F Rego, C. A. Teles, M. S. Prado, S. M. A. Matos, D. N. Santos, L. A. dos Santos and S. Cairncross (2007) 'Effect of city-wide sanitation programme on reduction in rate of childhood diarrhoea in northeast Brazil: Assessment by two cohort studies', *The Lancet*, vol 370, pp1622–1628

Bartlett, S. (2005) 'Water, sanitation and urban children: The need to go beyond "improved" provision"', *Children, Youth and Environments*, vol 15, no 1, pp115–137, www.colorado.edu/journals/cye/15_1/a6_Sanitation.pdf, accessed June 2008

Benidickson, J. (2001), 'The development of water supply and sewage infrastructure in Ontario, 1880–1990s: Legal and institutional aspects of public health and environmental history', Background paper for the Walkerton Inquiry, University of Ottawa, Ottawa, http://tspacetest.library.utoronto.ca:8080/bitstream/1778/4019/1/10294043.pdf, accessed June 2008

Brocklehurst, C. and J. G. Janssens (2004) *Innovative Contracts, Sound Relationships: Urban Water Sector Reform in Senegal*, Water Supply And Sanitation Sector Board Discussion Paper Series, no 1, January 2004, www.worldbank.org/html/fpd/water/pdf/WSS_Senegal.pdf, accessed June 2008

Browder, G., S. Xie, Y. Kim, L. Gu, M. Fan and D. Ehrhardt (2007) *Improving the Performance of China's Urban Water Utilities*, World Bank, Washington, DC, www-wds.worldbank.org/external/default/WDSContentServer/WDSP/IB/2007/09/25/000310607_20070925111156/Rendered/PDF/409640P0704130Public.pdf, accessed June 2008

Business News Americas (2007) 'Cities ministry pre-selects 669 sanitation projects for PAC funds', *Business News Americas*, 27 April 2007

CAG (Castalia Advisory Group) (2004) 'Sector note on water supply and sanitation for infrastructure in East Asia and the Pacific Flagship', Paper commissioned for the ADB–JBIC–World Bank East Asia Pacific Infrastructure Flagship Study, http://lnweb18.worldbank.org/eap/eap.nsf/Attachments/background+12/$File/EAPWaterandSanitation.pdf, accessed June 2008

Clarke Annez, P. (2006) 'Urban infrastructure finance from private operators: What have we learned from recent experience?', World Bank Policy Research Working Paper 4045, World Bank, Washington, DC

Crook, R. and J. Ayee (2006) 'Urban service partnerships, "street-level bureaucrats" and environmental sanitation in Kumasi and Accra, Ghana: Coping with organisational change in the public bureaucracy', *Development Policy Review*, vol 24, no 1, pp51–73

Engerman S. and K. Sokoloff (2006) 'Colonialism, inequality and long-run paths of development', in A. V. Banerjee, R. Benabou and D. Mookherjee (eds) *Understanding Poverty*, Oxford University Press, New York

Estache, A. (2006) *Infrastructure: A Survey of Recent and Upcoming Issues*, World Bank Infrastructure Vice-Presidency and Poverty Reduction and Economic Management Vice-Presidency, World Bank, Washington, DC

EU (European Union) (1991) Council Directive 91/271/EEC of 21 May 1991 Concerning Urban Waste-Water Treatment, Article 3, http://eur-lex.europa.eu/LexUriServ/LexUriServ.do?uri=CELEX:31991L0271:EN:HTML, accessed June 2008

Ferriman, A. (2007) 'BMJ readers choose the "sanitary revolution" as greatest medical advance since 1840', *British Medical Journal*, vol 334, p111, www.bmj.com/cgi/content/full/334/7585/111-a, accessed June 2008

Financial Times (2008) 'New Delhi poised to throw tax net on offshore deals', *Financial Times*, 28 February 2008

Government of India (2007) *Report of the Steering Committee on Urban Development for Eleventh Five Year Plan (2007–2012)*, Planning Commission, New Delhi, http://planningcommission.nic.in/aboutus/committee/strgrp11/str11_hud1.pdf, accessed June 2008

Gurría, A. (2007) *Water: How to Manage a Vital Resource*, Secretary-General, OECD, OECD Forum: Innovation, Growth and Equity, 14–15 May 2007, Paris, www.oecd.org/dataoecd/41/54/38583737.pdf, accessed June 2008

Hall, D. and E. Lobina (2007) *Water as a Public Service*, Public Sector International Research Unit (PSIRU), University of Greenwich, Greenwich, UK, www.psiru.org/reports/2007-01-W-waaps.pdf, accessed June 2008

Hall, D. and E. Lobina (2008) *Sewerage Works – Public Investment in Sewers Saves Lives*, Public Sector International Research Unit (PSIRU), University of Greenwich, Greenwich, UK, www.psiru.org/reports/2008-03-W-sewers.doc, accessed August 2008

Kapur, D. S. (2007) 'What is ailing the sanitation sector in India?', WaterAid country representative, India on the occasion of World Toilet Day, 19 November 2007, www.wsscc.org/fileadmin/files/pdf/For_country_pages/What_is_ailing_poor_sanitation_coverage_in_India.pdf, accessed June 2008

Mackenbach, J. P. (2007) 'Sanitation: Pragmatism works', *British Medical Journal*, vol 334, no 1, s17, www.bmj.com/cgi/content/full/334/suppl_1/s17?maxtoshow=&HITS=10&hits=10&RESULTFORMAT=&fulltext=Mackenbach+2007&searchid=1&FIRSTINDEX=0&resourcetype=HWCIT, accessed June 2008

McKinley, T. (2005) 'MDG-based PRSPs need more ambitious economic policies', Policy discussion paper, United Nations Development Programme (UNDP), www.undp.org/poverty/docs/sppr/MDG-based%20PRSPs%201-05%20Background%20Paper%20(New%20York).doc, accessed June 2008

MWSS (Metropolitan Waterworks and Sewerage System) (2008) *Water Service Performance*, Balara Quezon City, the Philippines, www.mwssro.org.ph/publication_water_service_perfomance.htm, accessed June 2008

New Straits Times (2007) 'Friday Indah water finances weak', *New Straits Times*, Bangsar, Malaysia, 22 June 2007

Obrist, B., G. Cissé, B. Koné, K. Dongo, S. Granado and M. Tanner (2006) 'Interconnected slums: Water, sanitation and health in Abidjan, Côte d'Ivoire', *European Journal of Development Research*, June, vol 18, no 2, pp319–336

OPP–RTI (Orangi Pilot Project Research and Training Institute) (2008) *Low Cost Sanitation Programme: Replication of the Low Cost Sanitation Programme Outside Orangi*, www.oppinstitutions.org/creplicationofspnd.htm, accessed June 2008

Pharasi, S. and C. A. Kennedy (2002) 'Reflections on the financial history of Toronto's urban water infrastructure', Annual Conference of the Canadian Society for Civil Engineering, Montreal, Québec, Canada, 5–8 June 2002, http://pedago.cegepoutaouais.qc.ca/media/0358894/wps/en/contenu/doc/confgen/pdf/GE062-PHARASI-Kennedy.pdf, accessed June 2008

Piketty, T. and N. Qian (2006) 'Income inequality and progressive income taxation in China and India, 1986–2015', Centre for Economic Policy Research (CEPR) Discussion paper no 5703, http://ssrn.com/abstract=922116, accessed June 2008

Sachs, J. (2007) *China's Lessons for the World Bank*, Project Syndicate, May 2007, www.project-syndicate.org/commentary/sachs129, accessed June 2008

Smith, L., S. Mottiar and F. White (2003) *Testing the Limits of Market-Based Solutions to the Delivery of Essential Services: The Nelspruit Water Concession*, Research Report 99, Social Policy Series, Centre for Policy Studies, Johannesburg, www.cps.org.za/cps%20pdf/RR99.pdf, accessed June 2008

UNDP (United Nations Development Programme) (2005) *Health, Dignity, Development – What Will It Take?*, www.unmillenniumproject.org/documents/WaterComplete-lowres.pdf, accessed September 2008

UNESA (United Nations Economic and Social Affairs) (2008) *Economic and Social Development*, www.un.org/esa/desa/, accessed September 2008

UNESCO (United Nations Educational, Scientific and Cultural Organization) (2006) *Water: A Shared Responsibility*, UN World Water Development Report 2, Paris, UNESCO, www.unesco.org/water/wwap/wwdr2/index.shtml, accessed June 2008

Whittington, D. (2006) 'Reflections on the goal of universal access in the water and sanitation sector: Lessons from Ghana, Senegal and Nepal', in *Liberalisation and Universal Access to Basic Services: Telecommunications, Water and Sanitation, Financial Services, and Electricity*, OECD Trade Policy Studies, OECD and World Bank, Paris, Chapter 5

WHO–UNICEF (World Health Organization–United Nations Children's Fund) (2006) *JMP Coverage Estimates Improved Sanitation China 2006*, Joint Monitoring Programme for Water and Sanitation, www.wssinfo.org, accessed June 2008

World Bank (2003) *World Development Indicators 2003*, World Bank, Washington, DC

World Bank (2004) *Global Monitoring Report: Policies and Actions for Achieving the Millennium Development Goals and Related Outcomes*, World Bank, Washington, DC, http://siteresources.worldbank.org/GLOBALMONITORINGEXT/Resources/0821358596.pdf, accessed June 2008

World Bank (2007) *Spending for Development: Making the Most of Indonesia's New Opportunities: Indonesia Public Expenditure Review 2007*, World Bank, Jakarta, http://siteresources.worldbank.org/INTINDONESIA/Resources/Publication/280016-1168483675167/PEReport.pdf, accessed June 2008

World Bank (2008) *Key Development Data and Statistics*, World Bank, Washington, DC, http://go.worldbank.org/1SF48T40L0, accessed June 2008

7

Interfaces and Inter-Sector Approaches: Water, Sanitation and Public Health

Léo Heller

Introduction

This chapter discusses water and sanitation services (WSS), exploring the interfaces between the technical and policy-oriented fields of knowledge, as well as between different public policy sectors involved in their organization and delivery. It argues that an interdisciplinary approach and the adoption of practical actions grounded on a transversal perspective that cuts across the different sectors of public policy involved are essential to the success of WSS initiatives. The chapter also starts from a double assumption:

1 Interdisciplinary and transversal approaches are not very frequent in public policy and management activities connected with WSS and tend to have a limited scope.
2 Policy interventions to improve WSS must be sensitive to the specific conditions characteristic of the target region or country, and especially to their level of socio-economic development.

Although the discussion is mainly based on the experience of developing countries, where the challenges are greater, it also draws lessons from, and contributes to, the WSS agenda of developed countries.

In assessing the historical development of WSS services in African, Asian or Latin American countries, it is possible to identify an important common factor: the legacy of a technological and managerial framework introduced by the colonizing developed countries. A glance at the history of the initiatives in the field of WSS reveals a high incidence of technological imports, mostly from Europe since the 19th century and, from World War II onwards, also from the US. In general, this technological framework was introduced in developing countries without a proper consideration of the local circumstances and systemic conditions that may affect the working of WSS, not least because the prevailing understanding of the time was that technical options and solutions were neutral, universally valid and independent of local contexts. Consequently, the physical, geographical, meteorological, urban, social and cultural characteristics, among others, of the recipient regions and countries were usually not properly taken into account in the design and development of WSS systems during this early period.

In fact, this traditional, supposedly context-neutral, technological framework still prevails today in developing countries, and very little has been achieved regarding technological perspectives that genuinely respond to the specific regional or national characteristics and conditions, despite the significant (but often problem-specific) efforts of the academic community. This state of affairs has worrying implications since different levels of socio-economic development among countries tend to also entail the adoption of often widely different agendas for WSS. This problem is particularly evident when we observe the interfaces between the different fields that intersect in the running of WSS, such as the inter-linkages between these services and broader economic-financial or environmental health issues, given that the WSS agendas of different countries, and most particularly those of developed and developing countries, often have different demand hierarchies, if not altogether different problems and demands. For instance, while the current technological agenda in developed countries reflects the existence of priorities characteristic of a situation where the universalization of coverage has been achieved long ago, in developing countries the agenda must be primarily oriented at finding an urgent solution to the protracted social debt in terms of unequal access to often unreliable WSS still affecting a large proportion of the population.

Likewise, the organization of WSS in less developed countries is frequently based on management models imported from developed countries without fully taking into account specific local conditions. As a result, it is possible to identify inconsistencies between the management models implemented, which have often been designed in the particular circumstances of developed countries, and the local organizational, legal, policy-institutional and cultural frameworks, and these inconsistencies are the source of inefficiencies and ineffectiveness. Furthermore, the still emerging practices of planning and evaluation tend to rank low in the scale of priorities set by the political and managerial apparatus in charge of WSS in developing countries, which contributes to maintain the status quo.

In this context, this chapter focuses on the deficiencies in management resulting from the fragmented perspective that prevails in WSS and related public policies. It is argued that such fragmentation between the different fields involved in WSS, and the continued lack of recognition of the relevant interfaces, becomes an impediment for achieving greater or better results in terms of human and environmental health, among other benefits, to be derived from these services.

Additionally, and independently of the level of socio-economic development, the growing complexity of the situations in which WSS operate requires interdisciplinary and transversal inter-sector approaches. Thus, in developed countries that have solved the problem of universalization long ago, current problems such as the new risks to human health derived from the development process or the increasing environmental concerns facing water management have also moved interdisciplinary and inter-sector approaches higher up the WSS policy agenda.

In this regard, this chapter argues that achieving success in the interventions in the interfaces between WSS and related fields requires the development of professional teams whose members have both the specific specialist knowledge in the relevant fields and a capacity for dialogue and coordinated action across the different disciplinary divides. Similar considerations apply to the managerial and institutional organization of WSS, which also require the introduction of mechanisms that facilitate greater dialogue and cooperation across the different sectors closely linked with WSS, such as water resources management, environment policy, public health, social policy, urban and regional planning, and public services, among others.

Within this framework, the first section of the chapter discusses some theoretical aspects of the interdisciplinary and inter-sector relationships of WSS. Next, it explores the specific interface between WSS and public health, and then the public health links of the different dimensions that must be considered in order to develop public policy and management activities for enhanced WSS. The chapter argues that there is a need to replace the currently prevailing paradigms dominated by the engineering approaches that give centrality to the means and the technical dimension of WSS, and to develop new interdisciplinary and inter-sector approaches that place the protection and promotion of human health as the central objective of WSS. Finally, the conclusion summarizes the main lessons.

Water and sanitation services and their interfaces

The interfaces connected with WSS can be discussed from a number of perspectives. By way of simple illustration, WSS could be seen either as a field of knowledge or as a public policy arena, both of which are characterized by a complex configuration of, respectively, interdisciplinary and inter-sector linkages.

Thus, looking just at WSS as a field of knowledge, it deals with the rigorous understanding of the technical requirements facing the provision of an adequate supply of drinking water and safe collection and disposal of sewage to the population, including the design and implementation of the relevant technological solutions. In terms of knowledge production, this traditional approach has made WSS predominantly a preserve of civil and public health engineering. However, over time it has become evident that there is a need to incorporate the contributions from other fields of knowledge in order to achieve a more complete understanding of the whole array of processes and factors involved in all stages of providing WSS.

For instance, the natural sciences are required for understanding the physical and chemical processes involved in the treatment of raw water in order to make it safe for human use, as well as in the treatment of wastewater to reduce its negative impact upon the natural and human environment. In addition, the human and social sciences offer indispensable tools for improving our understanding of how WSS actually work beyond their physical-natural and technical dimensions, which is necessary for enhancing the impact of WSS interventions upon the population, facilitating the appropriation and effective use of WSS systems by the users, better integrating these services within the urban and rural metabolism, identifying the historical factors of failure and success in the management of WSS, or achieving a more systematic comprehension of the relationship between WSS and the geographical features of a given location, just to mention a few examples. The social sciences can also make substantive contributions by challenging the very assumptions upon which the traditional approaches and technological frameworks for WSS are founded, and unearthing the reasons why, despite the significant scientific and technological advances made in this field of knowledge, the actual provision of essential WSS worldwide is largely unequal, and technological solutions often even reinforce rather than diminish the existing social inequalities. Last, but not least, the life sciences, including the health sciences, provide an understanding of the impacts of WSS upon human health in its various dimensions, which we consider later in more detail. Thus, given the multidimensional character of WSS, the contribution of different fields of knowledge has the potential to increase the benefits that they can bring to their users, which requires significant changes in the individual disciplinary fields in order to develop more integrated and interdisciplinary approaches.

Similarly, looking now at WSS as a public policy arena, the potential intersections between public sanitation policies and various other policy sectors are closely related to the specific institutional structures and organization characteristic of different political cultures. However, beyond these differences between countries and regions, in most cases it is possible to identify close similarities in the interfaces between WSS and health policy, environmental policy, regional development and town planning policy, water resource policy or social policy, to mention just a few examples, some of which I consider later.

The need to strengthen the interfaces between WSS and related spheres of public policy is increasingly recognized. The United Nations Human Settlements Programme (UN-Habitat), for example, explains the deficiencies of WSS in the urban areas of poor countries as the result of a combination of related, contributing and underlying causes, and stresses that sustainable solutions require taking into account the links between WSS and water resources management, demographic planning, and the urbanization and regularization of slums (UN-Habitat, 2003). There is also a growing recognition of the need for a systemic approach to WSS that incorporates their interfaces with the environment, water resources and social conditions to supersede the traditional narrow understanding of WSS as restricted to the activities of service provision (Lindqvist et al, 2001, p391). Other authors have also emphasized the importance of greater integration between WSS and other public policy fields such as public health, the environment, housing, energy, transport and urban planning (Moore et al, 2003). Similarly, empirical work carried out in Botswana pointed to the lack of coordination between urban planning and the planning and management of water resources as an obstacle for achieving the results envisaged by the WSS authorities (Toteng, 2002). Some authors have argued that this lack of coordination in the interface between closely related policy fields is often explained by the simplifying approaches adopted in relation to the impact of the local conditions where the initiatives take place – for example, approaches that conceive of the urban milieu as a physical environment to be modified, neglecting its social complexity (Stephens, 1995).

In this regard, it is possible to assess these interrelations between WSS and related policy fields by examining the Millennium Development Goals (MDGs) from the perspective of the role that WSS play in their achievement. In developing this reasoning, it becomes clear that WSS have an impact upon most of the MDGs, and not just upon goal 7, target 10, which specifically addresses these services:[1]

- *Eradicate extreme poverty (goal 1, target 1)*: improving the conditions of water and sanitation produces a better home environment, improving housing and contributing to breaking the cycle between poverty and the health–disease process.
- *Eradicate hunger (goal 1, target 2)*: improving environmental conditions has proven effects on combating infant malnutrition through reduction of the incidence of parasitosis and diarrhoea.
- *Achieve universal primary education (goal 2)*: improving health standards through sanitation means more children in school and the installation of toilets in schools has a significant effect upon girls' attendance since they can have privacy at school, thereby meeting their physiological needs.
- *Gender equality and women's empowerment (goal 3)*: water supply in homes, with greater accessibility, has the potential to liberate women, who are, in general, responsible for supplying water to their deprived families and for reproduction, valuing their role.
- *Reduce child mortality (goal 4)*: the effect of WSS initiatives on reducing infant mortality is widely proven.
- *Combat HIV/AIDS, malaria and other diseases (goal 6)*: the role of WSS initiatives in reducing levels of infectious parasitic diseases is widely accepted by the scientific community. Specifically, studies have shown the link between WSS and the illnesses cited in the MDGs: HIV/AIDS, since a lack of adequate sanitary conditions increases the risk of opportunistic infection, in addition to the fact that the illness increases individual susceptibility to water-related diseases (WHO and UNICEF, 2005); and malaria, because the vector is waterborne.

The multiple positive effects of introducing WSS initiatives, as shown by the MDGs, are sufficient to demonstrate both the importance of interdisciplinary and inter-sector approaches in the design and implementation of WSS public policy and management.

THE RELATIONSHIP BETWEEN WATER SUPPLY AND SANITATION (WSS) AND HEALTH

In assessing the inter-linkages between WSS, hygiene and public health, it is important to bear in mind the multiple dimensions of WSS, and that each dimension has a specific pattern of interfaces, as illustrated below (based on Prüss-Üstün et al, 2008):

- Water and sewage can be a means through which pathogens and toxic chemical substances are transmitted.
- WSS can be viewed as a service (that could be extended to include solid waste management, storm-water management and irrigation) whose absence could increase the risk of transmission of a number of diseases.
- Habits such as personal and domestic hygiene are determining factors in increasing the services' benefits to human health.
- The way in which water resources are managed may increase or decrease disease risks.

Some authors have expanded the scope of these interrelations by considering WSS to be a commodity, a saleable product, an approach that may have a significant health impact upon the target population. This issue has provoked intense debate between the advocates of WSS commodification and their opponents, who argue that WSS services represent a public and social good, a citizenship right or, more recently, even a human right that cannot be treated as a private good. Authors defending this latter position base their analysis on the theory that the management of WSS by private companies would prioritize the pursuit of private profit, thereby compromising the system's benefits to public health. Such reasoning has already inspired empirical research such as a study carried out in Germany suggesting that privatization of WSS could increase the risk of transmitting waterborne carcinogens (Fehr et al, 2003).

The effects of inadequate WSS provision on human health have been known since ancient times, and perhaps the most remarkable scientific expression of this old tradition was John Snow's investigation into one of the cholera outbreaks that affected London during the mid-19th century. Ever since, epidemiological studies supported by a number of health indicators continue to demonstrate the impact that an absence of adequate conditions of water supply or sewage collection and disposal can have upon human health (Esrey and Habicht, 1986; Esrey et al, 1991; Heller, 1998). Likewise, recent systematic analyses and meta-analyses show the positive impact that improvements in water supply, water quality, sewage collection and hygiene can have upon the occurrence of diarrhoea, which can experience reductions of between 25 and 37 per cent as a result (Fewtrell et al, 2005).

From a global perspective, it is estimated that the inadequate provision of water, sanitation and hygiene is responsible for 5.7 per cent of the world's burden of diseases measured in disability-adjusted life years (DALYs) (Prüss et al, 2002), although it is assumed that this is a conservative figure and that the real impact could be as high as 10 per cent of the world's DALYs (Prüss-Üstün, 2008). Within this context, one of the main factors is the incidence of diarrhoea, which accounts for a very large proportion of WSS-related health problems, and constituted the second greatest world's burden of disease in 1990 (Michaud et al, 2001). In fact, 88 per cent of the diarrhoea-related burden of disease is attributed to the inadequate provision of water supply, excreta disposal and hygiene, and the greatest impact of the disease is concentrated upon children from less developed countries (WHO, 2002).

From another angle, the positive economic impact of successful WSS initiatives has been similarly demonstrated. For instance, a recent study that applied the classification of the World Health Organization's Commission on Macroeconomics and Health[2] to explore the cost-effectiveness of public health interventions in developing countries showed that public WSS initiatives can be both 'cost-effective' and 'very cost-effective', based on the amount of the gross domestic product per capita spend for each DALY averted (Haller et al, 2007), and a similar conclusion was reached in a study on household-based interventions to improve water quality (Clasen et al, 2007). In addition to being cost effective, other authors found that such initiatives are also cost beneficial in that the return on a monetary unit invested could range between 5 and 46 in developing countries. The capital gained comes from:

- direct health-related economic benefits (e.g. costs avoided owing to fewer cases of diarrhoea);
- indirect health-related economic benefits (productivity gains); and
- benefits unrelated to health (e.g. time saved thanks to better access to WSS).

In economic terms, this last factor was found to have the greatest impact (Hutton et al, 2007).

As a note of caution, the analysis of the health and economic impacts of WSS interventions must take into account the heterogeneity of different contexts. Thus, in wealthy regions the hygiene and sanitation needs and the expected health impacts for WSS interventions are very different from those corresponding to slums or rural areas. In addition to this, crucial factors – such as the quality of the operation and maintenance of WSS, and even the technological conception of the systems – have a crucial impact upon the results. This is important because the quality of the services is usually more precarious in developing countries, where intermittent water supply, deficiencies in the quality of the water supplied, absence of sewage treatment and inadequate provision of waste disposal are widespread. Moreover, the often precarious domestic installations for water and sanitation characteristic of many homes in developing countries constitutes an additional and crucial risk factor for faeco-oral transmission.

In a broader perspective, theories on the relationship between WSS and health have to be considered within the wider framework of environmental health, which has been developed to take into account the health hazards inherent in different situations, including homes, workplaces or outdoor environments. In such conditions, the contamination of air, water and food can be considered to be the main hazard to human health, in addition to the risk of accidental injury. The interventions foreseen within the framework of environmental health include, in addition to WSS and hygiene, air pollution control, storm-water drainage, solid waste management, food management, building regulations, vector control, and workers' health (Cairncross et al, 2003). Placing the WSS–health relationship within the context of environmental health can afford a wider vision of the risks to human health posed by environmental factors and, in turn, make WSS initiatives more effective.

The following section examines the possible shifts in behaviour that can be expected when the design and implementation of public policy and management programmes for WSS are oriented by the potential health impacts of such interventions.

PUBLIC POLICY AND MANAGEMENT OF WSS FROM A PUBLIC HEALTH PERSPECTIVE

The incorporation of the public health perspective within the formulation and implementation of public policy and management programmes for WSS could trigger far-reaching transformations in the conventional framework of this sector and enhance the beneficial impacts of these services, a change that could be even more radical if the health benefits of WSS are adopted as the key driver for WSS policy decision-making. Such a transformation would imply a significant reordering of priorities from the currently prevailing conceptions in public policy and management, which give priority to the means by which WSS are delivered, to a more advanced framework that places more emphasis and a higher value on the ends of these services. Generally, the development of WSS policies has been mainly approached from the engineering perspective, placing a higher value on the technological dimension of the processes, often to the detriment of crucial considerations, such as the meaning and ends that should be the guiding principle for WSS in the first place. This is of utmost importance given that, in historical perspective, the success story of WSS in Europe and North America was largely driven by the adoption since the late 19th century of public policies that gave top priority to the universalization of these services, looking at the prevention of disease transmission and epidemics. Clearly, transforming the prevailing WSS paradigm by subordinating the means (the technological dimension) to the ends (achieving universal provision of safe WSS) has major conceptual, technological and political implications for the practices currently in use in the sector, including the need for inter-sector coordination, which require a more in-depth understanding and elaboration.

In this connection, we develop below a contribution to this debate that follows the main conceptual guidelines that informed the approach of this book, and which draws a distinction between the spheres of public policy and management and explores some of their interfaces in the sub-dimensions of:

- legal and institutional frameworks;
- planning, policies and programmes;
- administrative organization of the services and its implementation; and
- evaluation.

Legal and institutional frameworks

Evidently, the legal and regulatory apparatuses of a given country tend to be influenced by its distinctive political culture and practices, including the level of definition and detail applied to their laws, norms and regulations. However, leaving aside these specificities, the adequate provision of universal WSS presupposes the existence of legislation where the rights and duties of the parties involved in the organization and delivery of these services are clearly established. In particular, it is expected that the legal and regulatory instruments would define the roles of the different actors, including service providers, regulatory bodies, users and agents operating in the interfaces between WSS and other sectors, among others. These requirements are essential everywhere, but are particularly crucial in those countries where the universalization of these services has not yet been achieved, which includes a large majority of the world's countries.

In addition to the influence of the political traditions of different countries on the legal and institutional frameworks adopted for WSS, the different conceptions that exist with regard to the expected level of public intervention also have significant influence on the legal and institutional dimension of these services. Clearly, the traditions of thought that defend a minimal level of state intervention tend to favour more generic and less 'interventionist' laws and regulations, which in those cases where there is a close alignment with neoliberal conceptions of the role of the state may even leave the task of regulation to the 'market forces' that are supposed to self-regulate the adequate functioning of the services. Contrariwise, the defenders of greater state intervention through regulation, control and monitoring of WSS activities believe that there is a need for legal frameworks that provide more specific rules for the organization and running of WSS. There are many examples of the practical consequences of these confrontations between rival views of the role of legal and institutional frameworks in the WSS sector – among others, the recent debates in Brazil that eventually led to the new Water and Sanitation Law passed in 2007 (Heller, 2006).

Always keeping in mind the differences between countries in relation to such issues as the allocation of responsibility for WSS (jurisdiction), the place of these services in the scale of policy priorities (hierarchy) or the prevailing understanding of the state's role in this area, I argue that legal frameworks for WSS must include a health focus as an essential component in at least two respects:

1 In the allocation of roles to institutional agents in the field of WSS, it is essential to recognize the need for more attention at the interfaces with other sectors, and particularly with the health sector, and to establish concrete mechanisms for integration of activities within these interfaces.
2 There must be explicit planning for the differentiated allocation of functions and duties between institutional agents – for example, in the separation between, first, the control of water quality for human consumption, an attribution of the WSS suppliers, and, second, the surveillance of the standards of the water quality actually delivered for human consumption, an attribution of public health authorities.

Needless to say that it is also essential that the interface with WSS be also recognized in the health legislation, which should establish the co-responsibility of health authorities on policy planning and implementation in relation to these services.

From another perspective, the actual structure of public-sector WSS activities, from the national to the local level, is crystallized in institutional frameworks, which as already discussed often assume different characteristics according to each country's traditions and prevailing policy preferences. Whatever the case, it is crucial that the inter-linkages with public health become embedded in the institutional frameworks of WSS, given the strategic importance of institutional drivers (e.g. directives, norms, programmes, etc.) in the promotion of systemic change. Thus, there is a need for integrative mechanisms operating at the executive level to foster a closer interaction between the different sectors, and particularly between WSS and public health – for instance, by strengthening the focus on health prevention and promotion in the public health area, while tuning WSS policies to the main objective of enhancing the conditions for human health.

Additionally, it is important to better define and strengthen the responsibilities of the health sector in relation to WSS – for example, in the surveillance of water quality for human consumption or in epidemiological surveillance. In the latter case, surveillance of WSS-related diseases could have a significant role in planning and assessing the efficiency, efficacy and effectiveness of the actions implemented in the WSS sector.

Planning, policies and programmes

Planning at the level of WSS, something that is unfortunately carried out very rarely in the majority of developing countries, could potentially promote a convergence of efforts

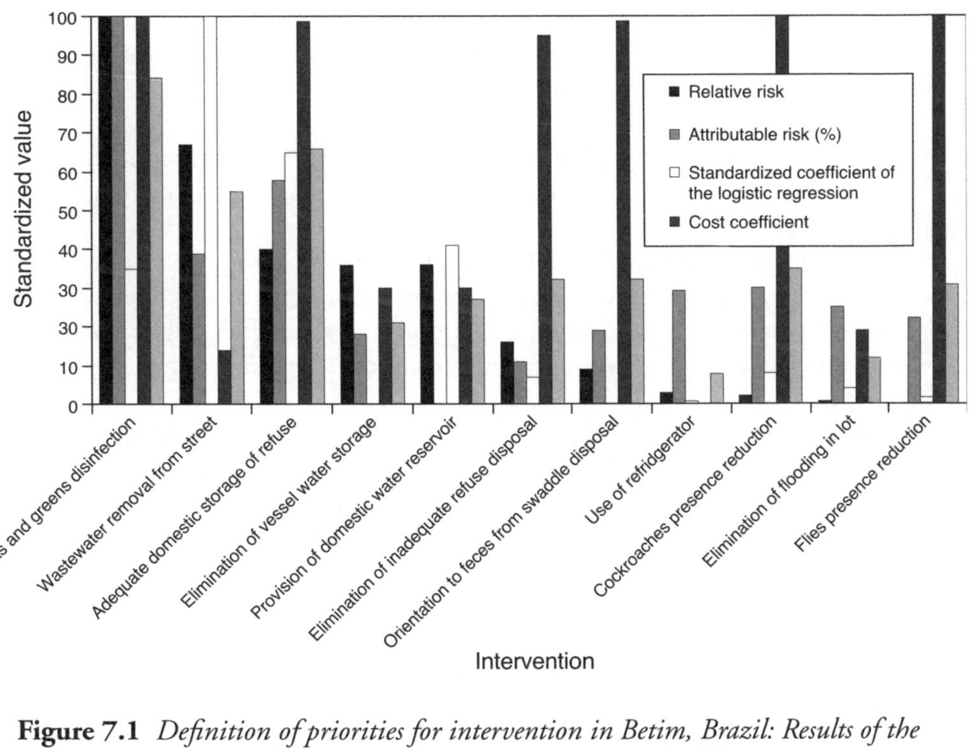

Figure 7.1 *Definition of priorities for intervention in Betim, Brazil: Results of the standardized coefficients of impact upon health*

Source: Heller et al (2005)

when focusing on adequately designed objectives. From a public health perspective, planning may be a crucial tool for establishing priorities for intervention aimed at health impact maximization.

By way of example, Figure 7.1 presents a comparison of WSS and hygiene initiatives prioritized according to their potential impact on health. This comparative model was developed as part of an epidemiological study looking at associations between WSS and hygiene and infantile diarrhoea.

One potentially effective tool for planning water services with a focus on protecting human health are the water safety plans (Davison et al, 2005). The adequate implementation of these instruments may facilitate the much needed shift from the currently prevailing approaches in water supply, which tend to be narrowly focused on the inner working of these services, to a more systemic perspective that encompasses the whole process involved in the delivery of water supply from the source through the user and back to the environment. Such a holistic approach would also place more emphasis on evaluating and managing risk, and identifying the critical points in the functioning of water supply systems.

Similarly, in an ideal situation, WSS policies should reflect the actions planned by the relevant authorities following the general directives adopted at the national level, which also ideally would normally be part of a multiannual planning strategy (e.g. with a five-year planning horizon). However, evidently the extent to which specific programmes promoting WSS may induce a greater or lesser public health focus will be influenced by the relevant institutional contexts where these services operate. Within this framework, WSS policies may include the following aspects:

- the practice of multiannual planning of WSS that target the improvement of public health conditions, including the adoption of methodologies for benchmarking performance;
- the incorporation of policy principles that are essential for the effective and inclusive organization of WSS in multiannual government planning (these would include such principles as universality; equality; the aim of being integral; public management; social participation and monitoring; inter-sector coordination; and quality) (Costa et al, 2006);
- the inclusion of mechanisms for allocating priority for public funding to situations where the public health impact of WSS interventions can be greater;
- the adoption of specific measures for strengthening local powers, given that it is at the local level that WSS actions can be better harmonized with the promotion of public health objectives.

Specific programmes, for their part, can really enhance the impact of focused WSS interventions on public health, which requires the application of appropriate methodologies to identify the most vulnerable situations, particularly those that are more likely to benefit from the initiatives.

The administrative organization of WSS and their implementation

As discussed in other chapters of this volume (see especially Chapters 5 and 8 to 10), the organization of WSS services may follow a variety of alternative management models (e.g. public, private, community led, etc.) and may be subject to different legal and institutional arrangements. However, although there exist comparative studies that have tried to assess the performance of these different models with regard to such aspects as, to give an example, commercial or managerial efficiency, there is a need for comprehensive and systematic evaluations of the comparative efficacy, effectiveness and efficiency of these different management models in relation to the public health benefits and impact.

Nevertheless, independently of the particular management model chosen or the juridical and institutional arrangements in place, WSS obviously need to be adequately organized to deliver a quality service to the entire population, which includes the necessary planning to address such issues as future needs of service expansion or other long-term challenges that may arise from unforeseen changes. In specific reference to the interface between WSS and public health, the organization of WSS requires addressing the following basic issues:

- the adequate operation of WSS to avoid health risk situations that may result from potential deficiencies in water quality, intermittency in the water supply, restrictions in the water volumes supplied for essential human uses, and exposure of the population to sewage, amongst other factors;
- permanent surveillance and monitoring of water sources to tackle situations that may imply health risks during water consumption;
- the adoption of tariff structures that are based on the principle of social solidarity and do not impose restrictions in the access to WSS on the basis of people's capacity to pay;
- permanent control of water quality, including the immediate communication to users and regulatory authorities of any potential health risks arising from water consumption;
- strengthening the participation of WSS users and local authorities as the central locus for the promotion of public health;
- coordinating WSS activities with the activities of epidemiological surveillance in order to reduce risk situations based on identification of the occurrence of water-related diseases;
- developing integrated programmes for sanitary and health education.

Clearly, achieving these objectives demands the adoption of interdisciplinary approaches to WSS and public health, given that addressing these questions only from the traditional engineering perspective, which continues to prevail in this sector of activity, has proved insufficient. This conclusion can be derived from the observation of the traditional ways in which engineering professionals involved with WSS continue to be trained, of the review of the main topics and approaches adopted in the scientific publications of this field, and from the actual practices that can be observed in the activities of planning, design, implementation and operation of these services. In particular, a closer interaction between engineers and health professionals could help to establish an 'epidemiological reasoning' in decision-making activities and to foster more interdisciplinary action in the field of WSS.

Moreover, with regard to the daily operation of the services, a closer working relationship between the fields of sanitation and public health is a fundamental requirement to enhance the role of citizen participation in decision-making activities, which is at the centre of the problem of water governance, a theme explored in several chapters of this book. In this regard, while in the health sector there exist important experiences of participatory and social monitoring activities in developing countries, albeit not without difficulty (Zakus and Lysack, 1998), as a general trend, in the WSS sector, social participation has been much more limited and less effective. The evidence of these comparatively poorer levels of participatory experiences in WSS is the absence of a more rigorous theoretical foundation and of empirical studies about citizen participation in the public policy and management of WSS.

In this regard, it is possible to envisage a number of possible interactions between public health and WSS activities that may promote more substantive citizen participation. From the inclusion of WSS issues in the agendas of participatory forums of the health sector to the articulation of public health and WSS activities aimed at stimulating citizen participation, there exist many possibilities for incentivizing greater beneficial integration for both sectors. The experience also shows that the success of such initiatives largely depends upon the conditions for substantive, not merely rhetorical, citizen involvement in decision-making, policy development, planning and monitoring of WSS actions. It is also crucial that citizens are given the conditions to understand WSS not as a separate and isolated sector of human activity, but rather as a complex field closely interlinked with a number of other spheres, and most particularly with the protection of human health.

Evaluation

The evaluation of WSS and their effects is a fundamental step for assessing the achievement of goals and, in particular, as a mechanism for feedback on planning. An important component of the process is that of assessing WSS impacts upon health. The incorporation of the practice of evaluation in the routine of the services has the potential to make the actions more effective. Evaluation of the impacts of WSS on health, in particular, however, is not carried out frequently, especially in developing countries. This is not only the case with WSS, given that, in general, the assessment of health impacts tends to be much more frequent in developed countries, which usually have less need of these types of evaluation (Erlanger et al, 2008).

The rationale behind assessing the impact of WSS upon health is that different approaches and initiatives in this field can lead to different effects on health, which highlights the need for an adequate and careful design of WSS interventions that take fully into account the social, environmental, economic, technological and cultural characteristics of the target areas. Furthermore, the results of evaluations can act as feedback for decision-making in the services.

Recent studies have evaluated the impact of WSS interventions on the health of the recipient population. For instance, two wide-ranging longitudinal studies were carried out in the city of Salvador, Brazil, aimed at evaluating the impact of a large sanitation programme that included the laying of more than 2000km of sewer pipes, the building of 86 pumping stations, and connection of more than 300,000 households to the sewerage network over eight years. The project sought to increase the coverage of household connections from 26 to 80 per cent. The epidemiological study, controlling the effect of several confounding factors, showed a reduction of 22 per cent in the prevalence of diarrhoea in children under the age of three (Barreto et al, 2007).

Other health indicators, besides the traditional use of diarrhoeal prevalence, are also recommended to evaluate the impact of WSS measures, including anthropometric indicators and enteroparasitoses, as well as specific pathologies such as cholera, typhoid fever, viral hepatitis and leptospirosis (Briscoe et al, 1986; Heller, 1998). Obviously, there must be compatibility between the health indicator and the epidemiological framework adopted in the case under assessment.

Another important procedure that can be implemented for evaluating health risks resulting from WSS interventions is the quantitative microbial risk assessment (QMRA) (Haas et al, 1999). This tool is simultaneously useful for evaluating and planning enhanced WSS interventions, and enables the quantitative assessment of the suitability of different technical solutions for controlling the risk of infection by pathogenic organisms, as well as comparing the relative risks of different technical alternatives. For instance, this technique has been used to verify the risk of infection by specific pathogens, especially protozoa and viruses, present in raw water sources or in water treated with different processes.

Another aspect of evaluation relates to the need for a critical approach to assess the logic of WSS plans and programmes. Although by their very nature these instruments include important political components, they are generally presented as 'professional' or 'technical' initiatives that are therefore naturally beneficial to the users and lacking in political content or intention. This relates to the evaluation of public policy in the field of WSS as distinct from the evaluation of the impact of these services.

CONCLUSIONS

As argued throughout this chapter, bringing WSS management into line with a public health perspective is not just desirable; rather, it is imperative to ensure that the implementation of WSS interventions result in the effective protection and promotion of human health. In other words, the incorporation of a public health perspective within public policy and management of WSS is a fundamental requirement to reorient the organization of these essential services in function of their ends, thus breaking with the hegemonic paradigm that gives priority to the technical means as the main drivers of management.

However, in order to promote the advancement of such new paradigms for WSS, there is much work to be done in the various levels of planning and implementation of public policy in order to introduce the concepts and mechanisms that may favour this process. As discussed earlier, the relevant areas of action are particularly the legal and institutional frameworks for the organization of WSS, the relevant policies and programmes, the activities of planning and administrative organization of the services and their evaluation, and the substantive participation of citizens in all stages of the processes, including policy design and monitoring of results.

There are major challenges ahead that will require significant efforts in this endeavour. In particular, there is first the need to facilitate the approximation of the conceptual frameworks of the professionals operating in the fields of WSS and public health in order to develop more interdisciplinary approaches in this interface. Second, there must be a greater convergence of perspectives in the public policies of both sectors of activity, which means breaking down the prevailing inflexibility and fragmentation of the existing bureaucratic apparatuses in order to develop the required inter-sector coordination.

NOTES

1 The following paragraphs on the MDGs have been adapted from UN (2008).
2 The WHO Commission on Macroeconomics and Health (CMH) defined as 'cost-effective' those health interventions where the ratio of the cost to the gross domestic product (GDP) per capita is less than three times (WHO–CMH, 2001). On this basis, the studies that apply the CMH classification define as 'cost-effective' those public health interventions that help to avert each disability adjusted life year (DALY) at a cost of between one and three times GDP per capita, and as 'very cost-effective' those interventions where the cost is less than the GDP per capita. The WHO (2008) defines DALY as:

> ... a health gap measure that extends the concept of potential years of life lost due to premature death (PYLL) to include equivalent years of 'healthy' life lost by virtue of being in states of poor health or disability. The DALY combines in one measure the time lived with disability and the time lost due to premature mortality. One DALY can be thought of as one lost year of 'healthy' life and the burden of disease as a measurement of the gap between current health status and an ideal situation where everyone lives into old age free of disease and disability.

REFERENCES

Barreto, M. L., B. Genser, A. Strina, M. G. Teixeira, A. M. O. Assis, R. F. Rego, C. A. Teles, M. S. Prado, S. M. A. Matos, D. N. Santos, L. A. Santos and S. Cairncross (2007) 'Effect of city-wide sanitation programme on reduction in rate of childhood diarrhoea in northeast Brazil: Assessment by two cohort studies', *The Lancet*, vol 370, pp1622–1628

Briscoe, J., R. G. Feachem and M. M. Rahaman (1986) *Evaluating Health Impact: Water Supply, Sanitation, and Hygiene Education*, International Development Research Centre, Ottawa, Canada

Cairncross, S., D. O'Neill, A. McCoy and D. Sethi (2003) *Health, Environment and the Burden of Disease; A Guidance Note*, Department for International Development (DFID), London

Clasen, T., L. Haller, D. Walker, J. Bartram and S. Cairncross (2007) 'Cost-effectiveness of water quality interventions for preventing diarrhoeal disease in developing countries', *Journal of Water and Health*, vol 5, no 5, pp599–607

Costa, S. S., L. Heller, L. R. S. Moraes, P. C. Borja, C. H. Melo and D. Sacco (2006) *Successful Experiences in Municipal Public Water and Sanitation Services from Brazil*, ASSEMAE, Corporate Europe Observatory (CEO), Transnational Institute (TNI), World Development Movement (WDM), Amsterdam

Davison, A., G. Howard, M. Stevens, P. Callan, L. Fewtrell, D. Deere and J. Bartram (2005) *Water Safety Plans: Managing Drinking-Water Quality from Catchment to Consumer*, WHO, Geneva

Erlanger, T. E., G. R. Krieger, B. H. Singer and J. Utzinger (2008) 'The 6/94 gap in health impact assessment', *Environmental Impact Assessment Review*, vol 28, pp349–358

Esrey, S. A. and J.-P. Habicht (1986) 'Epidemiologic evidence for health benefits from improved water and sanitation in developing countries', *Epidemiol Review*, vol 8, pp117–128

Esrey, S. A., J. B. Potash, L. Roberts and C. Schiff (1991) 'Effects of improved water supply and sanitation on ascariasis, diarrhoea, dracunculiasis, hookworm infection, schistosomiasis, and trachoma', *Bulletin of the World Health Organization*, vol 69, pp609–621

Fehr, R., O. Mekel, M. Lacombe and U. Wolf (2003) 'Towards health impact assessment of drinking-water privatization – the example of waterborne carcinogens in North Rhine-Westphalia (Germany)', *Bulletin of the World Health Organization*, vol 81, no 6, pp408–414

Fewtrell, L., R. B. Kaufmann, D. Kay, W. Enanoria, L. Haller and J. M. J. Colford (2005) 'Water, sanitation, and hygiene interventions to reduce diarrhoea in less developed countries: A systematic review and metaanalysis', *Lancet Infectious Diseases*, vol 5, no 1, pp42–52

Haas, C. J., J. B. Rose and C. P. Gerba (1999) *Quantitative Microbial Risk Assessment*, John Wiley and Sons, Inc, New York

Haller, L., G. Hutton and J. Bartram (2007) 'Estimating the costs and health benefits of water and sanitation improvements at global level', *Journal of Water and Health*, vol 5, no 5, pp467–480

Heller, L. (1998) *Saneamiento y salud*, Pan American Health Organization, Lima, Peru (in Spanish)

Heller, L. (2006) *Access to Water Supply and Sanitation in Brazil: Historical and Current Reflections; Future Perspectives*, Human Development Report 2006, Occasional paper, http://hdr.undp.org/en/reports/global/hdr2006/papers/heller%20leo.pdf, accessed June 2008

Heller, L, E. A. Colosimo and C. M. F. Antunes (2005) 'Setting priorities for environmental sanitation interventions based on epidemiological criteria: A Brazilian study', *Journal of Water and Health*, vol 3, no 3, pp271–281

Hutton, G., L. Haller and J. Bartram (2007) 'Global cost-benefit analysis of water supply and sanitation interventions', *Journal of Water and Health*, vol 5, no 5, pp481–502

Lindqvist, J., S. Narain and A. Turton (2001) 'Social, institutional and regulatory issues', in Č. Maksimovič and J. A. Tejada-Guibert (eds) *Frontiers in Urban Water Management*, IWA, Padstow

Michaud, C. M, C. J. L. Murray and B. R. Bloom (2001) 'Burden of disease: Implications for future research', *Journal of American Medical Association*, vol 285, no 5, pp535–539

Moore, M., P. Gould and B. S. Keary (2003) 'Global urbanization and impact on health', *International Journal of Hygiene and Environmental Health*, vol 206, nos 4–5, pp269–278

Prüss, A., D. Kay, L. Fewtrell and J. Bartram (2002) 'Estimating the burden of disease from water, sanitation, and hygiene at a global level', *Environmental Health Perspectives*, vol 110, no 5, pp537–542

Prüss-Üstün, A., R. Bos, F. Gore and J. Bartram (2008) *Safer Water, Better Health: Costs, Benefits and Sustainability of Interventions to Protect and Promote Health*, World Health Organization, Geneva

Stephens, C. (1995) 'The urban-environment, poverty and health in developing-countries', *Health Policy and Planning*, vol 10, no 2, pp109–121

Toteng, E. N. (2002) 'Understanding the disjunction between urban planning and water planning and management in Botswana: A challenge for urban planners', *International Development Planning Review*, vol 24, no 3, pp271–298

UN (United Nations) (2008) *End Poverty – Millennium Development Goals 2015: Make It Happen*, UN, New York, www.un.org/millenniumgoals/, accessed September 2008

UN-Habitat (United Nations Human Settlements Programme) (2003) *Water and Sanitation in the World's Cities: Local Actions for Global Goals*, Earthscan, London

WHO (World Health Organization) (2002) *The World Health Report 2002: Reducing Risks, Promoting Healthy Life*, WHO, Geneva

WHO (2008) 'Disability adjusted life years (DALY)', in *Health Statistics and Health Information Systems*, WHO, Geneva, www.who.int/healthinfo/boddaly/en/, accessed September 2008

WHO–CMH (World Health Organization–Commission on Macroeconomics and Health) (2001) *Macroeconomics and Health: Investing in Health for Economic Development, Report of the Commission on Macroeconomics and Health*, WHO–CMH, Geneva

WHO and UNICEF (World Health Organization and United Nations Children's Fund) (2005) *Water for Life: Making It Happen*, WHO and UNICEF, Geneva

Zakus, J. D. L. and C. L. Lysack (1998) 'Revisiting community participation', *Health Policy and Planning*, vol 13, no 1, pp1–12

8

The Market-Centred Paradigm

Michael Rouse

INTRODUCTION

The word 'market' is synonymous with the private sector, but the public sector also adopts market-type approaches to achieve improved performance. This chapter includes discussion on the use of the private sector; but first it covers the more general aspects of market forces. There is an important distinction to be made between private organizations that are in business to make a profit for their shareholders, but importantly have to deliver what the customers want or go out of business, and the use of market-type forces to improve the efficiency of water utilities. There are degrees to which the public sector emulates private operations. In The Netherlands, publicly owned water companies are required to operate to private-sector accountancy rules; but they do not pay dividends to their local authority shareholders. In parts of Australia, provincial states require state-owned companies to be the exact equivalent of private-sector companies. Such companies are exposed to full tax rules and required to pay dividends to the state.

The starting point for considering changes in the approach to water service delivery is that, generally, the municipal model of delivering water services, although originally successful in the developed world in establishing water supply and sanitation, is not the most effective way of achieving efficient and sustainable services in the current context characterized by ever more-demanding management conditions. What is the evidence of this? Unfortunately, there is little or no data available on the performance of the municipal model. This is partly the answer, as it suggests that there are no performance measures. Performance measures are important to reflect public policies and targets, such as universal access or improved drinking water quality, and to achieve these as efficiently as possible in the best interests of the community. In the developed world, the symptoms of failure are an increasing trend towards different models and the pleas of local government for 'federal' funds for refurbishment of ageing infrastructures. In the developing world, the symptoms of failure are a decline in previously good services due to inadequate investment and failure to extend systems to supply the extremities of growing cities. Although the developing world has greater difficulty in achieving the necessary investment, the principles of required reform are similar.

WHY IS THE MUNICIPAL MODEL INADEQUATE?

In principle, the municipal model for public services is ideal. Utilities 'belong' to the local community and there should be public accountability through elected officials. The reality is very different, as there are a number of characteristics that can contribute to the inadequacy of municipal-managed utility operations. First, the lack of separation between policy and delivery functions, with associated political interference in what should be wholly operational considerations. Politicians can interfere in staffing levels, for example, seeing public-service operations as a means of creating employment, rather than allowing operational managers to determine efficient resource requirements. The argument for such political interference is that it is legitimate public policy, but unfortunately it results in conflicting operational objectives. The excuse for interference can be subsidies. The subsidies arise as necessary top-ups to finance operations due to the failure by local government officials to sanction the water tariffs necessary to achieve sustainability. As subsidies tend to be inadequate to finance maintenance and refurbishment of infrastructure, there is a spiral of decline in service. Inadequate finances are partly responsible for poor public-sector pay, which does not attract high calibre people, and which can increase the temptation for corrupt practices. A combination of these factors can result in low morale, with resulting poor service and spiralling decline. The process of decline is illustrated by 'the vicious circles' in Figure 8.1.

There are examples of successfully developed municipal operations; the most striking examples are those that have established world-class operations through the application of 'market' principles by introducing internal contracts. This means that there has to be part of the organization that issues a contract and another part that is the contractor, thus achieving separation of policy and delivery. The internal contract provides a means of setting performance requirements and performance monitoring. It requires separate and accurate accounting. Reporting the performance results to the public provides for transparency, reducing the opportunity for unpalatable information to be hidden prior to elections. Internal contracts also provide for both incentive bonuses and time limits on management tenure, thus providing 'carrot and stick' incentives for improved performance. Seattle in the US and Brisbane in Australia are two examples of cities adopting an internal contract approach. These success stories are described in a book on *Institutional Governance and Regulation of Water Services* (Rouse, 2007). Although both of these examples are from the developed world, the approach could be applied anywhere. The main benefit is clear focus of responsibilities, well-defined objectives and targets and, most importantly, transparency.

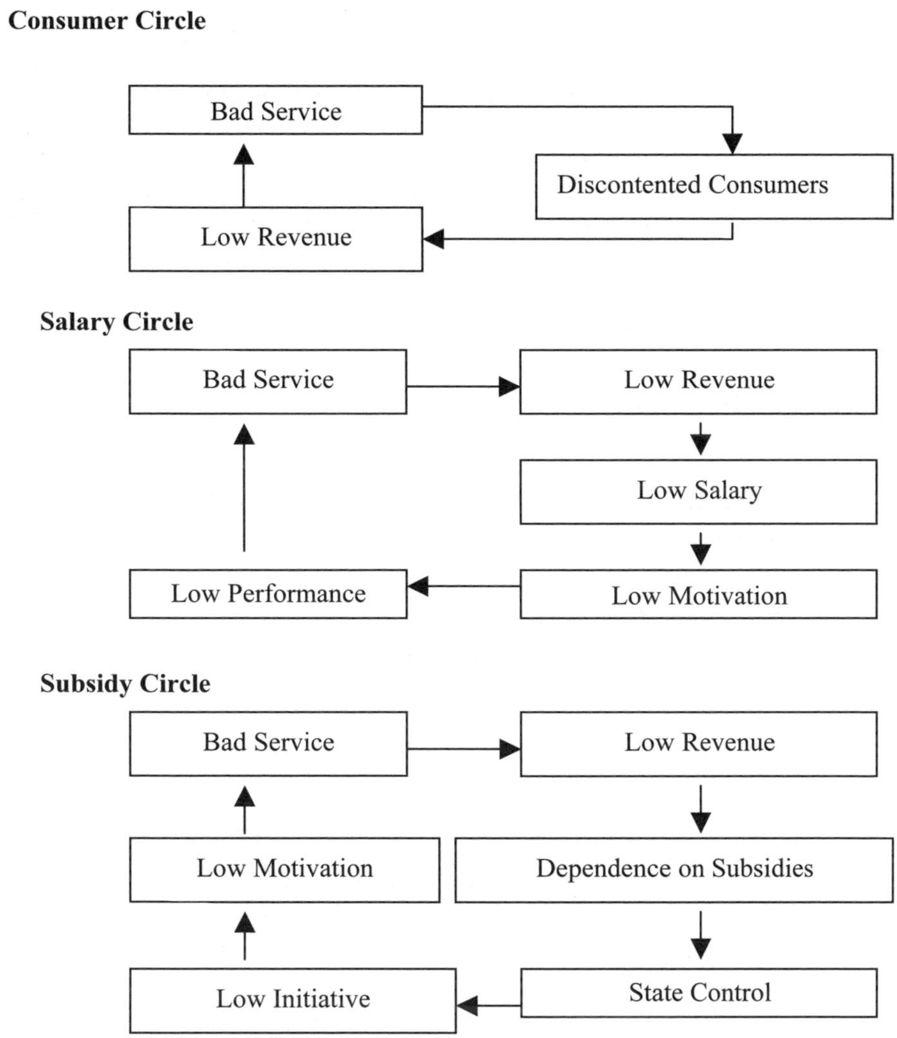

Figure 8.1 *The vicious circles of decline*

Source: Kontula (2004)

CORPORATIZED PUBLIC COMPANIES

It is much more common for countries and cities, in attempting to improve the performance of water services, to form public-owned water companies operating under private-sector laws of corporate governance: so-called corporatized public companies. These organizations are managed by private-sector type boards, can be required to pay taxes and dividends, and have to produce audited accounts. Due to the monopolistic nature of water services, prices cannot be determined by the market. There is an ongoing requirement for local authorities to take decisions on tariffs unless there is a separate economic regulator to manage that process.

The move to corporatized public companies is often accompanied by the amalgamation of many small water service operations into a few larger ones, as has happened in The Netherlands and in Australia. It is not suggested that the bigger the better, but that utilities have to be of a certain size to become viable operations. Companies have to be of a size able to attract good quality management either from the public or private sector, and able to afford necessary scientific support services. In The Netherlands, where there was a policy of integrating units to achieve a minimum size of 100,000 population served, the number of companies reduced from 94 in 1980 to just 13 in 2004. In the state of Victoria in Australia, as a result of widespread structural reforms during the 1990s, the reduction was even more pronounced, with 400 utilities being absorbed into 15 urban and 5 rural water authorities. Much earlier, in England and Wales, 1620 public-service bodies were incorporated within just ten river basin-structured Regional Water Authorities (RWAs). These RWAs, formed in 1974, effectively became corporatized public companies in 1982 when the local authority representative governing 'councils' were replaced with private-sector type boards. In Scotland, through a series of changes since the 1970s, there is now just one government-owned water and wastewater company.

These changes can be unpopular with local authorities who do not wish to lose responsibility for local water services, as was the case in England and Wales. In The Netherlands, although the local authorities relinquished management control, they became 'shareholders' in the larger companies, with collective responsibility for approving water service tariffs.

MONOPOLIES AND COMPETITION

Although currently there are attempts to introduce direct competition in England and Wales through common carriage (UK Government, 2003), water services are, generally, monopolies due to the dominant cost of the pipe networks. This limits the scope for direct market forces to drive efficiency. Instead, the concept of comparative competition has been introduced through the use of performance measures and league tables, often referred to as benchmarking. The approach is used as part of incentive-based regulation by the Office of Water Services (OFWAT), the economic regulator in England and Wales, to stimulate improved performance by the private sector companies (see OFWAT, 2007). Benchmarking is also used very effectively in The Netherlands, where involvement by water companies was initially voluntary, but has now been made compulsory. The success of both schemes in achieving significant performance improvements demonstrates that surrogate market forces can be applied equally to both public and private sectors.

MARKETIZATION IN CHINA

China uses 'marketization' in the development of a market economy. Initially, it applied to government exit from commercial industries; but its use has been extended to the non-commercial sector, such as municipal public utilities. The aim is to subject public utilities to market forces. A number of cities saw this as an opportunity to obtain private-sector money, and initially some of them entered into concession contracts with international private water companies without adequate preparation. A study of 17 cases was published in 2007 (Fu Tao and Lijin, 2007), which identifies the lessons to be learned in taking forward marketization by using the private sector. However, it is planned that marketization in China will be applied to the water sector as a whole through the development of benchmarking. This approach, embracing both public and private sectors, is important not only for China, but also generally as it is necessary to move away from the emotive hang-ups over the use of the private sector or the use of market forces. The public–private argument is non-productive and takes attention away from the task of extending the provision of water and sanitation services. This unproductive argument is recognized in the United Nations Development Programme (UNDP) *Human Development Report* of 2006 (UNDP, 2006) in discussing how to meet the Millennium Development Goals (MDGs) on water and sanitation.

THE PRIVATE SECTOR IS ALWAYS INVOLVED, SO WHY IS IT CONTROVERSIAL?

Although it is understandable and desirable that there is criticism of unsuccessful contracts, it is strange that the use of the private sector to manage water services is controversial, in principle, as the private sector is always involved to some degree. The private sector almost always designs and constructs new treatment plant; it provides information technology (IT) systems and tools and equipment, and in doing so either makes a profit or goes out of business. Yet, for some people, profit should not be associated with owning, financing or managing the service operation. It appears to be appropriate for government to finance water services, even if government has to borrow money (on which banks will make a profit), but not for there to be direct private-sector investment. For many it seems to be acceptable to have inefficient municipal-operated water services, but not efficient private-sector ones achieving a profit. Surely the question should be how to achieve efficient and sustainable water services.

One difficulty seems to be associated with water as a human right (WHO, 2002), a principle against which no one would argue; but it has been interpreted as meaning water should be free. Here it is necessary to differentiate between water itself being free and the need for payment for the provision of water services. One reason for resistance to the use of the private sector in managing operations arises from increases in water charges associated with private contracts. Often, prior to the letting of a contract, water charges under public-sector operations had been achieving a low level of cost recovery through charges, but with government removing the subsidy at the start of the private

Table 8.1 *Water charges in South Australia (1993 to 2000)*

Year	*Cost of first 136 kilolitres*			*Cost of average consumption of 250 kilolitres*		
	AUS$[a]	*Percentage increase since 1993*	*Percentage increase since 1996*	*AUS$**	*Percentage increase since 1993*	*Percentage increase since 1996*
1993	120	–	–	220	–	–
1996	172	43	–	275	25	–
2001	191	59	11	301	37	9
2006	–	–	–	335	52	22

Note: [a] Australian dollars (AUS$1 = US$0.96).
Source: author's elaboration based on tariff figures published in *Government of South Australia Gazettes* (1993–2006)

contract. In other situations, such as in Adelaide, South Australia, charges were increased substantially three years before the letting of a private contract. Later, the increases were incorrectly attributed to the private contract in an Australian *The Guardian* newspaper article (Mac, 2003). Table 8.1 gives the actual figures; an increase of 59 per cent took place between 1993 and 2001, but there was a 43 per cent increase prior to the start of the private contract in 1996. In the article, the increase of 59 per cent was attributed to the private contract.

Whether public or private, there should be full-cost recovery to achieve sustainable service. There is a good example in India in which a community-owned system achieved a sustainable unsubsidized system. In Kolhapur in the state of Maharashtra, four villages formed a joint water management organization to manage water supplies. In order to achieve sustainability, all of the operating and maintenance costs had to be recovered from charges, and virtually 100 per cent compliance with payments was achieved (DFID, 2000). The business principle of having to achieve a revenue surplus applied equally to a community-owned system in India as to a corporatized public company in the developed world.

SOME PRIVATE FAILURES: WHAT WENT WRONG?

In this section, examples in three countries are considered: Atlanta in the US, La Paz in Bolivia, and Chengdu in China.

Atlanta

One particular private contract that has received a great amount of publicity is in the City of Atlanta in the US. The city placed a 20-year contract with United Water Services, beginning in 1999, for the management of water supply, system maintenance, billing and consumer service. The annual fee to be paid to United Water Services for this work was US$21 million. The original annual operating cost by the city's own estimation had been US$48 million, so there was an estimated saving to the city of around US$ 400 million over 20 years. The contract was launched with a great deal of optimism; but by 2002 there was dissatisfaction in the city and following a change of mayor the contract was terminated in January 2003. What went wrong? Based on what is known now, there were unrealistic expectations. The city systems were in poor condition and needed significant investment, whereas the contract gave emphasis to cost savings to be used for other purposes. Table 8.2 shows that the state of the distribution system was significantly worse than allowed for in the contract. Although United Water Services did not achieve the anticipated annual savings of US$20 million, it did achieve US$10 million. However, this saving was not available to increase investment in the water infrastructure as the city took US$9.8 million in an annual franchise fee. Since the termination of the contract, the city has recognized that there needs to be significant investment in the infrastructure, largely for environmental improvements, but also including the water supply systems, and that water charges have to increase by 10 per cent per year until 2015 to pay for that investment.

Table 8.2 *Data showing the poor information on which the Atlanta contract was based*

Item	*Basis of contract*	*Found in practice*
Faulty water meters	1171	11,108
Mains breaks	101	279
Faulty fire hydrants	734	1633

Source: Moore (undated)

The lessons are clear. The use of the private sector is not a 'magic wand' – not a substitute for facing up to the need for investment in the infrastructure. In addition, contracts have to be based on sound information, and if the information is not available at the outset, the contract should include provision for obtaining sound data and for the contract to be reviewed accordingly. Had the Atlanta contract been established in a policy framework that recognized the need for investment in the infrastructure, and included a requirement to identify the scale of the problems in the distribution system, the private-sector company would have had the opportunity of making a significant contribution to delivering the required results.

La Paz

This is another example in which the contract was terminated by government. The La Paz contract was a 25-year concession designed to expand the service to the poor. Those bidding for the contract were required to state how many additional connections they would make for a specified tariff. The contract, which was placed with the Aguas del Illimani consortium, of which the major partner was Suez Environnement, included the requirement to install around 72,000 new connections in the poor neighbourhood area of El Alto. The government set a tariff increase of 35 per cent overall; but the tariff structure included a low social tariff for the first 30m^3 per month. The government, as part of the contract conditions, set a maximum connection charge of US$155; but following a pilot project working with neighbourhood associations, there was provision for the poor to pay for connection in US$5 instalments.

Initially, the contract was considered to be a success, and in 2001 it was proposed that the consortium expand the work to extend connections to communities outside the original area covered by the contract to cover 15,000 additional connections. Around that time, government took the decision to raise the combined connection charge for water and sewerage to US$450, which was half the annual minimum wage. Even with lower-cost condominial systems, in which pipes are laid under paths and across properties, rather than under roads, using local labour, the connection charge was only reduced to half. Later, the consortium was blamed for this high connection charge, but claimed that it had been unsuccessful in persuading the government to lower the charge (Hayward, 2005). There was also a dispute over the number of additional connections required to fully serve the poor area of El Alto. The consortium claimed that the original estimate in the contract was too high. Whatever the truth about the numbers, the contract resulted in significant improvements in the water service in La Paz. However, the contract became caught up in the political changes of the country, and in January 2006 the government announced that it would terminate the contract. This termination finally took place in January 2007 when President Morales said 'water cannot be turned over to private business; it must remain a basic service, with participation of the state so that water service can be provided almost for free' (*International Herald Tribune*, 2007).

What can be learned from the experience of La Paz? First, there is always a need for extensive public consultation, especially on water charges. There was good consultation initially, with neighbourhood associations able to influence connection charges and to make provision for an easy payment option for the poor. Later, connection charges were raised significantly, an easy payment option was not available and there was no effective consultation on these important changes. Second, institutional arrangements need to be in place to oversee private contracts to ensure that contract conditions are based on sound information, and that there is effective communication on output requirements and the cost of service improvements. This is best done by an independent regulator who is seen by the public to have the role of protecting consumers. This role is discussed later in the chapter under 'Success factors in the use of market-type forces'.

Chengdu

As mentioned earlier, China has looked to 'marketization' to energize its utility operations. There have been successes in some cities; but in other cities, one of which is Chengdu, according to Fu Tao and Lijin (2007), there have been problems. Chengdu entered into an 18-year build–operate–transfer (BOT) contract for the construction and operation of a water treatment plant. In this contract, the city agreed to pay the contractor for a fixed volume of water every year based on the demand predictions of 1999. The demand has not grown as predicted, and since the contract did not allow for a periodic review of the conditions, the city water company is obliged to continue to take the contracted amount of water. As a result, the city is unable to optimize through making maximum use of lower-cost water from its original treatment plants, with the consequence that water revenue no longer covers the costs of supply.

WHAT WAS DIFFERENT IN THE SUCCESSFUL USE OF THE PRIVATE SECTOR?

It is instructive to consider three examples, England and Wales, Chile and Ma'asham in China.

England and Wales

The various reforms have delivered sustainable water services with greatly increased investment in the infrastructure. Although privatization of water services in England and Wales took place in 1989, the earlier history is important. The full story is given in Rouse (2007); but those key elements of the history which were important to the success of privatization are listed below.

- The amalgamation of 1620 water and wastewater utilities into just ten Regional Water Authorities (RWAs) in 1974 created units of viable size.
- Importantly, there was no federal funding after 1974, which meant that the RWAs had to become financially self-sufficient.
- In 1982, the government decided that the RWAs should be more 'commercial' in approach and large local authority representative boards were replaced by smaller boards made up of people with business backgrounds.
- Twenty-six private statutory water companies existed at that time, so there was a history of the use of the private sector in the provision of water services.
- Privatization of the RWAs was achieved through flotation on the stock market, with some priority being given to staff and individual customers of the RWAs, achieving people ownership.
- The government recognized that privatization of a public service monopoly would need to be accompanied by strong regulation.
- The success of a five-year periodic review planning process coordinated by the economic regulator was noteworthy.
- There was openness and transparency, with all documents put into the public domain.

Chile

Privatization in Chile is regarded as one of the most successful exercises, and parallels are drawn with the UK experience. The relevant history begins in the 1970s when water services were virtually free, there were no incentives to use water wisely, and a lack of income for the operator had resulted in deterioration of the infrastructures. In 1977, the then state-owned companies were integrated within one national company, and tariffs were increased gradually to recover operational and maintenance costs. Later, the state company was split into one company per provincial state, but came under the overall direction of a holding company, Corfu. In 1990, the regulator, the Superintendent of Sanitary Services (SISS), was formed. Although the SISS had few powers initially and relied on Corfu for implementation, the regulatory system resulted in significant improvements in performance in some areas. The SISS had adopted 'a model company approach' to establish the extent to which performance could be improved. But the improvements were not universal, and in 1997, the Tariff Law was introduced to reinforce the need for sustainable cost recovery, with SISS being given the necessary powers of enforcement. To achieve the desired performance improvements and to increase the level of investment in the infrastructure, privatization was introduced in 1998 through a combination of sale of equity to large private investors and companies being floated on the stock market. Corfu retained 35 per cent of the equity, with government having a veto on transferring assets. Later, with a change of government, sale of assets ceased and privatization took the form of concession contracts.

The programme of privatization operating within a strong regulatory system has been very successful, producing world-class performance water utilities. The success can be attributed partly to the emphasis given throughout the reform programme, both in the public and privatization phases, to the needs of the poor. Local governments funded extensions to distribution systems to reach poor communities and the access charge could be paid in 60 monthly instalments over five years. There were also means-tested subsidies for the payment of water bills. These targeted subsidies, handled by local authorities, facilitated full-cost recovery and allowed the water utility managers to concentrate on running efficient water utilities without the conflicting requirement of being a social service. The Chilean success demonstrates that utilities can be operated in a market-type environment to benefit all citizens.

Too much attention is given to the futile arguments on public or private operation, and too little to creating the conditions, as in Chile (whether in the public phase or the private phase) for successful development of water services.

Ma'asham, China

The example of Ma'asham is important because it shows that successful use of the private sector can be 'home grown'. In 2002, the city of Ma'asham, in Anhui Province, east China, formed a joint-venture company with Beijing Capital Company to build additional treatment capacity. The relationship worked well, and in 2004 the joint-venture arrangement was extended to cover the whole water utility operation. The structure and a full description is given in the book on *Reform of China's Urban Water Sector* (Fu Tao and Lijin, 2007).

Ownership of the distribution system is retained by Ma'asham Water Company. The system is leased to the joint venture company operating under a concession contract. Investment is provided through the joint venture made up of Ma'asham Water Company and Beijing Capital Company. The arrangement has already delivered significant improvements in asset condition and operations. It will be interesting to follow the ongoing performance of this joint venture system. It appears to have some of the essential ingredients for success, with both the city and the private company having a joint financial and reputation interest. The involvement of Beijing Water Company brings in both investment money and management expertise. Beijing Capital Company is setting up a benchmarking scheme involving Ma'asham and its other joint venture operations, providing the opportunity for good experience to be shared.

SUCCESS FACTORS IN THE USE OF MARKET-TYPE FORCES

Governments and cities are rethinking how water services are provided. It is necessary to separate policy and delivery functions, with operational management allowed to manage water services without political interference. The role of governments, whether national or local, should be to develop effective policies on infrastructure investment, cost recovery and charging systems to accommodate the poor. Governments also have to find a means of obtaining the performance benefits that arise from a market approach. Essentially, due to the dominant cost of the pipe networks, water services are a monopoly, so it is necessary to simulate market forces through the use of comparative competition techniques such as benchmarking. League tables are effective as no manager wishes to be bottom of any league, and in Chile 'the model company' is used to determine performance targets. In The Netherlands, economic performance is driven by estimating the potential annual cost savings based on perfect operations and striving to obtain those savings (Jonkers, 2004). Over time, additional potential costs savings are identified and the targets 'recalibrated'.

This discussion is important because in a debate on the use of pseudo-market forces, the word privatization does not appear. The issue of public or private operations should not be the dominant question. The important requirements are those elements necessary to achieve sustainable water services (Rouse, 2007), and these include full-cost recovery and operational managers being given incentives to improve performance. Due to highly publicized failures and an effective anti-privatization lobby, most countries' governments are nervous about using the private sector. However, as shown in this chapter, with sound government policies and effective regulation to provide objective improvements and transparency, privatization has produced world-class water utilities: so how should governments approach the employment of the private sector? Consideration of successes and failures identifies some key factors.

LESSONS FROM PRIVATE-SECTOR SUCCESSES AND FAILURES

Most importantly, use of the private sector is not a 'magic wand'. Letting a contract to a private operator will not suddenly solve fundamental problems. It will not remove the need for substantial investment to refurbish infrastructures, which has to be paid for whether it is public or private investment. It is not a substitute for sound policies and should not be slotted into existing governance structures. What privatization does do is provide access to operational management with wider experience, and it should lead to better definition of objectives and greater transparency. It is a means of harnessing the energy of the commercial world to achieve better service and higher efficiency.

It is essential that conditions are conducive to use of the private sector and that the public are fully consulted on why privatizations, or any other changes, are being considered. For this reason, reforms, whether public or involving the private sector, should not be rushed; but time should be taken to develop effective communication with the public, with particular attention given to policy on water charging and provisions for the poor. The general public needs to be convinced that changes will bring stated benefits. If the public (not pressure groups who have different agendas) cannot be convinced of the benefits, then governments should take heed. The debate with the public should consider not just a privatization option, but other options and the pros and cons of each. In the debate, it is important for the public to recognize that improved services, and especially investment in deteriorated infrastructure, cannot be achieved without an impact upon water charges, and that charges will need to increase in either public-sector or private-sector options. Without such consultation, anti-market forces movements will have the opportunity of undermining even well-prepared developments by using information selectively. Public communication is one reason why a strong and transparent regulatory system is an advantage. It can avoid the problems experienced in Atlanta by ensuring that contracts are based on sound information and that the public are not given unrealistic expectations.

Strong regulation was introduced in England and Wales on privatization and in Chile ahead of privatization; but it is not common in structures involving some form of concession contract. I believe that this is a deficiency, as independent regulation can provide the basis for successful contracting-out. It can provide clarity on the conditions on which contracts are based, with due consideration where there is poor information on system condition; it can provide the system for integrated planning to avoid Chengdu-type situations; it can provide the framework for periodic reviews of contracts; and, as mentioned earlier, it can promote effective consumer consultation and transparency. It is often resisted by both cities and contractors as interfering in commercial transactions; but this should not be the case. It is not for a regulator to be involved in the contract details, but to ensure that the information on which contracts are based is sound. This suggests some overseeing of the organization letting a contract, which works very effectively in Wales, in which D r Cymru is the licensee but contracts out operations to the private sector. The role of regulators in concession contracts is discussed in more detail in Rouse (2007, pp26–28).

Regulation can also be very effective in assisting in the establishment of a market-based structure in public-sector operations, through managing integrated planning in order to determine water tariffs, to managing comparative competition systems. In particular, decisions on tariffs by regulators will be based objectively on need as there is poor experience of governments in setting sustainable cost-recovery tariffs. Two good examples of full regulation in public-sector operations are in Scotland and the state of Victoria, Australia. There are also mixed public and private situations. There has been success through the introduction of regulation in Zambia, where nine commercial units and 22 local authorities are regulated by the National Water and Sanitation Council (NWASCO, 2003). In this poor country, significant process has been made towards full-cost recovery and sustainable water services.

This gives emphasis to the most difficult aspect of water charging, especially as in most cities the starting situation is low-cost recovery. Whether operations are public or private, water services can never be truly sustainable without full-cost recovery from charges, including adequate provision for refurbishment of distribution systems. This requires a sound policy on the transition to full-cost recovery. General subsidies, which benefit the rich more than the poor, should be diverted to assisting the poor in getting affordable access and in providing assistance in paying bills, as in Chile. The public needs to be aware that this must happen anyway, and it should happen whether or not the private sector is involved. Strangely, since it is the 'rich' who will have to pay more (the poor should receive any subsidies), they are likely to be most vocal in opposition.

Having established the conditions under which communities can benefit from privatization, governments or cities need to decide upon the form of privatization. It could take the form of joint ventures, as in Ma'asham, which would include some 'sale' of equity as part of bringing in external investment, or more likely there would be some form of concession contract. Setting up effective contracts involves knowledge and experience not generally available in local governments. For some guidance on what constitutes effective procurement, see Rouse (2007, pp151–155).

CONCLUSIONS

The use of market forces has resulted in world-class performance water utilities providing a good service to all citizens. Market forces can be applied through privatization, or more commonly through providing the conditions under which public-sector management can operate in a commercial way. Whether public or private, this requires:

- separation of policy and operational functions, with no political interference in operational management;
- the replacement of general subsidies with water charges set at a level to provide for full-cost recovery in an efficient operation;
- subsidies targeted to the poor, with emphasis on easy payment systems;
- the introduction of some form of comparative competition;
- the application of effective procurement principles in setting up private contracts or in establishing internal local government contracts;
- some form of independent body (a regulator) to manage planning and tariff-setting processes, and to ensure realism in establishing contracts and, above all, transparency and public participation.

REFERENCES

DFID (UK Department for International Development) (2000) *Sustainable Community Management of a Multi-Village Water Supply Scheme in Kolhapur, Maharashtra, India*, Water and Sanitation Programme Field Note, SPI Series 2, DFID, London

Fu Tao, C. M. and Z. Lijin (2007) (eds) *Reform of China's Urban Water Sector,* IWA Publishing, London

Hayward, K. (2005), 'Bolivian government signals contract end', *Water*, vol 21, 25 January, www.iwapublishing.com/template.cfm?name=bolivian_government_signals_contract_end, accessed June 2008

International Herald Tribune (2007) 'Bolivia's Morales celebrates foreign water company's exit, plans more nationalization', *International Herald Tribune*, 4 January 2007, www.iht.com/articles/ap/2007/01/04/america/LA_GEN_Bolivia_Nationalization.php, accessed June 2008

Jonkers, P. (2004) 'The Dutch water industry – lessons learned', Presentation at the International Water Association/Japan Waterworks Association/Federation of Japan Water Industry Workshop, held in Tokyo, May 2004

Kontula, E. (2004), 'Finnish cooperation in Vietnam', Presentation at the World Bank/WaterAid Workshop on Modes of Engagement with the Public Sector Water Supply Providers in Developing Countries, London, August 2004, www.wateraid.org/documents/plugin_documents/publicsectorproviders.pdf, accessed July 2008

Mac, P. (2003) 'Water privatization the problem, not the solution', *The Guardian*, 14 May 2003, www.cpa.org.au/garchve03, accessed May 2008

Moore, A. (undated) 'Atlanta dissolves water privatization', *Water Industry News*, http://waterindustry.org/Water-Facts/atlanta-1.htm, accessed June 2008

NWASCO (National Water and Sanitation Council) (2003) *Water Sector Reform in Zambia*, www.zambiawater.org.zm/nwasco/reform.pdf, accessed May 2008

OFWAT (Office of Water Services) (2007) *Levels of Service for the Water Industry in England and Wales 2006–2007 Report,* OFWAT Publications, Birmingham

Rouse, M. (2007) *Institutional Governance and Regulation of Water Services*, IWA Publishing, London

UK Government (2003) *The Water Act 2003*, Her Majesty's Stationery Office, London

UNDP (2006) *Human Development Report*, UNDP, New York

WHO (World Health Organization) (2002) *Water for Health Enshrined as a Human Right: United Nations Committee on Economic, Cultural and Social Rights*, www.who.int/mediacentre/news/releases/pr91/en/, accessed May 2008

9

Complementary Paradigms of Water and Sanitation Services: Lessons from the Finnish Experience

Jarmo J. Hukka and Tapio S. Katko

THE PURPOSE OF WATER AND SANITATION SERVICES

The overall framework of this book is laid on two major assumptions – that water and sanitation services (WSS) constitute a social right of citizenship, and that there is a need to better integrate the public policy and management aspects of WSS within both analysis and practice. The book also recognizes that the particular forms of organization characterizing WSS worldwide are historically shaped and affected by multidimensional constraints and local conditions. The first assumption implies that although water supply and sanitation can be subject to a number of social forms of delivery, it is a state obligation to guarantee the universal access to these essential services. Of course, the acknowledgment of the right of access to WSS does not imply that these should be free of charge, as the costs of providing these services have to be covered in order to ensure their financial sustainability. The second assumption focuses on the need to promote a more balanced approach to the organization of WSS, an area of activity that has been, and is still largely, dominated by technical and managerial considerations. In order to overcome the grave problems affecting WSS worldwide, and particularly in less developed countries, it is crucial to bring together both the technical-managerial and the public policy aspects of the activity. Finally, the book also departs from recognizing the historical diversity of WSS organizational forms, also captured by the notion of 'insdiversity' (institutional diversity) discussed elsewhere in this volume (see Chapter 5). In this regard, the ownership and management of water and sanitation services is, in many respects, a mainly local issue.

It is heavily influenced by the scale, geography and history of the location, where social, cultural, political and economic aspects often have a paramount role, while the interplay between different levels of government (local, regional or federal, but increasingly also supra-national, as in the case of the European Union) may have a decisive impact upon the running of WSS.

Within this framework, this chapter aims at identifying and discussing various alternative or, rather, complementary paradigms that may be considered in the development of sustainable and viable water and sanitation services. The approaches presented here are the customer–citizen responsive paradigm, the multiple-stakeholder paradigm, the local government paradigm, the cooperative paradigm and the multiple producers paradigm. We have drawn on the Finnish experience of WSS, which, as explained below, could be considered an example of good practice in this field, but have also complemented our reflection with materials derived from research projects on the rest of Europe and the Americas.

CUSTOMER–CITIZEN RESPONSIVE PARADIGM

It is increasingly acknowledged worldwide that water undertakings – which are often also responsible for sewerage and proper wastewater management and, thus, ensure an integrated approach in water services production – must become more accountable in relation to their activities. This includes more transparency in a number of crucial areas where WSS utilities must provide accurate, regular and up-to-date information to their customers about such issues as compliance with water quality standards, but also other important aspects, such as pricing, investment decisions or changes in institutional arrangements (e.g. decisions about the introduction of new management models). However, in Europe and elsewhere, there is also a growing recognition that the accountability of WSS cannot be limited to providing information to customers. As illustrated in the debates that took place in Europe during the negotiations over the Water Framework Directive, there is acknowledgement that sustainable and democratic water management, including WSS, requires 'active', and not merely passive, citizenship, which implies developing a culture of dialogue, negotiation, subsidiarity and institutional strengthening (EC, 2001).

In this connection, WSS utilities have typically used the term water 'users' or 'consumers', while more recently the term 'customer' has been introduced, reflecting changes in the institutional environment (Seppälä et al, 2004, p87). If seen merely from the WSS utilities' point of view, this may be considered justified since they normally sign the supply contract with their customers. However, considering the ultimate purpose of community water supply and sanitation, it is obviously at least equally important and justified that WSS utilities consider people not just in their roles of customers, but also as citizens. Obviously, these two are not alternatives but, rather, different sides of the coin depending upon the viewpoint. Utilities have to involve citizens in the decision-making processes regarding key concerns that significantly affect their communities since they have a unique obligation to the public as providers of an essential service. Citizen involvement is also important since WSS provision has an impact upon many different aspects of community life, including social, political, health, ecological, economic and cultural considerations. These concerns are sometimes mutually contradictory – for instance, improving the ecological state of local water bodies or renewing ageing infrastructure might imply a significant increase in WSS fees, and these contradictions can be best resolved through dialogue between the involved stakeholders and the public (Seppälä et al, 2004, p87).

Customer and citizen orientation or, rather, responsiveness may involve various steps, as suggested in Figure 9.1. In short, it can be described by a chain from supply-driven to demand-driven management, and further from customer orientation to customer and citizen responsiveness. The first step is to offer customers a holistic solution to their needs. A few steps higher, value- and partnership-based approaches are used (Seppälä et al, 2004, pp87–90). Instead of customer orientation, we would suggest parallel customer and citizen responsiveness. According to Hadley and Young (1990), such a responsive public service model aims to make public service production more efficient and effective, and, particularly, to offer an alternative to the ultraliberal model of private-sector production that has been promoted worldwide since the 1980s. Customer- and citizen-responsive public services are enabling rather than dependency-creating, holistic rather than fragmented, and open to citizens' participation and influence.

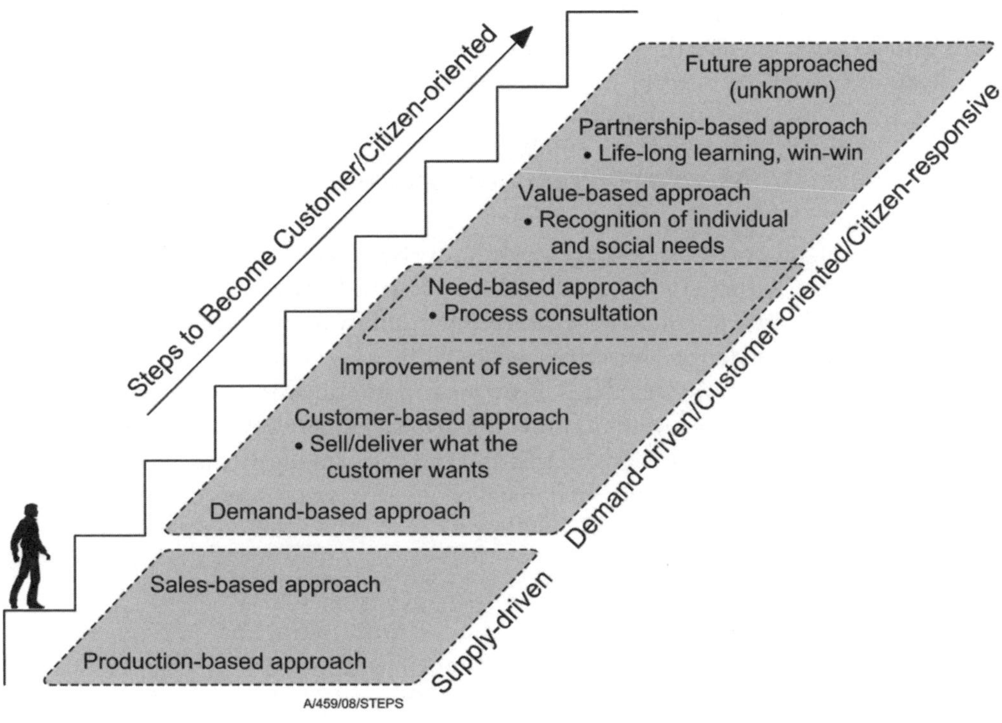

Figure 9.1 *Steps of customer and citizen orientation in a water utility*

Source: adapted from Seppälä et al (2004), reprinted with the permission of the *Journal of the American Water Works Association*

It is obvious that within WSS utilities, the management culture in many places may still be consumer oriented, while the roles of people as customers and citizens are not taken into consideration, particularly under the heavy stress posed by economic efficiency requirements. Whatever the case, customer and citizen responsiveness are integral components of the process of improvement in WSS management. On the one hand, local governments and WSS utilities may involve citizens through consultations about proposed policy decisions, such as organizing citizen workshops to discuss the desirability, acceptability and feasibility of a project to build a new treatment plant, or similar issues. This is done, for instance, in some Brazilian cities through the organization of citizen councils that have an active participation in the discussion of urban policy, including WSS, and whose decisions often inform policy decisions at the various levels of government (da Costa et al, 2006; see also Chapter 20). Similar processes have been successfully implemented elsewhere, including many cities in Africa, Asia, Europe and Latin America (e.g. Balanyá et al, 2005, quoted in Castro, 2009). On the other hand, in many places, citizens have not been passively waiting to be invited to participate in crucial decisions about WSS and have been rather proactive in the exercise of their citizenship rights. For instance, Castro (2007) has examined the influence of citizen movements in shaping policy decisions over WSS in Latin America, where the promotion of unregulated privatization of WSS utilities has fuelled significant social unrest since the 1990s.

However, it is not only people in their roles of customers and citizens that need to be taken into account, but also the perceptions and interests of other involved stakeholders. This leads us to consider next the multiple-stakeholder paradigm.

Multiple-stakeholder paradigm

The production and provision of water services involves many stakeholders with various roles and interests. Grigg (1996, pp14–15) has classified these stakeholders into four major categories. Service providers have direct responsibility for water management, while regulators are responsible for overseeing service rates, water quality and environmental impact, health issues or service standards. Planners and coordinators take care of additional functions, while support organizations provide a diversity of services and goods, including research, technical services, supplies, information and data, among others. In turn, Seppälä et al (2004) identified the following major stakeholders in water services:

- water utilities;
- consumers or direct users;
- indirect stakeholders;
- proactive stakeholders (e.g. environmental pressure groups);
- policy-making bodies;
- regulatory agencies;
- financing agencies; and
- water utility owners.

Whether these classifications are exhaustive or not, what clearly emerges from their analyses is the high complexity and diversity of interests and actors involved in public policy and management of WSS.

In this regard, the general institutional framework for WSS services sketched in Figure 9.2 may appear somewhat complicated as it involves a great number of stakeholders, potentially with conflicting interests. For example, in Finland the major organizations that have an impact upon water and sewerage services are the European Union (EU), the central government, and the local governments or municipalities (Juhola, 1995, in Hukka and Katko, 2003). Citizens as individuals and interest groups have their own perceptions and means of pushing for changes that advance their vested interests or prevent innovations that may undermine them. Experiences, for example, from planning regional water and sewerage systems tend to show that through proper partnership, involvement and dialogue among various stakeholders, possible contradictions or bottlenecks can be avoided or reduced. In the opposite case, years of delays may occur.

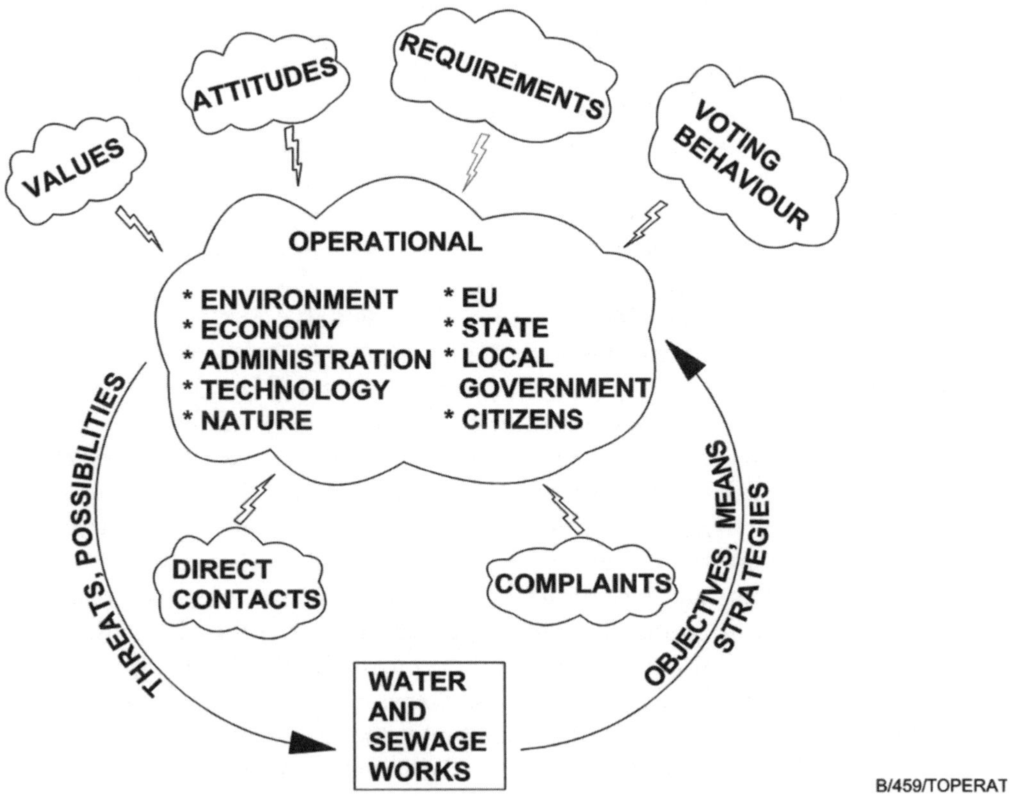

Figure 9.2 *The complex framework of water services*

Source: Juhola (1995), modified by Hukka and Katko (2003, p8)

In this context, Finland is presented as an example of water administration and legislation that draws on a variety of paradigms. One justification for this could be that the country has been ranked at the top in several water- and environment-related international comparisons. For instance, according to the Water Poverty Index (WPI) published in 2002, Finland is the highest-ranking country with a WPI of 77.9 points (Lawrence et al, 2002). The WPI evaluated performance in the following five categories: resources; access; capacity; use; and environmental impact. Even though the Finnish institutional arrangements may not be replicable as such, they present an example of an enabling environment for successful water governance (Hukka et al, 2006). However, we do not suggest that the Finnish experiences discussed here would be directly transferable or replicable to other conditions. Besides, in spite of citizens' perceptions of lack of corruption in the country, as reported by Transparency International (2007), we believe that there is much scope for achieving higher transparency in decision-making processes in Finland, and this also applies to the WSS sector. For instance, although, according to historical perspective, the diversity of management options has been one of the cornerstones in the overall evolution of WSS services in the country, the latest findings from several research projects suggest that this situation may have changed for the worse. In particular, our research shows that the EU's subsidiarity principle, which 'is intended to ensure that decisions are taken as closely as possible to the citizen and that constant checks are made as to whether action at Community level is justified in the light of the possibilities available at national, regional or local level' (EU, 2008) is not always adhered to in Finnish WSS policy and management. More alarmingly, we have also found that decision-makers do not always consider all potential options available, as they certainly should. Nevertheless, and although these and other worrying trends need to be addressed, we believe that there exist important lessons to be learned from the Finnish experience in relation to the development of multi-stakeholder paradigms for WSS.

In Finland, responsibility for WSS is in the hand of central, regional and, particularly, local authorities. The Ministry of the Environment directs Finnish environmental and housing policy, use of land areas, construction and preparation of legislative proposals. It is also responsible for land-use planning, preserving a good living environment, developing regional and community structure, culture and heritage. As for water, it deals especially with wastewater and environmental issues (Hukka and Seppälä, 2004). The Ministry of Agriculture and Forestry is charged with development of WSS, use and management of water resources, performance of facilities management and land surveying. The Ministry of Social Affairs and Health, for its part, oversees environmental healthcare. It issues instructions concerning, for instance, water quality, permitted chemicals and the minimum standards for untreated water. State provincial offices are organs of state regional governments that have supervisory powers over municipalities, which, again, are responsible for the provision of environmental healthcare. The framework is completed with the Finnish Environment Institute (FEI), which is a research and development centre for environmental issues. In cooperation with regional environment centres, the FEI maintains an environmental information system and takes responsibility for research and monitoring of environmental standards (Hukka and Seppälä, 2004).

In turn, the municipalities provide basic services for their inhabitants. They oversee land use and building within their jurisdictions and have responsibility for water, energy supply, waste disposal, street maintenance and environmental protection. Municipalities are in charge of taking measures to provide WSS services; but they do so normally in cooperation with other key stakeholders, including the Finnish Water and Wastewater Works Association, the Finnish Association of Local Authorities, employment and economic development centres, the Finnish Competition Authority and the National Consumer Administration.

In addition, WSS in Finland are also controlled by environmental permit offices, which are independent authorities granting various water-related permits according to the Water Act and the Environmental Protection Act. They also hold the power of decision over environmental permits for key enterprises, while less important permits are granted by regional environmental centres and municipalities.

A brief digression is warranted here to mention that Finnish regulations and laws related to WSS services can be categorized into four main areas: water services; health; water and environmental protection; and complementary legislation. The Water Services Act contains provisions on the development of water services, as well as the organization of water services and rates. A key principle of the act is that water and wastewater user charges should cover all of the investments and operating costs of an undertaking. Yet, subsidies to water services from the municipality, the state and the EU are possible. The act also stipulates that the charges may include 'a reasonable rate of return'. The properties located within the service area of a water and sewerage undertaking should, as a rule, be connected to its networks. This guarantees the economic viability of the water and sewerage undertaking (MAF, 2001).

According to the Water Services Act, municipalities have to prepare so-called water services development plans in cooperation with the water and sewerage undertakings within their territory and neighbouring municipalities. Municipalities are also to participate in regional water services planning (MAF, 2001, 2002). The municipal council makes decisions concerning the general bases for charges for municipal and other services (AFLRA, 2004; Finlex, 2006). Furthermore, all municipalities have to approve the operational areas of the water and sewerage undertakings.

This succinct revision of the organizational features of the Finnish WSS sector offers a good example of what we term the multiple-stakeholder paradigm. As already pointed out, these paradigms are not mutually exclusive and, in fact, coexist and interact very closely between them. This is particularly true in relation to the local government paradigm, which we consider next.

LOCAL GOVERNMENT PARADIGM

In the international context, the authors have noticed that too often the difference between central and local government goes unrecognized. However, in relation to water services, the use of the term 'government' may give a false impression since in many countries it is local governments that play the most important roles in WSS-related decisions. Of course, we should not forget the important role of the central, sometimes the federal, government as the sector's policy formulator. Nevertheless, it is a historical fact that the vast majority of WSS systems in the Western world, and probably elsewhere, have been developed under local government ownership while also procuring works, services and goods from the private sector. It is our understanding that, on the basis of the successful experiences with WSS Western countries during at least the past 200 years, capacity-building by both local governments and utilities and the local private sector – instead of promoting the involvement of multinational private water companies not familiar with local conditions – should be the long-term objective. This approach can be called the local government paradigm.

In international water policy discussions, it has been often noted that water management and, especially, WSS management is largely a local issue. This is closely related to the subsidiarity principle, one of the cornerstones of EU legislation, which as already noted 'is intended to ensure that decisions are taken as closely as possible to the citizen' (EU, 2008). In fact, already in 1992 the International Conference on Water and the Environment (ICWE) in Dublin adopted the principle that the management of services should be carried out at the lowest appropriate level (see Figure 5.1 in this volume), with full public consultation and involvement of users in the planning and implementation of water projects (UN, 1992). These principles place local governments at the very centre of decision-making and management activities in relation to WSS, and there are excellent examples of the importance of municipalities in the development and management of these services in Europe and elsewhere, as shown in this volume, particularly in Chapter 11 on France, Chapter 13 on the Nordic countries, Chapter 14 on European trends and Chapter 20 on Brazil.

In this regard, for instance in Nordic Europe, municipalities or local governments have traditionally been relatively stronger than the regional-level administration powers that exist in other countries. In the case of Finland, the first municipal companies created were the water supply and electricity utilities established during the 1880s by the largest towns after the Municipal Act was enacted in 1875 (Katko, 1997). One of the major reasons to establish piped water supply systems was the need to provide water for fire-fighting (Juuti and Katko, 2005, p220), while water quality, health and sanitation also played a role. After Finland gained her independence in 1917, the duties of municipalities were increased dramatically. This was partly due to the legislation that put new requirements on municipalities and, partly, to municipalities' own initiatives. Since the 1980s, the trend has been towards decentralization and more freedom for municipalities to decide how to organize and implement assigned tasks (Kettunen, 1999). More recently, however, the central government has been criticized for not funding sectors such as healthcare adequately. Yet, in water services, the central government has always funded less than 10 per cent of annual investments; it has provided financing primarily for groundwater investigations and promotion of inter-municipal cooperation. Compared to such countries as Sweden and the US, this share is remarkably low.

In Finland, municipalities have the responsibility to ensure that WSS services are provided while, in practice, they also produce them, either through their own municipal technical department, a public utility company, a joint-stock company owned by the municipality, or a supra-municipal system of some form. In areas beyond the official operational jurisdictions of water and sewerage utilities, water cooperatives owned directly by consumers often provide and produce the services. Municipalities may support the establishment of these cooperatives, and also offer operational assistance. The municipality may also set standards for the service quality of water cooperatives, which must be constructed to meet the requirements set for the municipality's own water system. This arrangement guarantees the quality of the operations, and a smooth transition, in case the municipality decides to take over the water system in the future. Several other options also exist, as pointed out by Takala (2007), such as buying services jointly or separately from other utilities or private companies.

In the broader European context, there is a very wide diversity of options for WSS services, as shown by recent research by Juuti and Katko (2005, pp235–237) on the evolution of these services in 29 European cities in 13 countries. Considering the large differences between cultures and traditions of urban planning, typologies of water resources management, legal and administrative families, cultures and organizations, and the variation in the roles of local governments in the European context, this was hardly surprising. The role of local governments in each society also largely explains how water services are, and can be, managed. For instance, in France there are as many as 36,000 municipalities, which fosters inter-municipal cooperation (see Chapters 11 and 14). This fact, combined with other policies, helps largely to explain the increasing roles of private operators in that country rather than, for instance, the direct financing arrangements of municipalities (Juuti and Katko, 2005, pp46–47). The other extreme of the spectrum is represented by England and Wales, where the role of local government has remarkably declined, particularly since the 1970s, which partly helps to explain the swift privatization of public water authorities in 1989, with little or no debate or local government resistance. The dramatic shift from local to regional management of WSS that took place in 1974, in particular, severely undermined the influence of local governments in the provision of utilities in England and Wales, a process that was further accentuated through additional reforms during the 1980s (for an analysis of the demise of local authorities in the provision of essential services in England and Wales, see Okun, 1977; Parker and Penning-Rowsell, 1980; and Maloney and Richardson, 1995, all quoted in Castro et al, 2003). From a different viewpoint, the relationships between central and local governments can be also explored on the basis of how local-level activities are financed. In the Nordic countries as well as in France, Austria and Spain, municipalities finance most of their activities with their own income – mainly from local taxes – while in England and The Netherlands the opposite is largely true: municipalities' own revenue forms only a minor part, and the majority of financing consists of grants from the central government (Pietilä, 2006, p67).

There are also excellent examples of the role of local authorities in the provision of WSS. For instance, in the US, water services are often produced by city-wide water and/or wastewater works, as well as various types of supra-municipal water, water conservation, groundwater or sanitation districts that may have various types of connections to local governments, counties and district courts (CWEF, 2004). In Brazil, da Costa et al (2006) reported a significant number of examples of successful municipal WSS utilities operating in a wide range of urban centres, large and small, rich and poor. In their assessment, the authors examined the performance of the municipal WSS utilities in relation to what they termed the nine 'principles of a public environmental sanitation policy':

1 universality;
2 equity;
3 integrality;
4 municipal responsibility;
5 public management;
6 participation and social control;
7 cross-sectoral integration;
8 service quality; and
9 access.

One of the successful local governments identified is that of Porto Alegre (1.5 million people), where WSS are provided by the Municipal Department of Water and Sewerage (DMAE). DMAE is the largest local authority provider of WSS in Brazil, and has already achieved universal coverage for water supply and 85 per cent coverage for sewerage. Porto Alegre's municipal government is well known for its use of participatory budgeting and other democratic reforms, which have been even praised by the World Bank as an example of sound public policy and management of essential services. The city of Alagoinhas (140,000 inhabitants) in the poor north-east of Brazil and Santo Andre (670,000 inhabitants) in the industrial region around São Paulo are also good examples of successful local government organization and operation of WSS highlighted in this research. The Brazilian case shows that public water utilities can achieve universal access even under very difficult socio-economic conditions. A strong political will and the implementation of mechanisms to allow citizens' participation and 'social control' over the process were identified as the key features helping to explain the success of these local governments in running WSS (da Costa et al, 2006).

In many cases, municipal WSS utilities are increasingly outsourcing planning and design, construction and maintenance activities to the private sector. Figure 9.3 shows the most widely used type of public–private cooperation, where core operations are performed by municipality-owned utilities and none-core operations are bought from the private sector. The tower symbolizes local government ownership and the division between core and non-core operations. The foundation of the tower of this imaginary multicultural community represents the core operations of WSS utilities and local government ownership. The core operations consist particularly of strategic planning and management, contracting-out based on competitive bidding, and overall administration of the systems. The central government – the roof of the tower – should control compliance with the set WSS policy with the support of regional authorities (Hukka and Katko, 2003, p119).

One of the most important trends in the last two decades has been driven by the policies of commercialization of local government services, including WSS, which, in many cases, are being reorganized on the basis of private-sector management principles.

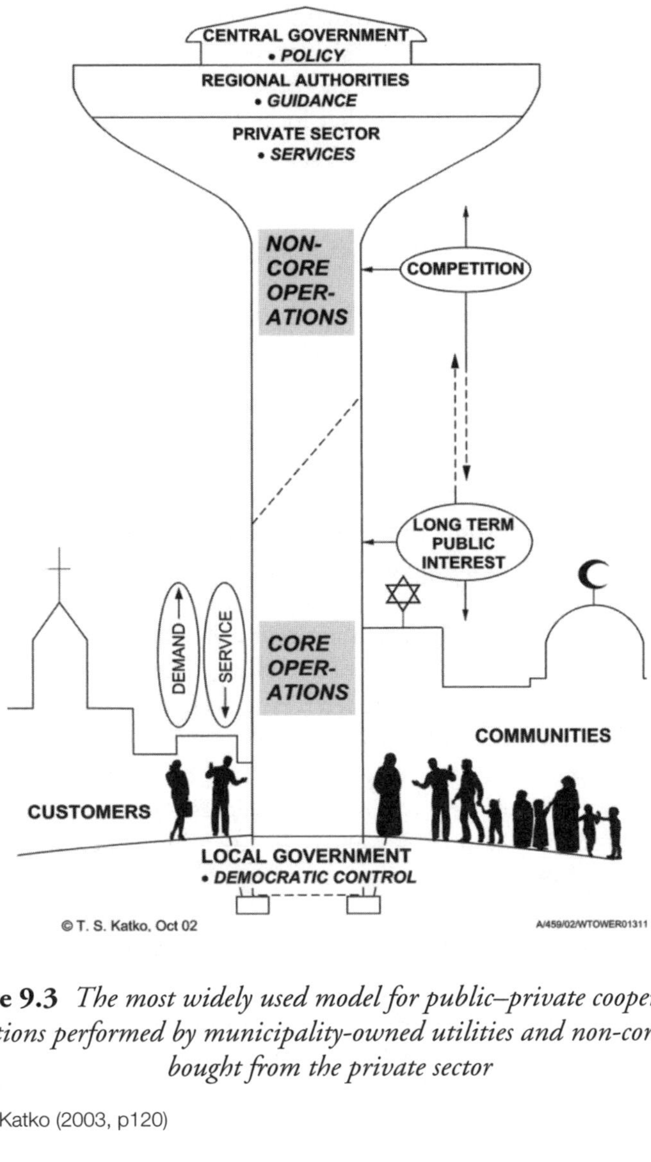

Figure 9.3 *The most widely used model for public–private cooperation: Core operations performed by municipality-owned utilities and non-core operations bought from the private sector*

Source: Hukka and Katko (2003, p120)

In relation to these trends, Windischhofer (2007) argues that municipalities have played the main role in promoting the commercialization of WSS, although this process has been often driven by central government policies that require the adoption of private-sector models and practices by local governments, often referred to as new public management. Windischhofer also criticizes the policies implemented by central governments promoting the reform of local public services by cutting the municipalities' access to financial resources and encouraging them to adopt private-sector managerialism and entrepreneurialism. On similar lines, Vinnari and Näsi (2008) explored the effects of introducing the new public management model in the public WSS sector, especially from the viewpoint of financial management and accounting. They found that the application of business-sector practices in the public sector does not necessarily guarantee transparency and accountability, but opens up possibilities for hidden taxation and other types of creative accounting by local government authorities. They conclude that local governments must also be subject to stricter regulation and clearer 'rules of the game'.

Cooperative paradigm

Water co-operatives owned by the users, and earlier also largely managed by them, are one option for arranging water services, particularly in rural areas. Finland has a long tradition of water cooperatives where users are the owners and decision-makers, which have successfully operated in rural and smaller urban settings for decades (Katko, 1992). According to Takala (2007), there are some 1500 water and wastewater associations of this kind in Finland. These user-owned systems are usually rural and small compared to municipal waterworks, while some of them may supply up to 10,000 people.

The formation of a water cooperative requires a champion or 'spark plug' (Katko, 1994), a board of directors and consumers/members (see Figure 9.4). Typically, a cooperative needs an active champion who initiates the project and is often involved in the management of the system. The role of authorities has been negligible, although in recent years water cooperatives have obtained financial support from the government and the municipality. The main responsibility, however, rests on the cooperative itself.

While some of Finnish water cooperatives have been dissolved or merged with larger works, new water associations have also been established in sparsely populated areas since the 1990s, lately also for sewerage. In townships, they are nowadays run on a commercial basis, while smaller and newer co-operatives in rural areas rely more on external support.

Considering the future, it is possible for associations to provide adequate water and sanitation services independently, in cooperation with other associations or municipal waterworks, or to rely on buying services from the private sector. One alternative is to merge waterworks into larger entities; but as suggested by a recent study, all feasible options should be considered in decision-making (Watertime, 2005). As for the other Nordic countries, Denmark has some 2500 water cooperatives, while Norway and Sweden formally have none. In Sweden, municipalities have to assume the responsibility for running services for any larger group of consumers (Takala, 2007). The US has various forms of small water supply arrangements that also draw on cooperative principles (Tamm, 1991).

There is also a longstanding history of WSS co-operatives in Latin America. For instance, in Bolivia, major urban water utilities are managed as cooperatives under customer ownership, such as the Saguapac Co-operative, which serves about 800,000 residents in the city of Santa Cruz. Experiences gained from both the utility performance and customer satisfaction perspectives have been good (Nickson, 1998). To the authors' knowledge, this is the world's largest water utility run as a cooperative. Another successful case of water cooperatives in Bolivia is Cosmol, a local service provider in Montero. In

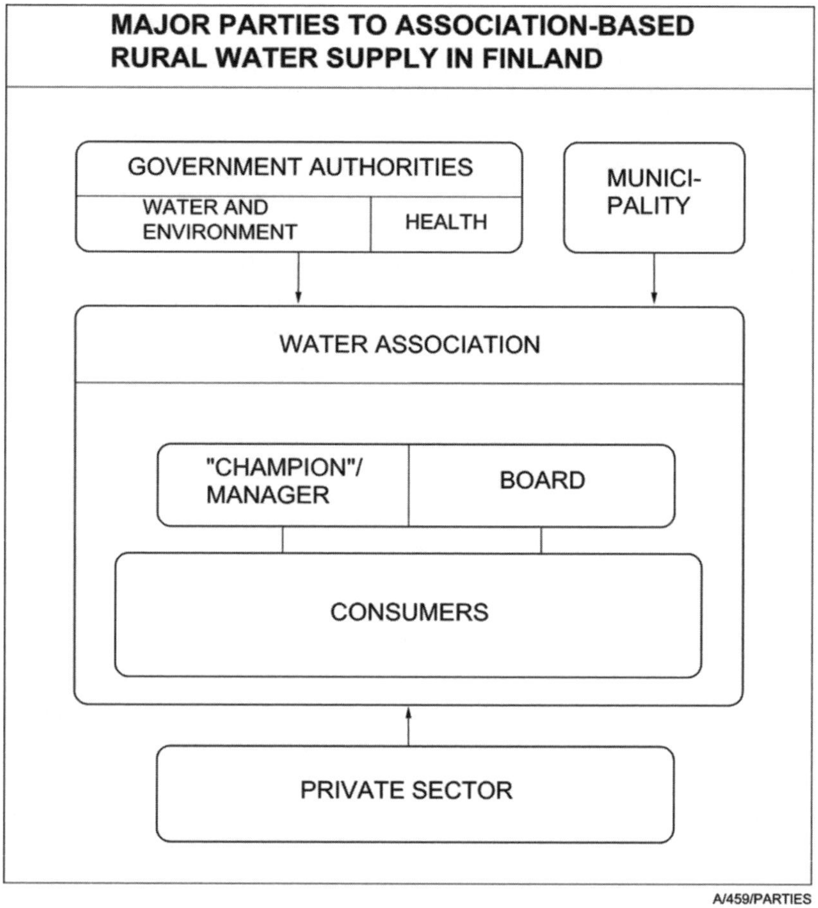

Figure 9.4 *Key stakeholders related to water cooperatives*

Source: Katko (1992); Hukka and Katko (2003, p104)

Brazil, a 'satellite type' of cooperative was introduced for rural water supply and sanitation during the 1990s. According to Heller (2006), it has proved successful. The key idea behind it is that rural localities group together geographically around a supra-municipal authority that has the technical and administrative capacity to provide the needed support. This form of cooperative is self-financed, with little state and federal government involvement. In Argentina around 1999, some 10 per cent of the population was served by cooperatives (Azpiazu et al, 2004). Among these cooperatives, WSS utilities is a case in the municipality of Moreno in the Buenos Aires Metropolitan Area, which is the object of a recent study (Hardoy et al, 2005). Through awareness-raising and a participatory assessment of water and sanitation provision, a local partnership-based management unit was formed. It is hoped that this will be institutionalized into a local water authority. Given that extending conventional water and sewerage services to the many un-served settlements is unrealistic in the short term, the study concluded that providing WSS to the poorest areas of Buenos

Aires is likely to happen only if all the actors involved are committed to working together towards a solution. To some extent, the analysis can be generalized to similar situations in the deprived urban areas of most developing countries.

A different form of cooperative arrangement was implemented in Wales, when in May 2001 the publicly owned, not-for-profit company Glas Cymru acquired the privately owned Welsh Water (Dŵr Cymru). Glas Cymru was owned by a group of 200 representatives of a broad spectrum of Welsh interests. Some of the aims of this acquisition were to reduce customers' water bills and to publish performance incentives. Although the British water industry does not necessarily recognize this as a significant shift away from privatization back to public ownership, it has been argued that the Welsh case shows that customers' interests cannot be met properly by a fully privatized water market (Seppälä et al, 2004). According to Drakeford (2002), commercial rather than social pressures led to the demise of Dŵr Cymru. He points out that the public debate in Wales revolved around one core question: 'How far is it possible to deliver publicly desirable goals such as social inclusion, community safety and well-being through a market system based upon profit-driven companies?' (Drakeford, 2002, pp29–30).

Water cooperatives and other user-owned and user-managed utilities have won high marks for customer satisfaction and operational performance worldwide (for excellent examples see Chapters 17 and 18 in this volume). Yet, it is possible that customers may expect less of a smaller utility and are thus more satisfied with the service they receive. In any case, smaller user-managed utilities can provide useful lessons for larger utilities for improving their customer strategies and responsiveness (Seppälä et al, 2004).

In addition to the different individual options discussed so far, there are various possible combinations of organization forms for WSS. We have termed these options the multiple producers paradigm, which we briefly introduce in the following section.

MULTIPLE PRODUCERS PARADIGM

At the global level, the public sector is in the driver's seat as owner and operator of WSS. It has been estimated that only about 6 per cent of the world's population covered by water supply services are served by the seven biggest private multinational companies, and the percentage for sewerage coverage is even lower. Unfortunately, the international debate on private-sector involvement in water services has been almost exclusively based on the role of private WSS companies as operators or owners of the services. Yet, worldwide, the vast majority of these undertakings are publicly owned and operated – in many, if not most, cases by local governments. As already discussed, these public undertakings buy their goods, equipment and services to a large extent from the private sector based on competitive bidding (see Figure 9.3). These mainstream roles of the public and private sectors have been largely neglected in the conventional debates.

However, the overwhelming dominance of public-sector WSS does not mean that the roles of other stakeholders – such as citizens or water services users, local, regional and central government authorities, or private enterprises – should be underestimated or neglected. On the contrary, their roles must be assessed and formulated so that societal goals and the requirements of sustainable development are met in the field of WSS production and provision. These are some of the key principles behind the multiple producers paradigm, which again can be illustrated with some examples from Finland. In addition to municipal WSS utilities and consumer-managed WSS cooperatives, supra-municipal systems of various forms are also found in the country. The most traditional form involves signing bilateral agreements between various systems, municipalities or larger agro-industries. The food industry has often been a shareholder and a financier of regional water utilities in need of good-quality groundwater. If the quantities of industrial wastewater in question are large, joint financing and implementation with the municipality has been used.

In other cases, a joint local authority (formerly a federation of municipalities) may be established to serve a whole region as a bulk water supplier, an operator or a joint-stock company. The bulk (wholesale) supplier company typically owns water intakes, treatment plants and trunk mains, while each municipality has its own distributing system. The operator company can be a joint local authority or a joint-stock company that leases the facilities from the owner municipalities. The weakest or, rather, the least binding form of cooperation is one-time informal cooperation, while the strongest or most binding form is the joint-stock company.

Discussion and prospects

When considering a conceptual and general development framework for WSS services, the starting point should be the basic aim of these essential services, which in our perspective implies the adoption of the not-for-profit principle. If there is a profit or, rather, benefits, these must be collective in the form of improved WSS services and a better environment, or by making access to the services more affordable. In particular, in the developing economies, improved services are also expected to enhance the time savings, economic productivity, convenience and dignity of individuals and families, such as by helping to increase girls' school attendance in the poorest countries.

In the selection of an implementation model or models, long-term experiences and the nature of the services in their particular circumstances should be considered, especially when decision-makers come under heavy pressure from international stakeholders or other actors who promote one-size-fits-all solutions. The experience emerging from recent research suggests that there is a need to examine the various options available in any given situation instead of uncritically adopting solutions often presented as 'innovations', such as long-term operational contracts, let alone full concessions promoted by multinational private companies, a model which is still supported by some international financial institutions in spite of accumulating evidence of their pitfalls.

Among other options, the regionalization of WSS utilities may be justified from several perspectives. However, one fundamental question should be asked: how far is it possible to justify the expansion of piped water supply and sewer networks when we take into consideration the political, economic, social, cultural, technological, environmental and legislative requirements (as in the PESTEL framework)? For instance, a large regional wastewater system is currently planned for the Tampere region in Finland, possibly to collect wastewaters from 15 municipalities; but the vulnerability risks involved in such a megaproject have not been considered. Such deterministic approaches to technology development are highly questionable in any long-term considerations; but the WSS sector worldwide is still largely dominated by technological reductionism and one-dimensional thinking.

Another example of current trends is the case of The Netherlands, where there has been a strong drive to merge municipally owned WSS systems to create public water companies (public limited companies, or PLCs), the now dominant model in that country (Blokland et al, 1999). After extensive mergers of municipal WSS utilities, their number has declined dramatically. This leads to the question of how far it is sensible to merge such systems without losing the possibility of making comparisons and applying benchmark monitoring. However, it is worth clarifying that according to the Dutch legislation in force, all WSS undertakings have to be under municipal ownership.

Still, there is much scope for institutional diversity despite these trends – for instance, in countries with large rural areas, we tend to find a large number of small systems managed as cooperatives or other forms of user-led management. In practice, then, we will probably always find that the production of WSS services of various forms takes place at least at three different scales: supra-municipal, municipal and small systems – although not forgetting that in many countries there is also the fourth level of on-site systems.

In the end, decisions about what kind of WSS paradigm is chosen depends upon a conscious value-bound selection – that is, what decision-makers may consider to be most viable and essential. From the perspective adopted in this book, the nature of WSS requires a long-term vision where the duties of the public sector as guarantor of the social rights to essential welfare services and goods must be the central consideration in decision-making. Therefore, the authors argue that the selling of public assets of natural monopoly such as WSS utilities should be the last option for decision-makers, if an option at all. We are concerned with the fact that some current policies being implemented in WSS, particularly leveraged buyouts and the use of hedge funds as investment vehicles or extensively long operational contracts, are increasing the vulnerability of public infrastructure and making the protection of critical infrastructure more complex and difficult. There is clear evidence showing that private investors and operators are neglecting the critical long-term issues that are at the centre of essential WSS in favour of the pursuit of short-term profits (Hukka and Katko, 2007). We believe that instead of aiming at short-term economic objectives, the WSS of human communities have to be seen as part of a country's critical infrastructure. Minimizing the vulnerability of WSS and adopting long-term and multidimensional thinking must become central to the activities of policy-making, planning, operation and management of these systems.

CONCLUSIONS

When assessing the various paradigms available for the organization of sustainable WSS, it is crucial to remember that the ultimate main purpose of these services is to serve the essential needs of communities. It is also important to emphasize that the variety of WSS management options included in the paradigms discussed in this chapter are complementary (not mutually exclusive) alternatives. These different options may actually support each other when we try to develop sustainable WSS policies, strategies and management options for various conditions. In practice, there are few situations where the conditions for WSS may be identical and facing similar problems, and therefore public policy and management of these services cannot be treated as being the same way everywhere. Furthermore, the approach proposed here poses an additional challenge for WSS professionals as they must go beyond merely technical concerns and join the task of persuading decision-makers, who are often disconnected from the realities of WSS on the ground, that the crucial objective of WSS is the secured satisfaction of the essential needs of human communities.

We conclude our analysis by suggesting that the best approach is adopting the *diversity paradigm* in public policy and management of WSS, avoiding the mechanical preference for particular options even in the case of a single country, let alone in the wider international context. When the strategic decisions regarding policy, governance, provision and production of WSS services are to be made, there is a need to analyse the strengths and weaknesses of the various paradigms from the point of local requirements and preferences.

ACKNOWLEDGEMENTS

The authors wish to thank the reviewers for their useful comments, partners of the PRINWASS, Watertime (EU) and Urban Water Conflicts (UNESCO) projects for their cooperation, and the Academy of Finland (nos 115397 and 210816) for its financial support.

REFERENCES

AFLRA (Association of Finnish Local and Regional Authorities) (2004) *Gateway to Local Finland*, Helsinki, Finland, www.kunnat.net/k_etusivu.asp?path=1;161;279, accessed 5 December 2007

Azpiazu D., M. Schoor, E. Crenzel, G. Forte, J. C. Marín and J. Rozé (2004) *D22 Strategic Country Report Argentina*, PRINWASS project, University of Oxford, Oxford, www.prinwass.org, accessed June 2008

Blokland, M., O. Braadbaart and K. Schwartz (eds) (1999) *Private Business, Public Owners: Government Shareholding in Water Companies*, International Institute for Infrastructural, Hydraulic and Environmental Engineering (IHE), Delft, The Netherlands

Castro, J. E. (2007) 'Poverty and citizenship: Sociological perspectives on water services and public–private participation', *Geoforum*, vol 38, no 5, pp756–771

Castro, J. E. (2009) 'Commons and citizenship: The contradictions of an unfolding relationship', in S. Helfrich (ed) *Genes, Bytes and Emissions: Common Goods and Citizenship*, Heinrich Böll Foundation Editions, Mexico City, in press

Castro, J. E., M. Kaika and E. Swyngedouw (2003) 'London: Structural continuities and institutional change in water management', *European Planning Studies*, vol 11, no 3, pp283–298

da Costa S. S., L. Heller, L. R. S. Moraes, P. C. Borja, C. H. de Melo and D. Sacco (2006) *Successful Experiences in Municipal Public Water and Sanitation Services from Brazil*, Transnational Institute, Amsterdam, www.tni.org/detail_pub.phtml?page=books_brazilwater&menu=05k, accessed May 2008

CWEF (Colorado Water Education Foundation) (2004) *Citizen's Guide to Colorado Water Law*, revised edition, http://cfwe.org/CitGuides/CG–Law2004.pdf, accessed June 2008

Drakeford, M. (2002) 'Providing water in Wales: Is there a third way? The Welsh experience with public and private utilities and the emergence of the not-for-profit mode', Documents from the second PRINWASS Research Workshop, University of Oxford, Oxford, 28 February 2002, www.prinwass.org/docs_Feb02.shtml, accessed May 2008

EC (European Commission) (2001) *A Sustainable Europe for a Better World: A European Union Strategy for Sustainable Development*, European Commission's proposal to the Gothenburg European Council, EC, Brussels

EU (European Union) (2008) 'Subsidiarity' in *Europa Glossary*, EU, Brussels, http://europa.eu/scadplus/glossary/subsidiarity_en.htm, accessed May 2008

Finlex (2006) *Valtion säädöstietopankki* [*The State's Legislative Data Bank*], Ministry of Justice, Helsinki, Finland, www.finlex.fi/fi/, accessed May 2008

Grigg, N. S. (1996) *Water Resources Management: Principles, Regulations, and Cases*, McGraw-Hill, New York

Hadley R. and K. Young (1990) *Creating a Responsive Public Service,* Harvester Wheathsheaf, Hertfordshire, UK

Hardoy, A., J. Hardoy, G. Pandiella and G. Urquiza (2005) 'Governance for water and sanitation services in low-income settlements: Experiences with partnership-based management in Moreno, Buenos Aires', *Environment and Urbanization*, vol 17, no 1, pp183–199

Heller, L. (2006) *Access to Water Supply and Sanitation in Brazil: Historical and Current Reflections; Future Perspectives*, Human Development Report Office, Occasional Paper 2006/24, United Nations Development Programme (UNDP), New York, http://hdr.undp.org/en/reports/global/hdr2006/papers/heller%20leo.pdf, accessed 5 December 2007

Hukka, J. and T. Katko (2003) 'Water privatisation revisited – panacea or pancake?', *Occasional Paper Series*, no 33, International Water and Sanitation Centre (IRC), Delft, The Netherlands, www.irc.nl/page/6003, accessed May 2008

Hukka, J. J. and T. S. Katko (2007) *Vesihuollon haavoittuvuus* [*Vulnerability of Water Services and Critical Infrastructure Protection*], Foundation for Municipal Development, Helsinki, Finland, available in Finnish at www.kaks.fi or www.polemiikki.fi/files/1133-25459_TutkJulk58.pdf, accessed May 2008

Hukka, J. J. and O. T. Seppälä (2004) *WaterTime Project Report: National Context Report – Finland*, University of Greenwich, Greenwich, www.watertime.net/wt_cs_cit_ncr.asp#Finland, accessed May 2008

Hukka, J. J., E. M. Vinnari and P. E. Pietilä (2006) 'A quest for effective water governance: Look who's leading the WPI', Paper presented to the 32nd Conference of the Water Engineering Development Centre (WEDC), Loughborough University, Colombo, Sri Lanka, 13–17 November 2006

Juhola, P. (1995) *Vesihuoltolaitos yhdyskuntien palveluorganisaationa* [*Water and Sewage Utility as a Service Enterprise*], A5 Licentiate thesis (in Finnish), Tampere University of Technology, Institute of Water and Environmental Engineering, A5, Tampere, Finland

Juuti, P. and T. Katko (eds) (2005) *Water, Time and European Cities: History Matters for the Futures*, www.watertime.net/Docs/WP3/WTEC.pdf, accessed May 2008

Katko, T. (1992) *The Development of Water Supply Associations in Finland and Its Significance for Developing Countries*, Discussion paper series no 8, UNDP and World Bank, Water Supply and Sanitation Programme, New York and Washington, www-wds.worldbank.org/servlet/main?menuPK=64187510&pagePK=64193027&piPK=64187937&theSitePK=523679&entityID=000009265_3961002175639, accessed May 2008

Katko, T. (1994) 'The need for "champions" in rural water supply', *Waterlines*, vol 12, no 3, pp19–22

Katko, T. (1997) *Water! Evolution of Water Supply and Sanitation in Finland from the Mid-1800s to 2000*, Finnish Water and Waste Water Works Association, Helsinki, Finland

Kettunen, P. (1999) 'Local government services', *Finnish Local Government Studies*, vol 27, no 3, pp332–341

Lawrence, P., J. Meigh, and C. Sullivan (2002) *The Water Poverty Index: International Comparisons*, Keele Economics Research Papers no 19, Keele University, Keele, UK, www.keele.ac.uk/depts/ec/wpapers/kerp0219.pdf, accessed May 2008

MAF (Ministry of Agriculture and Forestry) (2001) *Water Services Act (119/2001)*, Unofficial translation, Helsinki, Finland, www.finlex.fi/en/laki/kaannokset/2001/en20010119, accessed May 2008

MAF (2002) *Guidelines for the Water Services Act 119/2001* (in Finnish), Helsinki, Finland, www.mmm.fi/julkaisut/julkaisusarja/MMMjulkaisu2002_1.pdf, accessed May 2008

Nickson, A. (1998) 'A water co-operative for a large city: Does it work?', University of Sussex, Institute for Development Studies, Brighton, UK, www.id21.org/id21ext/2ban1.html, accessed 30 November 2007

Pietilä, P. (2006) *Role of Municipalities in Water Services*, PhD thesis, no 617, Tampere University of Technology, Tampere, Finland

Seppälä, O. T., R. P. Rajala and T. S. Katko (2004) 'Customer-responsive water and sanitation services', *Journal of the American Water Works Association*, vol 96, no 6, pp83–93

Takala, A. (2007) *Vesiyhtymien toiminnan kehittäminen* [*Operational Development of Water and Wastewater Associations*], MSc thesis (in Finnish), Tampere University of Technology, Tampere, Finland

Tamm, G. (1991) *Institutional Framework of Small Community Water Supply Systems in the United States: A Review of Experience and Lessons for Developing Countries*, Discussion Paper Series no 7, UNDP and World Bank, Water Supply and Sanitation Programme, New York and Washington, DC, www-wds.worldbank.org/external/default/WDSContentServer/WDSP/IB/1991/08/01/000009265_3961001205423/Rendered/PDF/multi_page.pdf, accessed May 2008

Transparency International (2007) *Corruption Perceptions Index 2007*, Transparency International, Berlin, www.transparency.org/policy_research/surveys_indices/cpi/2007, accessed May 2008

UN (United Nations) (1992) *The Dublin Statement on Water and Sustainable Development*, International Conference on Water and the Environment (ICWE), UN, Dublin, www.gdrc.org/uem/water/dublin-statement.html, accessed March 2008

Vinnari, E. M. and S. Näsi (2008) 'Creative accrual accounting in the public sector: "Milking" water utilities to balance municipal budgets and accounts', *Financial Accountability and Management*, vol 24, no 2, pp97–116

Watertime (2005) *WaterTime Final Results: Good Practice Recommendations and Participatory Decision Support System*, University of Greenwich, Greenwich, www.watertime.net/, accessed 24 March 2008

Windischhofer, R. (2007) *Municipal Entrepreneurialism and the Commercialization of the Finnish Water Sector*, PhD thesis, Acta Electronica Universitatis Tamperensis 641, Tampere University of Technology Tampere, Finland, http://acta.uta.fi/pdf/978-951-44-7029-5.pdf, accessed May 2008

10

Community Organization and Alternative Paradigms for Improving Water and Sanitation in Deprived Settlements

Gordon McGranahan and Martin Mulenga

Introduction

There is agreement on the need to reduce the share of the world's population with inadequate water and sanitation, but not on how this should be achieved. A centuries old argument over whether water and sewerage utilities should be privately or publicly operated was revived during the 1990s. Such debates do a good job of pitting different vested interests in the water sector against each other, and a bad job of focusing attention on measures likely to improve provision in deprived settlements. First, as described below, whether a utility works efficiently and equitably is only contingently related to whether the operator is private or public. Second, utilities are often not even present in low-income settlements, while cooperation among residents and between community groups and the organizations that work with them is almost always critical.

Water and sewerage networks and indoor piping are not just designed to provide services, but also to shift responsibility to engineers and 'service providers', ensuring that people do not have to cooperate with their neighbours to secure adequate water and sanitation. With these networked technologies – the norm in affluent settings – even the most selfish neighbours will not take away your household's water supplies or expose you to their faeces. Without such technologies, however, even mildly selfish behaviours can create major public health risks in the community and beyond. Even simplified sewerage systems, designed to be affordable in low-income settings, cannot function well without a level of community participation unnecessary with conventional networked systems (Paterson et al, 2007).

More specifically, in settlements where household water and sanitation services are lacking, much depends upon the ability and willingness of community members to adopt sanitary practices that protect their neighbours as well as themselves; to agree on what sorts of shared water and sanitation facilities are affordable and acceptable, and where they should be located; to put forward effective representatives that can lead self-help initiatives and negotiate with governments or private providers; and, if necessary, to develop the skills and levels of trust needed to handle finance and manage projects. The ability and willingness of community members to do such things can be influenced but not prescribed from the outside.

It can be difficult for overburdened and underpaid people to cooperate, particularly where there is a history of conflict and mistrust. The ideal community, with members working together in perfect harmony, is just as much an illusion as the perfect market or the perfect plan. On the other hand, changes that increase the opportunity or capability for communities to pursue collective improvements can be critically important.

Better organized communities cannot only help themselves, but can be central to getting markets or governments to play a more positive role. In economic terms, they can help to correct public and market failures.

Public failures in water and sanitation provision often arise from poor relations between local authorities, utilities and the residents of deprived communities. Even ostensibly pro-poor policies can be subverted, as when government authorities impose unrealistically high standards for household water and sanitation provision, creating not a goal for poor communities to aspire to, but a stick to beat them with. This is often part of a broader process by which large settlements become 'informal' and exist in contravention of the formal regulatory system (UN-Habitat, 2003). In such conditions, well-organized communities are in a better position to support realistic norms and to help secure government support in achieving them.

Perhaps the most relevant market failures result from sanitation being, in large part, what economists term a 'public good', and from both bad water and sanitation creating 'externalities'. To the extent that water and sanitation problems affect groups rather than individuals, and that the related health risks each resident faces result primarily from the practices of others, markets responding to individual demands will not provide adequate supplies. If the group affected can combine their individual demands for water and sanitation, and create a collective demand, such market failures can be overcome. While, in practice, the groups affected do not form well-defined communities, community groups are better placed than individuals to overcome these market failures.

It is often argued that sanitation is more of a public good than water because you benefit yourself and your family when you secure better water supplies, but you benefit the rest of the community when you use better sanitation facilities. This sort of difference helps to explain why individualized markets drive water provision more easily than sanitary provision – and why sanitary improvements tend to lag behind water improvements. Somewhat perversely, given this widely perceived difference, it has been with (rural) water rather than sanitation that community management has most often been promoted. In effect, the fact that latrines can be constructed privately on a house-by-house basis, while securing water often requires a collective effort, has obscured the more important fact that improved sanitation is a more public good.

It is easy to exaggerate these differences. Inexpensive water improvements, such as public wells and public taps, are not easily 'sold', and only yield their full benefits if people use them in a cooperative fashion. Benefits are easily dissipated by excessive queuing, monopoly pricing or groundwater contamination. Small private water vendors often play an important role, but this too can be undermined by a lack of local cooperation (Kjellén and McGranahan, 2006; McGranahan et al, 2006). Even good indoor piping in your home will not protect your children when they play and drink other water in neighbours' homes. More important, where provision is poor, water and sanitation problems are closely bound up with each other, with hygiene problems and with a range of other localized environmental health problems (McGranahan et al, 2001). The resulting combination of hazards creates what has been termed a syndemic, where multiple diseases and conditions combine synergistically to create a large excess burden of disease (Singer et al, 2006).

The remainder of this chapter is divided into three main sections. The first examines some of the dangers of idealizing the perfect plan, the perfect market or even the perfect community. The second illustrates how central community organization is to recent successes in rural sanitation improvement. The third turns to urban water and sanitation, and examines successful examples of community involvement. The conclusion is not that a community-driven approach should replace market and government-driven water and sanitation provision, but that community groups do often hold the key to better provision.

THE DANGERS OF IDEALIZING MARKETS, PUBLIC PROVISIONING OR COMMUNITY MANAGEMENT

The 1990s and early 2000s saw a concerted attempt to promote more private-sector involvement in the water sector and beyond. In 'developing' countries, this was led by international development agencies, including most notably the World Bank (Finger and Allouche, 2002). The promotion of more private participation (also known as privatization) was not based on evident successes of private water and sanitation enterprises operating in these countries. It was part of a broader neoliberal agenda, spurred on by the fall of communism, and designed to reduce the role of government and expand that of private enterprise and markets. Proponents contrasted the messy reality of public water bureaucracies with idealized visions of competing private operators.

Not surprisingly, the reality of private operators was not in line with these ideals. Nor, however, was it the unmitigated disaster predicted by many critics. Indeed, it is difficult to discern whether private-sector participation had any effect at all on either efficiency (Braadbaart, 2002) or coverage (Clarke et al, 2004; Kirkpatrick et al, 2006). The envisaged private investments were not forthcoming, and many of the problems that afflicted public utilities also came to afflict privately operated utilities (Budds and McGranahan, 2003). While privatization and its reversals were highly significant in terms of national politics, they were not of much consequence for those without access to water or sewerage networks.

Community-based approaches have also been promoted, albeit on a much smaller scale and primarily in areas where utility operated systems could not succeed. WaterAid, the only major international non-governmental organization (NGO) focusing specifically on improving access to water and sanitation in deprived areas, has long emphasized the importance of community participation in its own work. Similarly, the International Water and Sanitation Centre (IRC), the major international research institute focusing on water, sanitation and hygiene conditions, has emphasized the importance of community-based improvements for decades. However, such approaches do not really threaten or serve vested interests in the sector, or fit within any of the dominant ideologies, and attract little attention outside of the water and sanitation sector.

It is possible but not desirable to idealize perfect community provision in much the same way that perfect market provision and perfect public provision have been idealized (McGranahan et al, 2001, Chapter 5). If the perfect market can be said to translate individual preferences into economically efficient outcomes through the invisible hand, and the perfect plan to translate public interest into state provision through bureaucratic rationality, the perfect community can be said to translate group decisions into collective action through voluntary or obligatory association. However, this sort of idealization represents much of what is wrong with the ways in which market and state solutions have historically been promoted, and is hardly a good model to extend to community-driven improvements. In any case, the success of community-driven approaches often lies not in replacing markets or state provisioning, but in making them more effective, efficient and equitable.

Where community management of water and sanitation has been promoted by international agencies, there has been a tendency to both idealize and standardize the approach. Indeed, the same development agencies that over-promoted the privatization of water and sanitation utilities were often simultaneously overselling certain community-based or community-driven development models (Mansuri and Rao, 2004). As already indicated, in the water and sanitation sector this applied especially to isolated rural water provision. Not surprisingly, perhaps, an evaluation of the impacts of community-based water services in Sri Lanka found that while projects that gave residents more say in project design were more likely to be successful, their impact was also greater where communities were already comparatively well organized (Isham and Kahkonen, 2002). Similarly, a review of community-managed water supply projects in Africa found that they often failed to come to terms with the complex realities of actually existing communities, and often applied an overly simple model of how and when community management can function effectively (Harvey and Reed, 2007).

An idea being promoted as a new urban paradigm is providing water and sanitation to groups of households (Mara and Alabaster, 2008). Local residents would be required to engage with the provider as a group, and to cooperate in managing selected components of the local water and sanitation systems. The authors claim this would bring costs down to the point where it would become realistic to meet international water and sanitation targets. Organizationally, the challenge is to ensure that:

- the formal providers provide the required services;
- local residents cooperate and make their required contribution.

As the authors note, since community cooperation has been achieved in very difficult circumstances (see the example from Karachi below), there is no reason to believe that the community side of this type of arrangement is inherently unattainable. On the other hand, there is no reason to believe that it will be achieved simply by changing the operating procedures of formal water and sanitation providers.

THE SUCCESS OF COMMUNITY-LED TOTAL SANITATION (CLTS) AND WHAT IT REVEALS ABOUT THE POTENTIAL ROLE OF RURAL COMMUNITIES

A conventional state-led approach to rural sanitation involves designing acceptable latrines and then promoting their construction and use. Despite many decades of experience, in conditions of poverty it is still difficult to get latrines constructed with this approach, and even harder to ensure that they are used and maintained. Subsidies are required, and even so coverage is often partial and maintenance poor. Private enterprises or social marketing are often involved; but it is difficult to market latrines of acceptable quality. Efforts have also been made to engage communities in constructing and managing improved latrines, with mixed results.

Among rural practitioners, it has long been recognized that community-led water and sanitation can perform as well in the right circumstances (Lockwood, 2004; van Wijk-Sijbesma, 2004). One of the biggest challenges is how to scale up community managed sanitation – part of the broader challenge of scaling up community-driven development (Binswanger and Aiyar, 2003). Without profits, there is no economic incentive for providers to expand community-led systems even if they are successful. On the other hand, without profits to cover, there can be a greater economic incentive for other communities to imitate successful initiatives, provided they are well informed.

Perhaps not surprisingly, in many of the more successful approaches, capital, operations and maintenance costs are not subsidized. First, not only is it difficult and often costly to manage subsidies, but the need for a subsidy makes success costly for the sponsor and can provide perverse incentives even within government agencies. Second, spontaneous expansion across communities is precluded unless the subsidy programme itself is demand responsive. On the other hand, without any subsidy the 'unacceptably' poor are unlikely to find 'acceptable' water and sanitation options affordable.

Community-led total sanitation (CLTS) is very explicit about both creating community demand for better sanitation and avoiding subsidies (Kar and Pasteur, 2005). It starts with a facilitator working in the community who reveals an unarticulated collective demand for sanitary improvement, and then finds a technology that meets this collective demand (Kar and Pasteur, 2005; Robinson, 2005). Rather than starting with the latrine, a 'good' that is private or shared among a few households, CLTS aims directly for the public good – faeces-free public space. With conventional participatory tools, the community confronts its open defecation problem as a group, seeking out the faecal material, mapping it and calculating its quantities. In successful cases, this leads to the realization that open defecation not only should, but can, be stopped if everyone works together. The health threats resulting from open defecation provide an additional justification for immediate action. When the discussion turns to technology, rather than providing information on a suite of technologies that may not be affordable, the facilitator gets people to come up with their own designs that they know they can afford.

CLTS has been very successful in its initial scaling-up, particularly in Bangladesh, where many claim it started, and in neighbouring India (Kar and Pasteur, 2005; Deak, 2008). By 2006, CLTS had helped thousands of villages in both Bangladesh and India to achieve open defecation-free status (Kar and Bongartz, 2006). It has spread to parts of Africa, as well as to Bolivia, Yemen and several other Asian countries (Deak, 2008, p9). There will be times when CLTS is misapplied and places where it is inappropriate; but it illustrates how community-driven approaches can be successful.

ILLUSTRATIVE EXAMPLES OF COMMUNITY ENGAGEMENT IN URBAN WATER AND SANITATION

The following examples all involve community-driven water and/or sanitation improvements that have been supported by organizations outside the water sector, and have been going for long enough to demonstrate their sustainability. Despite their urban locations and very different practices, some of the key principles are analogous to those applied with CLTS.

The first example involves two community-based networks and the NGO who supports them, working together to find affordable solutions to sanitation problems in Mumbai and Pune. The communal toilets that these groups have developed are very much part of a broader process of community organizing, centred on savings groups and extending well beyond sanitary improvement.

The second involves an NGO who chose to start with sanitation after it was identified as a priority by residents of the informal settlements of Karachi. One of its central innovations has been to build on the residents' own attempts to use the natural drainage system, with communities taking responsibility for financing, managing and even building the latrines and neighbourhood sewers (with non-financial assistance from the NGO), and the government taking responsibility for the trunk sewers and treatment.

The third involves an NGO in Buenos Aires who has also focused on water and sanitation, and the need for community organization. Despite the fact that, in this case, universal coverage with sewered toilets and piped water managed by the utility is a more realistic goal, innovations have also involved new forms of community organization.

While these three examples are from Asia and Latin America, somewhat similar examples can be found in Africa. Indeed, the approach developed in Mumbai and Pune has provided one of the inspirations for the water and sanitation efforts of Shack/Slum Dwellers International, with activities in Cape Town, Durban, Nairobi and Kampala (see www.sdinet.org/rituals/ritual7.htm). More generally, community organizations can play a critical role in many urban centres in Africa (see Kyessi, 2005, for details on communities organizing for water and sanitation in Dar es Salaam). Research in South Africa, Zambia and Zimbabwe indicates that residents are often unaware of the operations or nearby community organizations, limiting their ability to improve services on a large scale (Mulenga et al, 2004), but this also often applies in Asia and Latin America. Simplified sewerage systems, somewhat analogous to those developed for Karachi, have also been applied on a limited scale in Africa. In Ethekwini (Durban), South Africa, for example, a shallow sewer system on a model adapted from La Paz, Bolivia, was piloted. A study evaluating this pilot was very optimistic about the possibilities of using shallow sewer technologies and involving communities in managing the neighbourhood sewers (Eslick and Harrison, 2004). Finally, as in Buenos Aires, NGOs in Africa have been involved in helping to negotiate extensions of utilities' water networks, even if coverage in most of sub-Saharan Africa remains far lower than in Argentina.

Organizing residents around community toilets in Mumbai and Pune[1]

Many attempts to improve sanitation in urban India have been based on working with communities. Organizations of the urban poor have often focused on sanitary improvement as part of the efforts to improve conditions in the slums, sometimes with considerable success. The example described here is grounded in an alliance between the National Slum Dwellers Federation (NSDF), Mahila Milan (MM: a network of savings groups formed by women 'slum' and pavement dwellers) and the Mumbai-based NGO SPARC (Society for the Promotion of Area Resource Centres).

During the late 1980s and early 1990s, this alliance built and managed public toilet blocks in areas where provision was particularly poor or non-existent. The construction of each toilet block was usually preceded by a community-managed 'slum' survey to document the inadequacies – and local savings groups from the 'slum' often helped to design, implement and manage the toilet blocks. As with CLTS, collective discussions of sanitary deficiencies helped to create community demand and engagement, which, in turn, drove the siting, design and management. The toilets gave women more privacy, made queues work better (e.g. separate queues for men and women since with one queue, men push in), ensured a constant supply of water for washing and made better provision for children (many had separate children's toilets at the front so that children did not have to queue). Community management ensured that they could be maintained through user charges, with charges for use being lower than conventional 'public toilets' (typically, families paid a monthly charge giving all family members unlimited use). Caretakers and cleaners were identified from the local community and often lived on site as the new toilet complexes included accommodation for them.

Local governments ignored or discouraged these efforts until it became clear that they could discharge some of their own public responsibilities at a low cost through such schemes. The municipal commissioner in Pune (a city with over 2 million inhabitants) and then other city authorities ended up supporting this NSDF–MM–SPARC alliance in building community toilets; over 500 toilet blocks have been built to date – mostly in Pune and Mumbai. A major new toilet construction programme is now under way in the urban centres that are part of the Mumbai Metropolitan region, with support from the Mumbai Metropolitan Region Development Authority.

People sharing a toilet with people with whom they do not already cooperate on other matters are inclined to renege on their responsibilities and blame others for the 'filth'. Government authorities in India have often simply built public toilet blocks, perhaps with provision for emptying, but no arrangements to ensure that residents will cooperate in using, let alone cleaning and maintaining, the toilets. The result is all too predictable, and as a result of their many failures, communal latrines are treated as 'unimproved' facilities in the statistics monitoring progress towards the Millennium Development Goals (WHO and UNICEF, 2000).

Part of the success of the NSDF–MM–SPARC toilets has been that they provide such a visible and symbolic contrast to the dirty, smelly and often dysfunctional public toilets so often found in slums. Such toilets are a source of shame for local residents and come to symbolize the futility of trying to improve local conditions. This symbolism can be reversed by a community-managed toilet that is clean and well serviced. If the negative symbolism is reversed, the toilets can come to symbolize the moral strength and solidarity of local residents.

NSDF–MM–SPARC have numerous advantages over government agencies when it comes to developing community toilets that work. Organizations of the urban poor, constituted locally, have knowledge of local social and sanitary conditions that government agencies and public utilities cannot access. They have fewer layers of hierarchy and bureaucracy to contend with. Equally important, the organizational skills required to set up and run women's savings groups and to organize residents to negotiate collectively for land and housing rights are very relevant when it comes to organizing other collective challenges such as sanitation. Most of the skills learned by public officials and utility staff are not. Moreover, the sorts of exchanges through which groups within Mahila Milan and the National Slum Dwellers Federation regularly learn from each other are also effective for learning about community-based sanitation strategies – reports, memoranda and other instruments of government are not.

As this example makes clear, government authorities can work with organizations of the urban poor and the NGOs who support them. The capital costs for most of the communal toilets being developed through NSDF–MM–SPARC are funded by government authorities, and their latest contract is for the construction of between 10,000 to 20,000 toilet seats in ten municipalities in the Mumbai Metropolitan region. It will be important to monitor both failures and successes as this approach is scaled up. Community organizing can be a slow process and has its own rhythms. Initially, toilet construction was undertaken in places with strong representative community organizations. With the expansion in scale, NSDF–MM–SPARC have to work in communities without such organizations, creating tensions between the governmental and community processes. For instance, some local governments refuse to allow the local Mahila Milan groups to manage the maintenance and charge users to cover the costs of doing so. It is important to strengthen the role of communities in extending sanitation in India; but nobody familiar with the difficulties of working in low-income communities and with poorly funded government bureaucracies would suggest that it will be easy.

Neighbourhoods taking responsibilities for 'internal' sewers in Karachi[2]

The Orangi Pilot Project (OPP) was established in 1980 to help residents improve conditions in *katchi abadis* – the unplanned settlements where a large and growing share of Karachi's low-income population lived. Participatory research identified sanitation as a major problem. OPP (or now its Research and Training Institute, OPP–RTI) developed an approach to sanitation that was once considered a radical alternative to government, but has influenced city and national agencies so profoundly that many of the procedures and principles have become official policy (if not always official practice).

Whereas the toilet blocks in Mumbai and Pune described above were inherently communal, the sewers developed with OPP's assistance involved household connections, with communities organized to take responsibility for the last stages of the sewers. OPP staff had to work hard to get costs down to affordable levels and to help communities overcome the technical barriers (by providing maps, tools, designs, training and supervision). This involved numerous innovations, building on pre-existing drainage systems developed by local residents, which tended to follow natural water paths. It also involved numerous struggles against large loan-based sanitation projects. This included a successful effort to convince the governor of Sindh to cancel a US$70 million loan from the Asian Development Bank that would have been used to build a sewerage system ignoring pre-existing drains. OPP proposed instead to upgrade the pre-existing drains, link them to box trunks, and place treatment plants where the trunk sewers meet the sea. While these technical and economic innovations have been central to OPP's success, this section is more concerned with their approach to community organization, which has been equally innovative and central.

Two of the first barriers to improvement that the OPP identified were the psychological barrier of convincing the communities that not only the homes, but also the streets belonged to them, and the social barrier of getting the local residents on the streets to come together and organize (Pervaiz et al, 2008). Without overcoming these barriers, the OPP's economic and technical work would have been to no avail.

The OPP–RTI divides sanitation into 'internal' and 'external' development, corresponding roughly to community-controlled and state-controlled components. Internal development includes sanitary latrines in the house, an underground sewer in the lane and neighbourhood collector sewers. External development consists of trunk sewers and treatment plants.

The results in Orangi and elsewhere show that communities can finance, manage and build internal development provided that they are organized and provided with technical support and managerial guidance. As with CLTS, subsidies of the community-owned and community-operated technologies have not been necessary. Indeed, subsidies are felt to undermine the sense and the reality of local ownership of the 'internal' development. Instead, by building the 'externals', the public agency ensures that when one community solves its sanitation problem, it does not displace it onto the next community. While accepting this role is difficult for a public agency steeped in the conventional public service ethos, OPP–RTI is now viewed as a partner by many government officials responsible for sanitation provision, and OPP–RTI's maps, designs and recommendations are central to Karachi's current sanitation plan.

The OPP Research and Training Institute has helped the government agencies to convert natural drains into sewers and to develop drainage plans for most of Karachi. With the assistance of the OPP, or by adopting the same approach, communities have developed 'internal' sanitation that now serves almost 90 per cent of Orangi, amounting to over 865,000 people living in nearly 7600 lanes. These communities have invested about US$1.7 million of their own money in building and maintaining the 'internal' infrastructure, while the OPP has provided technical assistance and the local government has been building and maintaining the 'external' systems (see Pervaiz et al, 2008, p14).

The OPP's success is perhaps unique; but it has not been unqualified, and a number of attempts to replicate the approach have failed (Hasan, 2007; Pervaiz et al, 2008; Tayler, 2008). Nevertheless, almost 50,000 households in 11 other towns in Pakistan have built internal sanitation systems following the OPP model, and internationally the OPP's approach has informed many similar approaches. Moreover, many of the principles developed by the OPP have also emerged in other successful community-based sanitation systems.

Negotiating community water and sanitation improvements in Buenos Aires – IIED–AL[3]

The International Institute for Environment and Development – America Latina (IIED-AL) has been working since the mid-1980s with residents and community groups in Barrio San Jorge, a small unplanned and low-income settlement on the outskirts of Buenos Aires (Schusterman and Hardoy, 1997). Initially engaged to help construct a community mother and child centre, the IIED–AL developed long-term relationships within Barrio San Jorge, working on community initiatives ranging from land regularization to water and sanitation provision. While the IIED–AL responded to priorities expressed by local residents – including water and then sanitation provision – it has played a lead role in community organization, fundraising, and project planning and implementation (Schusterman et al, 2002). When water was first identified as a priority, the staff of IIED–AL worked with the community to plan and initiate a piped water system distributing the brackish groundwater supplies. However, when the opportunity arose to try to connect the barrio's pipes to the city's piped water network, this was seized. The opportunity came at the time when the principal Buenos Aires utility was being operated under contract by a private water consortium, Aguas Argentinas.

The contract gave Aguas Argentina an incentive to connect more homes; but the challenges of extending provision to poor and unplanned settlements were daunting. Indeed, a report by Lyonnaise des Eaux, the lead company in the consortium, made it clear that connecting settlements on unregistered land should be based on explicit public-sector approval, and that where there are conflicts, as in the case of squatting, this 'must first of all be settled at the political level, on the basis of comprehensive economic and legal studies' (Suez Lyonnaise des Eaux, 1998, pp14–17). While understandable, the problem with this approach is that settling land irregularities at the political level can take decades

– in many municipalities being 'in the process of being regularized' has become a long-term condition. In practice, after negotiations orchestrated in large part by IIED–AL, Aguas Argentinas, the local government and community groups agreed on a procedure for extending coverage. This approach then provided a basis for Aguas Argentinas to extend coverage to numerous other communities, and while it would be misleading to portray this as a community-driven process, it clearly relied on community organization (Schusterman et al, 2002).

Indeed, community organization was central to the water and sanitation improvements in Barrio San Jorge from the start. The first water projects were the joint initiatives of IIED–AL and the local co-op, Nuestra Tierra (Our Land). The elected neighbourhood commission that preceded the co-op had been created to negotiate the regularization of the settlement. Delays undermined residents' confidence in the organization; but residents nevertheless began to turn to the co-op to address a range of community issues, including water supply. Working together, IIED–AL and the co-op were making slow, piecemeal progress. They had failed to convince the public utility to extend services, and were developing small systems using the brackish groundwater. They then heard that Aguas Argentinas was connecting a neighbouring settlement, and with the agreement of local government began another long and difficult process of negotiation.

Even before it was finally agreed that Barrio San Jorge would be served by the utility's piped water supplies, IIED–AL and the co-op had gone a long way towards creating a local piped network. Responsibilities were divided in a manner somewhat similar to that described for OPP in the previous example, with 'street' and 'common' sections replacing 'inner' and 'outer' development: IIED–AL provided the technical support and materials; the co-op acted as an intermediary and helped to organize residents; the residents selected delegates for their 'street', paid charges via their delegates and contributed labour and a contractor built the 'common' sections.

Barrio San Jorge was the first un-regularized settlement to be connected to the formal Buenos Aires piped water network, and it set an important precedent. Aguas Argentinas was persuaded to connect Barrio San Jorge in large part by the local government, and this was evident in the timing – water began to flow two days before local elections. Securing community cooperation was critical to convincing Aguas Argentinas to connect the settlements. It remained a challenge, however, and issues of community organization continued to affect the water delivery system after Aguas Argentinas took responsibility. Thus, the community organizations never developed to the point where residents truly trusted their delegates with payments. This was a problem when the pipes were being laid, but continued to be a problem when Aguas Argentinas asked the community to organize collective payments. Residents refused, fearing that they would suffer from other residents' failure to pay, or from the community organizers taking a cut of the money. Aguas Argentinas, on the other hand, did not want to be in charge of collecting payments and to engage in disputes over disconnections. As a result, payments were individual, enforcement was lax, non-payment was common, and this continued until Aguas Argentinas pulled out in the wake of Argentina's economic crisis of the early 2000s.

Community organization was also critical to other low-income and un-regularized barrios trying to become connected to Aguas Argentinas's network, although the organizational forms and issues varied (Schusterman et al, 2002).

CONCLUSIONS

The examples provided here only begin to illustrate the varied ways in which community organization influences water and sanitation provision. They clearly demonstrate the importance of community organization, but do not imply that there is a community-driven model that can replace government- and market-driven models. Indeed, better organized communities are found to be critical to getting the government and markets to function effectively.

Where conventional piped systems are unaffordable, local cooperation, organization and involvement are particularly critical. External agencies cannot assume, however, that cooperation will simply emerge through participatory processes, particularly if that participation is part of a project cycle. Successful community-led water and sanitation efforts not only adapt to local conditions and build on local strengths, but are grounded in a tacit understanding of community politics that is more sophisticated than a conventional project manager can be expected to acquire, or conventional project guidelines can be expected to describe.

In rural areas, community management of improved water supplies received far more attention in past decades than did community management of improved sanitation. This is ironic since private incentives are typically more capable of driving water improvements than sanitary improvements. More recently, with the successes of community-led total sanitation, on the one hand, and disappointment with simple models of community-managed water supplies, on the other, the weight of attention has shifted.

In urban areas, there is more scope for conventional public service provision in the form of networked water pipes and sewerage. Only a minority of currently deprived urban households in low- and middle-income countries are likely to receive such services in the foreseeable future, however. For the rest, improvements are likely to depend heavily upon community involvement. In all of the urban examples provided in this chapter, government agencies have also been involved and, in most cases, so have private enterprises. Success, however, has been heavily dependent upon how communities have driven their part of the process.

NOTES

1 This section draws heavily on Burra et al (2003) and National Slum Dwellers Federation, Mahila Milan and the Society for the Promotion of Area Resource Centres (1997).

2 This section draws heavily on Hasan (2006, 2007, 2008) and Pervaiz et al, (2008).

3 This section draws heavily on Almansi and Tammarazio (2008); Hardoy and Schusterman (2000); and Schusterman et al (2002).

REFERENCES

Almansi, F. and A. Tammarazio (2008) 'Mobilizing projects in community organizations with a long-term perspective: Neighbourhood credit funds in Buenos Aires, Argentina', *Environment and Urbanization*, vol 20, no 1, pp121–147

Binswanger, H. P. and S. S. Aiyar (2003) *Scaling Up Community Driven Development: Theoretical Underpinnings and Program Design Implications*, Policy Research Working Paper 3039, World Bank, Washington, DC

Braadbaart, O. (2002) 'Private versus public provision of water services: Does ownership matter for utility efficiency?', *Journal of Water Supply Research and Technology-Aqua*, vol 51, no 7, pp375–388

Budds, J. and G. McGranahan (2003) 'Are the debates on water privatization missing the point? Experiences from Africa, Asia and Latin America', *Environment and Urbanization*, vol 15, no 2, pp87–113

Burra, S., S. Patel and T. Kerr (2003) 'Community-designed, built and managed toilet blocks in Indian cities', *Environment and Urbanization*, vol 15, no 2, pp11–32

Clarke, G. R. G., K. Kosec and S. Wallsten (2004) *Has Private Participation in Water and Sanitation Improved Coverage? Empirical Evidence from Latin America*, Policy Research Working Paper 3445, World Bank, Washington, DC

Deak, A. (2008) *Taking Community-Led Total Sanitation to Scale: Movement, Spread and Adaptation*, IDS Working Paper 298, University of Sussex, Institute of Development Studies (IDS), Brighton, UK

Eslick, P. and J. Harrison (2004) *Summary of Lessons and Experiences from the Ethekwini Pilot Shallow Sewer Study*, Water Resource Commission, Pretoria

Finger, M. and J. Allouche (2002) *Water Privatisation: Trans-National Corporations and the Re-Regulation of the Water Industry*, Spon Press, London

Hardoy, A. and R. Schusterman (2000) 'New models for the privatization of water and sanitation for the urban poor', *Environment and Urbanization*, vol 12, no 2, pp63–75

Harvey, P. A. and R. A. Reed (2007) 'Community-managed water supplies in Africa: Sustainable or dispensable?', *Community Development Journal*, vol 42, no 3, pp365–378

Hasan, A. (2006) 'Orangi Pilot Project: The expansion of work beyond Orangi and the mapping of informal settlements and infrastructure', *Environment and Urbanization*, vol 18, no 2, pp451–480

Hasan, A. (2007) 'The sanitation program of the Orangi Pilot Project–Research and Training Institute, Karachi, Pakistan', in A. M. Garland, M. Massoumi and B. A. Ruble (eds) *Global Urban Poverty: Setting the Agenda*, Woodrow Wilson Center, Washington, DC, pp117–150

Hasan, A. (2008) 'Financing the sanitation programme of the Orangi Pilot Project–Research and Training Institute in Pakistan', *Environment and Urbanization*, vol 20, no 1, pp109–119

Isham, J. and S. Kahkonen (2002) 'Institutional determinants of the impact of community-based water services: Evidence from Sri Lanka and India', *Economic Development and Cultural Change*, vol 50, no 3, pp667–691

Kar, K. and P. Bongartz (2006) *Update on Some Recent Developments in Community-Led Total Sanitation*, University of Sussex, Institute of Development Studies (IDS), Brighton, UK

Kar, K. and K. Pasteur (2005) *Subsidy or Self-Respect? Community Led Total Sanitation – An Update on Recent Developments*, IDS Working Paper 257, University of Sussex, Institute of Development Studies (IDS), Brighton, UK

Kirkpatrick, C., D. Parker and Y.-Z. Zhang (2006) 'An empirical analysis of state and private-sector provision of water services in Africa', *The World Bank Economic Review*, vol 20, no 1, pp143–163

Kjellén, M. and G. McGranahan (2006) *Informal Water Vendors and the Urban Poor*, Water Discussion Paper 3, International Institute for Environment and Development (IIED), London

Kyessi, A. G. (2005) 'Community-based urban water management in fringe neighbourhoods: The case of Dar es Salaam, Tanzania', *Habitat International*, vol 29, no 1, pp1–25

Lockwood, H. (2004) *Scaling up Community Management of Rural Water Supply*, International Water and Sanitation Centre (IRC), Delft, The Netherlands

Mansuri, G. and V. Rao (2004) *Community-Based and Driven Development: A Critical Review*, Policy Research Working Paper 3209, World Bank, Washington, DC

Mara, D. and G. Alabaster (2008) 'A new paradigm for low-cost urban water supplies and sanitation in developing countries', *Water Policy*, vol 10, no 2, pp119–129

McGranahan, G., P. Jacobi, J. Songsore, C. Surjadi and M. Kjellén (2001) *The Citizens at Risk: From Urban Sanitation to Sustainable Cities*, Earthscan, London

McGranahan, G., C. Njiru, M. Albu, M. Smith and D. Mitlin (2006) *How Small Water Enterprises (SWEs) Can Contribute to the Millennium Development Goals: Evidence from Accra, Dar Es Salaam, Khartoum and Nairobi*, Loughborough University, Water Engineering Development Centre (WEDC), Loughborough, UK

Mulenga, M., G. Manase and B. Fawcett (2004) *Building Links for Improved Sanitation in Poor Urban Settlements*, University of Southampton, Institute of Irrigation and Development Studies, Southampton

National Slum Dwellers Federation, Mahila Milan and Society for the Promotion of Area Resource Centres (1997) *Toilet Talk*, no 1, SPARC, Mumbai, India

Paterson, C., D. Mara and T. Curtis (2007) 'Pro-poor sanitation technologies', *Geoforum*, vol 38, no 5, pp901–907

Pervaiz, A., P. Rahman and A. Hasan (2008) *Lessons from Karachi: The Role of Demonstration, Documentation, Mapping and Relationship Building in Advocacy for Improved Urban Municipal Services*, International Institute for Environment and Development (IIED), London

Robinson, A. J. (2005) *Lessons Learned from Bangladesh, India, and Pakistan: Scaling-Up Rural Sanitation in South Asia*, Water and Sanitation Program South Asia, The World Bank, New Delhi

Schusterman, R. and A. Hardoy (1997) 'Reconstructing social capital in a poor urban settlement: The integral improvement programme in Barrio San Jorge', *Environment and Urbanization*, vol 9, no 1, pp91–119

Schusterman, R., F. Almansi, A. Hardoy, G. McGranahan, I. Oliverio, R. Rozensztejn and G. Urquiza (2002) *Public Private Partnerships and the Poor: Experiences with Water Provision in Four Low-Income Barrios in Buenos Aires*, Loughborough University, Water Engineering Development Centre (WEDC), Loughborough, UK

Singer, M. C., P. I. Erickson, L. Badiane, R. Diaz, D. Ortiz, T. Abraham and A. M. Nicolaysen (2006) 'Syndemics, sex and the city: Understanding sexually transmitted diseases in social and cultural context', *Social Science and Medicine*, vol 63, no 8, pp2010–2021

Suez Lyonnaise des Eaux (1998) *Alternative Solutions for Water Supply and Sanitation in Areas with Limited Financial Resources*, Suez Lyonnaise des Eaux, Nanterre Cedex, France

Tayler, K. (2008) 'Urban sanitation lessons from experience', *Waterlines*, vol 27, pp30–47

UN-Habitat (United Nations Human Settlements Programme) (2003) *Slums of the World: The Face of Urban Poverty in the New Millennium?*, UN-Habitat, Nairobi

van Wijk-Sijbesma, C. (2004) *Scaling up Community-Managed Water Supply and Sanitation Projects in India*, International Water and Sanitation Centre (IRC), Delft, The Netherlands
WHO and UNICEF (World Health Organization and United Nations Children's Fund) (2000) *Global Water Supply and Sanitation Assessment 2000 Report*, WHO and UNICEF, Geneva and New York

Part II

Country and Regional Experiences

11

Decentralization and Delegation of Water and Sanitation Services in France

Christelle Pezon

Introduction

Decentralization is the new watchword addressed to developing countries in the latest attempt to bring them on track for meeting the Millennium Development Goals (MDGs) for water and sanitation services (WSS) by 2015. The previous slogan was privatization. The two concepts are aimed at redesigning different aspects of water service management: decentralization targets the political authority responsible for WSS delivery, whereas privatization indicates which management option the public authority should follow to fulfil its mission. With decentralization, local authorities are entrusted with the responsibility of providing WSS under national regulation and policies that often limit the scope of their choices to just two possibilities: public or private management.

In France, decentralization and delegation have always been interrelated. Since the 1982 Decentralization Act, delegation of urban water services to private operators has become the rule. In a few years, local authorities serving a total population of around 8 million people switched from public to private management, which increased the market share of delegated urban WSS from 50 up to 75 per cent in terms of population covered. How different it was one century earlier, when the 1884 Local Authority Act marked the demise of private concession contracts and triggered a long period of municipal public management for urban water services. The share of delegated WSS increased only slightly from the 1950s to the mid-1980s in urban areas due to expanding networks that progressively reached a scale and extension that were beyond the administrative borders of cities. Urban communes failed to merge into larger organizations capable of operating WSS through public management. In order to adapt to the technical constraints of expanded networks, cities instead delegated their water services to big operators organized at a regional scale, which then prevented local authorities from amalgamating their water services into inter-communal utilities. In the case of urban communes, decentralization favoured the delegation model by preventing municipalities from developing local public–public partnerships; the existence of a few private water companies operating services nationwide made delegation an attractive option.

These private companies were born in large cities during the second half of the 19th century; but they grew up in rural France, after the socialist movement of 1900 to 1920 swept them away from urban communes. In theory, decentralization in water delivery also applies to rural communes. In practice, universal water delivery in France was made possible thanks to a strong state policy: few of the 32,000 communes with a population below 2000 inhabitants would have chosen to provide round-the-clock water services if they had had to develop the infrastructure on their local taxes or with their own citizens' financial resources. In rural communes, delegation was rooted in a centralized public policy that aimed to provide urban water to rural folks. Financial resources were secured at the national level and then channelled to the *départements* (the intermediate administrative level between communes and the state), where state engineers implemented the national water infrastructure plans (i.e. they defined the framework for the organization and management of rural water services). State engineers opted mainly for the creation of inter-communal arrangements that would then grant delegation contracts to private water companies that were technically capable of developing the water systems. Until the 1982 Decentralization Law, the system was subject to administrative regulation by *départements*. Like in urban areas, the high fragmentation of rural local authorities favoured delegation. Unlike in urban areas, the delegation of WSS in rural France was centrally implemented and monitored.

The 1999 Chevènement Law was a turning point in the organization of local authorities. Following the objective to rationalize the national territory by strengthening local authorities, the law has succeeded in reducing by six the number of local authorities effectively in charge. Many neighbouring communes that were unable to establish cooperation through inter-communal arrangements have decided to agree on adopting a single development policy that may include water and sanitation services. In many cases, the institutional functions allocated to the *communautés* allow them to develop the required capabilities (technical, administrative, legal, financial and managerial) to directly run their WSS in their jurisdiction, which had often been delegated to private companies for years. The law has now allowed a fairer competition between public and private management of water services.

This first section of this chapter presents an overview of the organization of WSS in France, followed by a section focusing on the rules and actors involved in the management of these services prior to and after the Chevènement Law. The final section provides the outlines for a prospective analysis of the situation of WSS in the country.

The organization of drinking water services: From communal fragmentation to community integration

The organization of drinking water services in France is very diverse. The first reason for this is the very large number of communes, which are the units in charge of organizing these services. The second reason is that communes have the possibility of choosing different options for organizing the production and distribution of drinking water. The 1999 Chevènement Law is the most recent attempt to rationalize the structure of local government in France through the promotion of new entities, the *communautés* (or communities). We consider these processes in the following subsections.

The legacy of communal division in the organization of water services

There are about 12,000 drinking water utilities in France, which is an unfortunate outcome of the extravagant number of communes that make up the republic, although the number has been slightly reduced over time from 44,000 in 1792 to about 36,500 today. This large number of communes has often been denounced as inappropriate for implementing policies and developing an effective bureaucracy; but the attempt to reduce the number of communes by amalgamating them has been clearly lost and the most promising solution accepted by the promoters of a more rational administration has been inter-communal cooperation (EPCI).[1]

Historically, after the failure of the authoritarian introduction of the municipal townships by the French Constitution of the Year III (Napoleon's Calendar), the basis for inter-communal cooperation was truly introduced by law on 22 March 1890. It established the figure of the 'single-purpose inter-communal syndicate' (*syndicat intercommunal à vocation unique*, or SIVU), which until 1959 was the only type of organizational arrangement allowing communes to transfer one (and only one) of their competences, water supply being a possibility.

On 5 January 1959, an ordinance introduced important transformations for the inter-communal syndicates. From then on, the syndicates were allowed to accumulate competences under the figure of the 'multi-purpose inter-communal syndicate' (*syndicat intercommunal à vocation multiple*, or SIVOM), and could also combine communes, syndicates and '*départements*' in mixed syndicates (SMs). The same ordinance instituted the formula for the creation of urban districts (DUs) to provide urban areas with a level of organization that could help them in coping with and managing their increasingly complex development. From 1966 onwards, the towns were also allowed to combine into urban communities (CUs), where the transfer of the competences in drinking water and sanitation are mandatory. The 1999 Chevènement Law provisionally closed the already long list of organizations likely to undertake responsibility for water services. The law created two new types of supra-communal entities, the community of communes (CC) for rural communities (*communauté de communes*) and the community of agglomeration (CA) for small towns and villages (*communauté d'agglomération*), and reserved the category of urban community (CU) for urban centres with more than 500,000 inhabitants.

The first syndicate for water services was created in 1907 in the Ardèche. It regrouped four communes, which allowed them to share the management of the water supply system and the maintenance of a reservoir. This was the first inter-communal production service for drinking water. By 1936, there were 290 drinking water syndicates constituted by 1641 communes, compared to some 1674 electricity syndicates that grouped 21,011 communes (Leydet, 1936). After the nationalization of the energy sector in 1946, drinking water services became the most important driver of inter-communal cooperation until the late 1980s.

In 1988 there were more than 15,700 water utilities, of which 3901 were inter-communal organizations formed by 25,400 communes (see Table 11.1). Ninety-four per cent of these organizations were syndicates with competence for the provision of drinking water only (SIVUs). This means that until as recently as the late 1980s, the large majority of the French water utilities (11,800, or 75 per cent of the total) were still organized at the communal level, and that a majority of communes (68 per cent) had transferred their responsibility for water services to specialized inter-communal organizations, rather than, as expected by the central bureaucracy, to the new supra-communal entities created at the level of districts and urban communities. Contrary to urban vocational organizations, syndicates allowed the communes to keep their control over water distribution and to limit the transfer of responsibility to the production of bulk water supply. Over half of the communes organized the activities of production and distribution of water in different forms. In some cases they would transfer the production of water to one syndicate and manage the distribution directly, in other cases they would invert the terms; in yet others, they would transfer both services to two different syndicates.

Table 11.1 *Communal and inter-communal services for drinking water in 1988 (France)*

Organization type	*Communes*	*SIVU*	*SIVOM*	*SM*	*District*	*CU*	*Total*
All inter-communal organizations[a]		12,200	2280	750	165	9	15,404
Drinking water services organizations[b]	11,800	3375	455	No data	62	9	15,701

Source: [a] Ministry of Home Affairs (2007); [b] Delamarre et al (1992)

The 'silent revolution'[2] *of the communities*

Eight years after the Chevènement Law was passed, 33,413 communes (91 per cent of the total) representing 54.2 million inhabitants (89 per cent of the population) voted for the creation of 2588 communities. Thus, to a certain extent, the number of local authorities has dropped from 36,700 communes to less than 5800 entities, including 2588 communities and about 3200 communes. All French cities, except Paris, which is a *département*, now belong to either a *communauté d'agglomération* or a *communauté urbaine*. In rural areas, the 2400 *communautés de communes* bring together, on average, 12.5

Table 11.2 *Drinking water service organizations in 2007 (France)*

Organization	*Commune*	*CC*	*CA et SAN*[a]	*CU*	*SIVU*	*SIVOM*	*SM*	*Total*
Number of services	8074	207	85	14	2949	374	143	11,846
Number of communes		2281	1242	358	19,762	3003	2060	36,780
Population (millions)	17.8	2.75	9.9	6.25	17.9	3.6	5.4	63.6[b]

Notes: [a] new agglomeration syndicate.
[b]The figure for population is inflated because of double-accounting in some communes, when services have been transferred by cascading into several EPCI.
Source: Canneva (2007)

communes and more than 11,000 inhabitants. The 14 urban communities, representing more than 6 million people, 85 agglomeration communities (50 per cent of the total) and less than 9 per cent of the communities of communes (207) are responsible for directly delivering drinking water services.

In 2007, two thirds of the drinking water utilities are still organized at the commune level, which concerns 28 per cent of the country's population (see Table 11.2). One third of the services (less than 4000) are organized through EPCIs. Three-quarters of these EPCIs have exclusive competence for delivering drinking water. More than 55 per cent of the communes (representing nearly one in three French people) organize their drinking water services under the single-purpose inter-communal syndicate framework (SIVU).

Although the traditional form of service organization (commune, SIVU) conserves its leadership (93 per cent of water services, 75 per cent of communes and 56 per cent of the population), it is declining: 3800 communal services (34 per cent) and more than 400 SIVU (13 per cent) disappeared during the last 20 years. In 2007, the communities served a population equal to that being supplied by the communal services or by the SIVUs (about 17 million people). Approximately 10 per cent of the communes transferred their drinking water responsibility to 306 community organizations who provided drinking water supply to nearly one in three French people.

The dynamic emergence of the communities introduced two major ruptures. The first was historical: for the first time the number of drinking water service units dropped (a reduction of 20 per cent in six years). The second rupture was organizational; the actual practices of drinking water provision were very different according to the type of organization involved. In the case of communes or syndicates, their involvement is often limited to either production or distribution, although they may also do both. In the case of communities, they have an integral responsibility for water services, from extraction at the source to metered distribution. Always in the case of communities, all decisions ranging from the conservation and improvement of the infrastructure or the management options to the quality and pricing of water is under a single authority: the community. This authority is inherited from the water utilities that were previously in charge of the services in its jurisdiction, and this includes the ownership of all their assets, and responsibility for existing debts, contracts and staff.

Managing drinking water services: Rules and actors

Like in the case of the organizational arrangements, the actual management of drinking water services in France is also very diverse. First because today management of water services is subject to different rules according to their size, and second because public and private management are both eligible for any service. We consider each of these aspects in the following sections.

The balanced budget rule

Historically, the number of communes that are authorized to finance their water services through taxation has tended to fall since the late 1960s. Today, the communes with less than 3000 inhabitants and the EPCIs where each of the commune members has less than 3000 inhabitants are still free from the constraints imposed by the balanced budget rules. These rules were established in 1995 and require, with few exceptions, that water utilities balance their books without contributions from/to the communes' general budget – that is, balancing their spending with revenue received from customers, whether the management of the services is public or private.[3] A total of 7015, or 89 per cent of, communal water services have less than 3000 inhabitants, representing 3.8 million people, or about 19 per cent of the population supplied by a communal service, and they are exempt from the new rules. The rest of the communes have to comply with the new rules.[4]

Table 11.3 shows that there are 2742 EPCIs, of which 2378 are SIVUs that are not subject to the balanced budget rule, irrespective of their type of management (public or delegated). They correspond to 17,723 communes and nearly 8.5 million inhabitants. If we consider together the communal and inter-communal utilities, a total of 9757 units or more than 80 per cent are not subject to the balanced budget rule. Jointly, they correspond to 24,738 communes with more than 12.3 million people, around one French person out of five. In other words, less than 20 per cent of the water utilities have to balance their spending through revenues paid by their customers, irrespective of their type of management.

A close examination of the EPCIs involved shows that nearly 81 per cent of the SIVUs are exempt from balancing their WSS accounts, while, in contrast, half of the rural communities are subject to the balanced budget rule. By definition, all of the agglomeration and urban communities are subject to the rule, disregarding the type of management they have chosen.

An important consequence of the reorganization of drinking water services in communities is that the choice of management type is not neutral: the drinking water services previously exempted from full-cost charges are now offset by the category 'grand' services, where tax-financed services, even partially, have become an exception.

Delegation of drinking water services: A public–private compromise

Communes and EPCIs can both directly manage their services (local public management, or LPM) or delegate it to a private operator (delegation of public services, or DSP). The

Table 11.3 *Distribution of the inter-communal cooperation organisms (EPCIs) according to the balanced budget regulation (France)*

Type of authority	*Population*						*Balanced-budget rule status*	
	<400	*400–1000*	*1000–3000*	*3000–10,000*	*10,000–20,000*	*20,000–50,000*	*Total not subject*	*Total subject*
Communities of communes	2	6	23	70	6	0	107	100
Communes	6	25	184	776	85	0	1076	1241
Population	533	4896	45,675	404,881	67,945	0	523,930	2,261,643
SIVU	201	484	908	693	86	6	2378	570
Communes	454	1462	4124	6580	1761	256	14,637	6584
Population	54,418	326,238	1,707,126	3,571,896	1,104,856	170,306	6,934,840	13,377,647
SIVOM	10	26	110	91	17	3	257	117
Communes	28	84	560	968	291	79	2010	1645
Population	2481	18,012	210,297	482,991	219,217	65,211	998,209	3,309,212
Total EPCIs							2742	787

Source: Ministry of Home Affairs (2007)

DSP is the norm, and it concerns 75 per cent of the population and of the volume of water supplied, while over one commune out of two are served by private operators. In terms of water utilities, the private operators control 4790 units (BIPE–FPEE, 2006) compared to more than 7000 in the hands of LMP, which are mostly communal and rural units, but also include community and urban cases, such as Strasbourg's CU, which serves drinking water to 15 million people.

To understand French drinking water management, DSP should not be equated with privatization, but rather considered as a compromise between the public and private sectors. In fact, DSP is a very widely practised leasing contract and it relies on public financing. The private leaseholder is 'delegated' the operation of the services on behalf of a commune or an EPCI, on average for an 11-year period. The private operators bill customers for the total charges and retain an amount covering their exploitation costs and profit margins, as defined in their lease contracts. They pass back to the delegating authorities any investment costs that they claim having financed. Today, funding for French water services is still overwhelmingly public, and private funding accounts for only 12 per cent of the investment destined to WSS services (see Table 11.4).

Table 11.4 *Finance for drinking water and sewerage services (France)*

	Finance for services (million Euros)			*Percentage*
	Drinking water	*Sewerage*	*Total*	
Agents			576	12
Local authorities – service budget	Unknown		2370	49.4
Local authorities – general budget	100	41	141	2.9
Water agencies	106	1054	1161	24.0
Départements	122	245	367	7.6
State	72	49	121	2.5
Regions	0	64	64	1.3
Total			4800	100

Note: = via the National Fund for the Development of Water Sources (FNDAE), which is now closed.
Source: BIPE–FPEE (2006)

The highest financial contribution comes from the local authorities (nearly 50 per cent), while the rest is provided by the 95 *départements* in the case of drinking water and by the water agencies for sewerage. Two types of financing can be observed: revenues from service charges and subsidies. The funding provided by the water agencies, the central government and, to a certain extent, the local authorities is mainly derived from revenues. The cost of subsidies is spread across the *départements* (122 million Euros) and the local authorities (100 million Euros) in the case of drinking water, while for sewerage it is distributed among the *départements,* the 21 regions and local authorities.

Leasing out management is radically different from privatization because it is an agreement over basic infrastructure that is public property and relies upon public financial investment, and it is a system that has even proved to be compatible with fiscal finance, and also works in small utilities that serve less than 3000 inhabitants.

A PROSPECTIVE ANALYSIS OF WATER AND SANITATION SERVICES

The relative market for DSP/LPM has not changed much over the last 15 years, despite the institutional evolution of water services following the Chevènement Law in 1999. This is surprising. In fact, community services should, in terms of harmonization, provide the population with quality and value for money WSS, notably by unifying the types of management. But the reality is different: the transitory period that lapses between the creation of the community and the achievement of equality goals in the delivery of public services often turns out to be longer than expected. When a community is established and takes on responsibility for drinking water services, it inherits the variety of service management arrangements that were in place before. The DSP contracts are not precarious and the community becomes tied to them until they run out. It is true that there are other options available for the communities, such as in the case of the CU created in Nantes, which deliberately opts to maintain a double system of management across its territory. In fact, although the evolution of WSS shows no noticeable modifications in terms service management choices – historical operators are confirmed in 90 per cent of the cases (Pezon and Bonnet, 2006) – there are several factors that invite us to entertain the hypothesis that this stability may only be apparent.

Evolution towards 'départementalisation'

We should first observe that the drop in the number of water utilities has also brought with it an increase in the number of mixed syndicates capable of delivering drinking water services.[5] These mixed syndicates – comprising communes and/or EPCIs and *départements* – are, in fact, contributing to a more general movement of *départementalisation* of rural water services (Grandgirard, 2007). This movement is not exclusive only to the responsibility for drinking water: the number of mixed syndicates has literally exploded since 1999, although the other types of syndicates have seen a total reduction.

Table 11.5 *Evolution of the number of mixed syndicates (1962 to 2006) (France)*

	1972	*1980*	*1988*	*1992*	*1999*	*2007*
Mixed syndicates	153	439	750	975[a]	1454	2749
SIVU	9289	11,664	11,967	14,596	14,885	12,149
SIVOM	1243	1980	2076	2478	2165	1501

Note: [a] = 1993.
Source: Ministry of Home Affairs (2007)

One possible effect from the double movement of *départementalisation* of rural services and organization of urban community services could be improved sector regulation thanks to the development of sufficient management capabilities at the *département* and community scales. Achieving critical size and management capability places communities and mixed syndicates in a good position to either have better control over the running of DSP contracts or have more leverage in their negotiations with private operators at the end of their contracts as the possibility of changing the services to LPM management becomes more credible. So, should we expect a kind of revolution in the management arrangements for drinking water services? If the majority of communities were to move to the same type of management for drinking water, nothing at the present time indicates what their preference would be, whether LPM or DSP. Two factors make prediction particularly difficult. The first is due to the uncertain character of a triggering element: up until 1999, the share of communes adopting LPM or DSP had experienced changes owing to the implementation of the balanced budget rule. In addition, as long as the communes were free to determine whether they would follow the rule or not, they had opted for LPM, which allowed them to reduce tariffs below cost for the services delivered. The expansion of the DSP option has progressively reduced the number of communes that could apply such tariff reductions. The passing of the 1999 Chevènement Law did not modify the rules governing the management options (LPM or DPS); but it has changed the configuration of the utilities through the reorganization of the communes, which makes it difficult to anticipate future decisions about the type of management preferred. A second factor of uncertainty is due to the fundamentally political nature of decision-making processes in the councils of French *départements* and communities. Elected public officers are now showing leadership in an area that was previously the preserve of technicians who used to carry out the work under syndicates. If the recent evolution of water prices continues, after decades of being already very high, they could regain a new impetus in water services that previously did not have budget constraints, but that in the current conditions have to adjust their prices to balance their costs. Similarly, although today the political debate about management options (LPM versus DSP) is less prominent, it may gain a new momentum in urban areas, as suggested by the commitment expressed recently by Bertrand Delanoë, Paris's re-elected mayor, to re-municipalize WSS in the capital on 31 December 2009 when the current delegation contracts expire. If this change were to happen in the French capital, it may, indeed, have national, if not even global, significance for WSS policies.

Towards the integration of water and sanitation services

In the majority of cases, the current drive to integrate the responsibilities for drinking water and sewerage to better manage the water cycle is leading, in practice, to the transfer of these responsibilities to the newly created supra-local communities. Thus, in 2007, 142 rural communities of communes (CC) and 76 communities of agglomeration (CAs, for small towns and villages) have followed the example of 14 large urban communities (CUs), where the communes have transferred both responsibilities, although in 653 CC (27 per cent) and 129 CAs (76 per cent) they have opted to transfer just one of the services: either drinking water or sewerage.

In the cases where communes opt for transferring just one of the responsibilities, they generally prefer to retain sewerage rather than drinking water. Apart from the cases where communes receive fiscal incentives to transfer the management of sewerage services to the community, there exist three elements that explain this trend. First, the *départements* are not as deeply involved in sanitation as they are in drinking water services. In the case of sanitation, the financial aspects are cared for by the six river basin agencies. At the local level, this situation makes the transfer of sanitation services to the community more vital, as communities are more likely to reach a minimal critical size. A second explanatory factor is the existence of efficient mixed drinking water syndicates associating communes and *départements*. In these cases, the communes frequently want to avoid disrupting the existing organization by transferring their responsibility to a community (enforcement of the substitution principle).[6] Lastly, the majority of sewerage services operate under PLM and it is easier to reorganize these services at the community level if they are run under the same type of management.

In the case of sanitation services, their organization at the community level helps to keep them under PLM management. In this regard, the experience of drinking water services is instructive. These services were generally under PLM management until the end of the 1960s, but have suffered difficulties in managing operations at the inter-communal scale. By becoming syndicates, essentially for technical reasons, the utilities have distanced themselves from the local centres of political decision, and the elected politicians have reacted to the relative loss of power they experience by 'externalizing' their responsibility for WSS through DSP. The communities are today the major political actors at the local level. Their size and their capability allow them to maintain essential services such as sanitation under public control. However, a strong uncertainty exists as the sanitation sector has not reached the same maturity as drinking water. In fact, the sewage collection network and the treatment stations are not up to standard: only 20,000 communes offer a full sanitation service. A total of 2500 communes have sewerage networks that completely or partially cover their territory but do not treat their effluent, and 14,120 communes do not have sewage collection services (IFEN, 2005). In sanitation, pipes are not as important as they are for distributing water. Opening up to the techniques of decentralized sanitation (septic tanks) gives rise to the multiplication of non-collective sanitation services, where management technologies and capabilities are new for both private operators and local authorities.

Conclusions

We have seen that the singularity of French drinking water services is largely explained, first, by territorial fragmentation and the large numbers of communes and, second, by the particular way of making private service management compatible with public finance and even with public investment. The capacities of local authorities responsible for water delivery and the degree of direct contribution from users towards financing the costs are unquestionably the key elements for analysing the organization and management of drinking water and sanitation services. The standard rural water service is the result of post-war voluntaristic action plans developed by the state in alliance with the private operators. From a management perspective, the main consequence of the technical choice made for networked water services in rural areas was to free up the services from the restricted boundaries imposed by the communes. In practice, centralization of rural water development led to the strengthening of nationwide private water companies, but also to the generalization of an unsustainable model of water delivery as most rural water services still rely on subsidies to meet their financial investment requirements (Ernst & Young, 2004). In France, the mechanisms of solidarity are still beneficial for rural users. But questions arise from the success of communities, on the one hand, and the emerging poverty trends, on the other. An increasing number of rural water services fall under the balanced budget rule as a consequence of the reorganization of communes in communities. If the technical option of networked water services was made viable through a public system of cross-subsidies from urban to rural water consumers, how much can rural water bills increase? For how long can urban–rural solidarity last when, in France today, the poverty is mostly urban?

In practice, decentralization led to networked water only in urban areas. Urban water services were developed at a time when innovative decentralized technologies for delivering water made it possible to adjust the technical perimeter of the services down to the local level. In 1884, the implementation of the 'one man, one vote' principle for local elections unquestionably helped to focus the attention of a large number of decision-makers on water services. The water services network was universalized on the basis of fiscal finance for investment in infrastructure: users began to be fully charged for the services two or three generations after connection to the network had become universal. Both in urban areas through local tax and in rural areas through a national funding mechanism, networked water services were universalized through public funding.

The French experiences of decentralization and delegation may be of some interest for the many developing countries engaged today in decentralization processes. For urban authorities, both in developed and developing countries, the continuity and further expansion of water services relies on local mechanisms of solidarity. In developing countries, there is a need to establish mechanisms based on solidarity between long-established users and the large population still deprived of a safe and reliable access to water. Most of the un-served population live in peri-urban areas, and the responsibility for providing WSS in those areas is often shared by several municipalities whose political interests may or may not coincide. At the scale of urban agglomerations, the lack of correspondence between political jurisdiction and the actual water needs of the people leads to territorial exclusion. The inability of neighbouring urban communes to cooperate reduces their managerial capacity and paves the way for the delegation of WSS to private operators. However, in the rural areas of developing countries, networked water services are unlikely to be developed unless central governments show determination and adopt clear policies to this end. Rural water services require decentralized techniques to deliver water that could be community managed and financed. Interestingly, this is happening in France for sanitation services, where rural communes are left on their own and users are now charged on a full-cost recovery principle: networked sewage collection is left aside as an unsustainable option and decentralized techniques are developed at the family level, under municipal regulation.

NOTES

1 For information for the period prior to the Chevènement Law, see Régismanset (1897); Fayolle (1908); Leydet (1936); Singer (1956); Hourticq (1959); Verdun (1961); Bloch-Lainé (1962); Bourjol (1963, 1975, 1993); Trorial and Astier (1966); Bèze (1968); Bernard (1969); Roussillon (1972); Terrazzoni (1975); Maurice (1976); Le Lamer (1986); Novarina and Martin (1988); and Perrin (1994).

2 This is borrowed from Philippe Dallier, Senate representative for Seine Saint Denis, in his *Rapport d'Information au Sénat*, no 193 (2005–2006), 1 February 2006.

3 According to the latest Accounting Application Instruction (M49), 1994.

4 Law no 88-13, 5 January 1988. Case law 9 April 1999, commune of Bandol versus two subscribers, Conseil d'Etat. See Pezon (2002).

5 In 2004, 4 per cent of the communes were regrouped into EPCI mixed syndicates, representing 3 per cent of the population. By 2007, the figures had increased to 7 per cent of the communes and 11 per cent of the population, respectively.

6 According to the substitution principle, when a commune that long ago had transferred its drinking water production responsibility to a SIVU now adhered to a community, the exclusive responsibility for drinking water services passes to the community.

REFERENCES

Allain, L. (1895) 'Note sur un procédé rapide et simple de stérilisation à froid des eaux de rivière destinée à la boisson', *Bulletin de la Société Scientifique Industrielle de Marseille*, 2nd quarter

Baudant, A. (1980) *Pont-à-Mousson (1918–1939) stratégies industrielles d'une dynastie lorraine*, PhD thesis in history, Publications de la Sorbonne, Paris

Baudin, L. (1907) 'Comment une ville défend ses eaux de source', *Revue Pratique d'Hygiène Municipale*, January

Bernard, P. (1969) *Le Grand Tournant des Communes de France: Des Communautés Nouvelles à l'Épreuve de l'Équipement*, Librairie Armand Colin, Coll. Science Administrative, Paris

Bèze, F. (1968) *Une formule de regroupement communal: Le SIVOM*, PhD thesis in law, Université de Toulouse, Toulouse

BIPE–FPEE (Fédération Professionnelle des Entreprises de l'Eau) (2006) *Les services collectifs d'eau et d'assainissement en France: Données économiques, sociales et techniques*, BIPE–FPEE, Paris

Bloch-Lainé, F. (1962) 'Pour une réforme de l'administration économique', *Revue Économique*, no 6, pp859–885

Bourjol, M. (1963) *Les Districts Urbains*, Berger-Levrault, Paris

Bourjol, M. (1975) *La Réforme Municipale*, Berger-Levrault, Paris

Bourjol, M. (ed) (1993) *Intercommunalité et Coopération Intercommunale*, Librairie Générale de Droit et de Jurisprudence, Paris

Buffet, B. and R. Evrard (1950) *L'Eau Potable à Travers les Âges*, Editions Soledi, Liege

Burel, J. (1912) *La régie directe considérée du point de vue de l'hygiène dans les villes, la question à Lyon*, PhD thesis in law, Editions A. Rousseau, Paris

Cahen, E. (1882) 'Moyen de se procurer partout de l'eau alimentaire de qualité parfaite et en quantité illimitée par le système Rouby', *Notice sur les Sources Artificielles*, November, Paris

Canneva, G. (2007) *Vers un système d'information des services d'eau et d'assainissement*, Report for the Medd, Laboratoire Gestion de l'Eau et de l'Assainissement, Montpellier

Coudevylle, A. (1983) *Les Concessions de Services Publics des Collectivités Locales*, Sirey, Paris

Courmont, J. and L. Lacomme (1907) 'La stérilisation par l'ozone des eaux urbaines', *Hygiène Générale et Appliquée*, November

Covo-Dahan, P. (1980) *Sous-traitance d'un service public communal. Stratégie d'implantation sur le marché de l'eau*, PhD thesis in economics, Université Paris-Dauphine, Paris

De Passy (1880) *Communication sur la filtration naturelle de l'eau dans les villes*, Stenographic copy, International Congress of Civil Engineering, 5–14 August 1878, Paris

Debauve, A. (1906) *Assainissement des Villes:Distributions d'Eau*, vol 3, H. Dunod and E. Pinat, Paris

Delamarre, A., F. Auriac, F. Durand-Dastès and P. Brossier (1992) *Les Services de Réseaux en France. Intercommunalité et Mode de Gestion*, Groupement d'Intérêt Public (GIP) – Réseau d'Etude des Changements dans les Localisations et les Unités Spatiales (RECLUS), Montpellier

Ernst & Young (2004) *Etude relative au calcul de la récupération des coûts des services liés à l'utilisation de l'eau pour les districts hydrographiques français*, Report for the Ministry of Environment and Sustainable Development, Paris

Fayolle, L. (1908) *Les Syndicats de Communes dans leurs Applications Pratiques*, Editions A. Rousseau, Paris

Frick, P. (1919) *Considérations sur l'Établissement des Projets de Distribution d'Eau Potable dans les Communes*, Dunod, Paris

Gaultier, L. (1915) *Stérilisation des eaux par les ultra-violets applicables aux armées en campagne et aux agglomérations*, Communication to the Paris section of the Association Générale des Hygiéniste et Techniciens Municipaux, 8 June 1915, Paris

Givaudan, A. (1978) *La Question Communale*, Editions de la RPP, Paris

Goubert, J.-P. (1984) 'La France s'équipe: Les réseaux d'eau et d'assainissement 1850–1950', *Les Annales de la Recherche Urbaine*, nos 23–24

Goubert, J.-P. (1987) *La Conquête de l'Eau: L'Avènement de la Santé à l'Âge Industriel*, Hachette Pluriel, Paris

Grandgirard, A. (2007) *De la gestion intégrée comme doctrine à l'intégration comme défi de gestion*, PhD thesis in management sciences, Ecole des Mines de Paris, Paris

Guichard, P. (1894) *L'Eau dans l'Industrie – Purification, Filtration, Stérilisation*, J.-B. Baillière et Fils, Paris

Guillerme, A. (1984) 'Capter, clarifier, transporter l'eau en France (1800–1850)', *Les Annales de la Recherche Urbaine*, nos 23–24

Guillerme, A. (1986) 'L'émergence du concept de réseau 1820–1830', *Groupe Réseaux, Cahier*, no 5, July

Guillerme, A. (1991) 'Réseau: Genèse d'une catégorie dans la pensée de l'ingénieur sous la Restauration', *Flux*, no 6, October–December

Hourticq, J. (1959) 'Le nouveau statut des syndicats de communes', *Départements et Communes*, April, p112

IFEN (French Institute for the Environment) (2005) 'Les collectivités locales et l'environnement', Volet Eau, La gestion de l'eau potable en France en 2001, Survey results, *Collection Etudes et Travaux*, no 44, IFEN, Paris

Le Lamer, C. (1986) 'L'évolution de la gestion communale en France depuis 20 ans. Fausses pistes et vrais problèmes', *Annuaire des Collectivités Locales*, CNRS, GRAL, Paris

Leydet, V. (1936) *Le Syndicat de Communes*, Librairie Technique et Économique, Paris

Loosdregt, H. B. (1990) 'Services publics locaux, l'exemple de l'eau', *Actualité Juridique-Droit Administratif*, vol 11, 20 November, pp768–778

Loriferne H. (ed) (1987) *40 Ans de Politique de l'Eau en France*, Economica, Paris

Marchadier, A. L. and H. Guinedeau (1910) *Projet d'épuration d'une eau de rivière destinée à la consommation publique*, Mans Water Services, Mans

Maurice, R. (1976) *Le Syndicat de Communes*, Editions Masson, Paris

Ministry of Agriculture, General Direction of Rural Engineering and Agricultural Hydraulics (1959) *Trois Enquêtes sur les Services Publics Ruraux en France*, vol 1, 'Inventaire des distributions rurales d'eau potable en France au 1er janvier 1954', Imprimerie Nationale, Paris

Ministry of Home Affairs, General Direction of Local Collectivities (2007) *Database of public institutions for inter-communal cooperation (BANATIC) Paris*, Ministry of Home Affairs, Paris

Monsarrat, G. (1920) *Contrats et Concessions des Communes et des Établissements Communaux de Bienfaisance*, Bibliothèque Municipale et Rurale, Paris

Murard, L. and P. Zylberman (1996) *L'Hygiène dans la République. La Santé Publique en France ou l'Utopie Contrariée 1870–1918*, Fayard, Paris

Novarina, G. and S. Martin (1988) 'La décentralisation', *Syros Alternatives, Décentralisation et Intercommunalité*, vol 11

Perrin, B. (1994) *La Coopération Intercommunale: Bilan et Perspectives*, Berger Levrault, Paris

Pezon, C. (2000) *Le Service d'Eau Potable en France de 1850 à 1995*, CNAM, Paris

Pezon, C. (2002) 'La dérégulation discrète de la distribution d'eau potable en France et l'émergence d'un nouvel acteur collectif, les abonnés', *Flux*, no 48–49

Pezon, C. (2003) 'Water supply regulation in France from 1848 to 2001: A jurisprudence based analysis', Paper presented to the Annual Conference of the International Society for New Institutional Economics, 11–13 September, Budapest, Hungary

Pezon, C. (2006a) 'The public private partnership French model for water services management: Genesis and key factors of success', *Urban Water Conflicts Working Series, Phase VI*, UNESCO–IHP, Paris, pp25–38

Pezon, C. (ed) (2006b) *Intercommunalité et durabilité des services d'eau potable et d'assainissement en France, en Italie et au Portugal*, Research Report, Ministry of Environment and Sustainable Development, Programme Politiques Territoriales et Développement Durable, Paris

Pezon, C. (2007) 'The role of users' cases in drinking water services development and regulation in France: A historical perspective', *Utilities Policy*, vol 15, no 2, pp110–120

Pezon, C. (2009) 'Public–private partnership in courts: The rise and fall of concessions to supply drinking water in France (1875–1928)', in B. Barraqué and I. Vlachos (eds) *Urban Water Conflicts*, UNESCO, Paris (in press)

Pezon, C. and F. Bonnet (2006) *Déroulement des procédures de délégation de services publics d'eau et d'assainissement. Analyses des procédures 2004*, Report for the Ministry of the Environment, Paris

Pezon, C. and S. Petitet (2004) 'Histoire de l'intercommunalité en France (1890–1999): La distribution d'eau potable en question', Communication at the Colloquium Les Territoires de l'Eau, Réseau Développement Durable et Territoires Fragiles, Université d'Artois, 26 March, Artois

Régismanset, H. (1897) *Les Syndicats de Communes: Étude sur la Loi du 22 Mars 1890*, Editions A. Rousseau, Paris

Roussillon, H. (1972) *Les Structures Territoriales des Communes: Réformes et Perspectives d'Avenir*, Bibliothèque de Science Administrative, Librairie Générale de Droit et de Jurisprudence, Paris

Singer, J. (1956) *L'Intervention des Collectivités Locales en Matière Économique*, Editions Aframpe, Paris

Terrazzoni, A. (1975) *Les fusions et regroupements de communes dans l'arrondissement d'Avranches*, PhD thesis in administrative sciences, University of Rennes, Rennes

Trorial, J. and H. Astier (1966) 'La réforme communale et l'aménagement du territoire; où en sont les regroupements de communes', *Le Moniteur des Travaux Publics et du Bâtiment*, no 15, pp17–34

Vanneufville, M. (1958) 'L'alimentation en eau de Paris', *L'Eau*, no 9, p201

Verdun, L. G. (1961) *Le groupement de communes en France*, PhD thesis in law, Université de Bordeaux, Imprimerie Bière, Bordeaux

Viguier, J. (1992) *Les Régies des Collectivités Locales*, Economica, Paris

Villard, G. (1885) *Etude d'un service d'eau pour la ville de Lyon. Principes généraux d'alimentation des villes en eau potable suivi de l'étude d'un tarif rationnel des eaux ménagères*, Paris

Villard G. (1887) *L'eau dans les villes. Ses fonctions diverses (alimentation, hygiène, industrie, distribution de force motrice à domicile)*, Ville de Lyon

12

The State of Urban Water Supply and Sanitation in Spain: Issues, Debates and Conflicts

David Saurí, Jorge Olcina and Antonio Rico

Introduction

In this chapter we will examine the current situation of water supply and sanitation services in Spain, with a special emphasis on the Mediterranean coast where the debate on water is more acute and where privatization processes are rapidly expanding in a context also of rapid urban and tourist development. As stated in the Introduction to this volume, the privatization of water supply and sanitation may affect universal rights of access to these services and introduce elements of inequity along income, gender, ethnic or market positional divides. In particular, the provision of water in sufficient quantity and quality and the safe disposal of wastewater constitute a human and social right as stated in the United Nations Millennium Development Goals (UN, 2000, 2002), as well as in the *European Declaration for a New Water Culture* signed in 2005 (EUWATER, 2005).

According to some sources, Spain occupies one of the first positions of the world in per capita water consumption (more than 400 litres per person per day, or lpd). This figure, however, may be misleading since it includes all of the water used in irrigated agriculture, which roughly constitutes 80 per cent of the total demand. Strictly speaking, urban water demand, on average, is usually situated below 200lpd and, in some cities such as Barcelona, it may be as low as 110lpd, one of the lowest figures for European cities. Having said this, it is also true that recent changes in the process of urbanization and the increasing urban sprawl in the peripheries of Madrid, Barcelona, Seville, Valencia and other Spanish cities has led to an expansion of outdoor water uses (irrigated gardens, swimming pools, etc.) that may be pushing the level of water consumption closer to North American or Australian figures (Domene and Saurí, 2006).

During the last decades, urban water in Spain has been subject to the pressures of privatization for two main reasons: first, the need of local councils to capture new financial resources for their exhausted coffers, and, second, the growing interest of private capital in entering new areas previously insulated from the markets in a process that David Harvey has called 'accumulation by dispossession' (Harvey, 2003). Privatization is part and parcel of wider political and economic processes or 'systemic conditions' surrounding water that may interact with physical constraints to produce what we could define as 'hybrid scarcities'. In the Mediterranean (as well as in other areas characterized by precipitation constraints, but also endowed with other resources), water is hardly a natural element, but is, rather, a resource appropriated at the source, diverted, transformed and (more so in modern times) also degraded. To use popular parlance, 'water flows uphill to money'. In other more sophisticated theoretical accounts, 'flows of water are flows of power' (Swyngedouw, 2004).

Water supply and sanitation in Spain are still largely dominated by 'technological fixes' – that is, water is brought to cities from reservoirs, intra- and inter-basin transfers, and, since very recently, by desalinization plants as well. Once the basic drinking source for many urban areas, groundwater has lost its primacy due, in many cases, to over pumping and pollution, especially near the coast where aquifers are affected by saltwater intrusion. Sanitation services also depend upon extensive sewerage networks that empty into fluvial courses or into the sea, sometimes with the sewage still untreated (see below). In recent years, demand management alternatives to curb consumption have been implemented in several cities. For the most part, these policies are still in their infancy with, possibly, the exception of economic instruments such as prices and taxes (Saurí, 2003). Generally speaking, the principles of universality and equity are followed in the provision of water and sanitation, but always at a price that varies substantially (roughly between 0.50 and 2.50 Euros per cubic metre in 2007) depending upon the city. Usually (although not always), urban water is more expensive where it is supplied by private companies, such as in the cases of Barcelona or Murcia. In some areas, however, the quality of drinking water, while being potable in sanitary terms, may be so deficient according to the users that they tend to buy bottled water at a cost considerably higher than that of tap water.

In the chapter we present, first, a general description of water supply and sanitation services in Spanish cities. Second, we discuss one specific case of social conflict related to urban water: the so-called 'Barcelona water war of the 1990s', in which a number of families (close to 90,000) refused to pay what they saw as an unfair rise in water taxation. Third, we focus on wastewater treatment and water reuse as particularly important aspects of the water cycle not yet well solved in Spain. The sluggishness characterizing Spanish wastewater policies contrasts with the dynamism of urban water supplies, now already in the beginning of a new era based on desalinization. In the conclusion we emphasize the fact that urban water supply and sanitation services in Spain remain strongly subject to the dynamics of urban growth.

Water supply and sanitation services in Spain: A brief account

As in many other countries, the provision of water and sanitation services in Spain remains under the responsibility of local councils (some 8100 in the whole country at the beginning of the 21st century). Water supply and sanitation are essentially a local endeavour, although this may change in the case of large agglomerations (e.g. Madrid and Barcelona) where supra-local or metropolitan authorities (the *Comunidad de Madrid*, or the Metropolitan Area of Barcelona) have been created to manage basic services such as water, public transportation or waste disposal.

Historically, the provision of water and sanitation services in Spanish cities, and especially in the big capitals, faced a similar challenge to that of large European and North American urban centres, which needed to accommodate health and sanitary concerns with urban growth and policy reforms. These were the times of the 'bacteriological city' (Gandy, 2004) in which local councils ensured the provision of water supply and extended the sewerage networks, sometimes as in Barcelona, after dramatic outbursts of infectious diseases (especially cholera). These reformist policies followed the philosophy of *tout a l'ëgout*, made possible thanks to the connections between the new buildings and the new sewerage networks in the expanding areas (*ensanches*) of the main Spanish cities (Gómez-Ordoñez, 1987) in the mid- to late 19th century. While all of these efforts took the spectre of disease away from the cities, the resulting (untreated) wastewater polluted rivers and aquifers, especially since the mid-20th century when urban growth accelerated in Madrid, Barcelona, Valencia, Seville and other large cities.

As already said, local councils were responsible for water provision and sanitation. Nevertheless, in some cases these services had been transferred to private companies under a concessionary regime designed after the French model. The most important of such cases is Barcelona, where water supply has been managed by private operators for 140 years, while sewerage has remained in public hands to date. This city therefore represents a peculiar case of very early privatization of urban water supply. The Compagnie des Eaux de Barcelone was created in Liège with Belgian and French capital in 1867. In 1882, the Compagnie was taken over by French shareholders and changed its name to Sociedad General de Aguas de Barcelona (SGAB). Further expansion occurred in 1897 when SGAB acquired the Empresa Concesionaria de Aguas Subterráneas del Río Llobregat, which pumped water from aquifers in the Llobregat River (Masjuán et al, 2006). During the 20th century, the company suffered various changes in ownership (including a brief public tenure during the Spanish Civil War), and from the end of the 1970s onwards, several banks sold their shares to a new holding formed by the French company Lyonnaise des Eaux and the Spanish Savings Bank La Caixa, which currently controls about 47.5 per cent of the new company known as Aigües de Barcelona (AGBAR). AGBAR is a multi-service company with interests not only in the water sector, but also in other activities such as healthcare and quality certification. Water, however, remains by far and large the main activity of the company, which has water business interests in Spain as well as in the UK, Algeria and several Latin American countries (Chile, Colombia and Cuba) and looks for expansion in Asia and North America.

Taking advantage of the disarray affecting many municipal water companies and of the financial needs of city councils, from 1980 onwards, AGBAR began to expand its operations in Spain, gaining the concession for supplying water to an increasing number of cities (alone or in partnerships). At the end of the 1990s, the holding supplied water to more than 750 municipalities and 11 million people (some 25 per cent of the Spanish population). At the same time, and as mentioned earlier, AGBAR invested in the Latin American water market (with some drawbacks such as the recent crisis with Aguas Argentinas in Buenos Aires, where the private concession was cancelled by the Argentinean government in 2006) and also in the European (the UK company Bristol Water was acquired in 2006) and African markets.

Nevertheless, in Spain, urban water supply and sanitation is also served by public companies, the most important of which are the Canal de Isabel Segunda (providing water to more than 5 million people in Madrid) and the Empresa Metropolitana de Abastecimiento y Saneamiento de Aguas (EMASESA) in Seville (with more than 1 million customers). Contrary to the case of Barcelona, where water supply is controlled by a private company, but sewerage is in public hands, in Madrid the public company is also in charge of sewerage and wastewater treatment. Overall, in the year 2000, 49 per cent of Spanish municipalities were served by a private company; 32 per cent by a public company; 12 per cent by public–private partnerships; and the rest by other types of consortia (del Romero, 2006).

As an example of the new privatization winds blowing in the Spanish water sector, we present the situation of the so-called 'Arco Mediterráneo' (Mediterranean Fringe) – that is, the coastal autonomous communities of Catalonia, Valencia and Murcia, which stretch along the Mediterranean coastline from the French border in the north to the Andalusian coast in the south, and also include the Balearic Islands. Madrid aside, this region is the most dynamic in Spain in terms of population and economic growth. It now concentrates about one quarter of the Spanish population (but more than 40 per cent of immigrants, both Europeans and non-Europeans) and about 35 per cent of the national gross domestic product (GDP). It is also where the penetration of water privatization has gone deeper. Data for the year 2000 show that while for Spain, as a whole, municipalities served by private companies were just below 50 per cent of the total, this figure rose to 77 per cent for those municipalities located in the Mediterranean region (del Romero, 2006).

The Spanish case therefore illustrates a variety of situations with regard to water supply and sanitation. First, there is no absolute privatization or full divestiture such as in Chile or the UK; nevertheless, private companies are increasing their presence in the water sector, especially the AGBAR group. Usually, privatization has involved a relatively substantial increase in prices, which, on the average, tend to be higher in the municipalities where the service is provided by a private company than elsewhere. As an interesting issue, privatization of water services has not produced noticeable social conflicts in Spain beyond some localized (and, for the most part, ephemeral) complaints about price rises. Rather, as we will see in the next section, water conflicts have stemmed from citizen opposition to public policies, especially water taxation and large-scale hydraulic projects. In this respect it could be argued that while privatization of water supplies has proceeded somewhat quietly, there have been public water projects that have raised a great deal of social controversy, which we will illustrate below.

Urban and territorial struggles for water supply and sanitation

During the last decades, urban water supply and sanitation issues have not provoked much upheaval in Spain despite the steady increase in privatization of these services. Rather, conflicts (or at least the most serious ones) have surfaced as a consequence of public decisions related to projects directed at increasing supplies (especially large inter-basin water transfers) and also to raises in water taxes. In this sense, the 'Barcelona water war' that we present next was a popular revolt against a taxation system perceived by the users as unfair.

The Barcelona water war of the 1990s: Equity and water taxes

Barcelona and its metropolitan area have always suffered the more expensive water of Spain and also one of the worst in taste and odour. Between 1987 and 1993, the price soared from 0.59 to 1.23 Euros per cubic metre, which is an increase of 108 per cent in just six years. This steep increase was not inspired by any privatization process, as the water supply is already privatized in the city, but by a new bundle of environmental taxes aimed at financing the construction of wastewater treatment plants, regional water supply schemes, flood control works and other hydraulic infrastructure. The European Wastewater Directive of 1991 was particularly germane to this increase in water prices since most of Catalonia and, especially, the Metropolitan Area of Barcelona lacked such facilities. In fact, river pollution at the end of the 1980s was so high that, in a famous court sentence, a judge ruled that a company was not liable for discharging waste into a stream near Barcelona because the river was already so contaminated that a 'little more pollution did not matter'. These new taxes (implemented almost overnight and without prior information to consumers) came to represent more than 55 per cent of the total water bill, whereas previously they constituted about 40 per cent, on the average. This sudden increase mobilized urban community groups in several districts of Barcelona, as well as in nearby cities. At the end of 1991 these groups agreed on a particular strategy of opposition: they would pay the part of the bill corresponding to the cost of the service (i.e. the portion charged by the water supply company) and put the money in a special bank account, but they will not pay the part corresponding to taxes. This line of action was chosen in order to avoid termination of the service for non-payment, which, for legal matters, could not proceed in the case of non-payment of taxes. At the end of the 1990s, almost 90,000 families in Barcelona and the Metropolitan Area followed this particular form of fiscal disobedience. The revolt was particularly strong in the working class neighbourhoods, who had also been important foci of struggles during the Franco dictatorship. These sectors reacted against the perceived unfairness of the impact of the water tax with respect to other areas in Catalonia and Spain, and especially with respect to the industrial sector, whose taxes were around fourfold lower than those charged to domestic consumers despite the fact that industries are thought to contribute much more to water pollution. The unfairness was also perceived with regard to the impact of the tax on some of the well-off sectors of the Barcelona Metropolitan Area. For instance, urban growth in some of the affluent coastal towns north of Barcelona had to face the problem of the important and negative effects of urbanization on the local hydrological cycle. The presence of many streams prone to flash flooding (and flash flood episodes have increased as a consequence of urbanization) forced towns to build costly flood defence works that were financed with the water taxes paid by all. Thus, in some respect, low-income families perceived that they could be financing, with their taxes, the flood protection works demanded by the higher-income groups populating the hills near the sea north of Barcelona. Finally, another equity issue raised by the community groups was related to the block-rate structure of the water bill. Because of its progressivity (the more you spend, the more you pay per unit of water consumed), this type of tariff tended to penalize large households even though their per capita consumption might be low. In 1995, four families went to court denouncing what they saw as an unfair treatment of large families, and two years later, the Higher Court of Catalonia ruled that prices should be modified to take into account household size. At the other end of the spectrum, inequity issues appear when small or single-member families pay higher prices per cubic metre of water because of a fixed service fee. This variant of inequity appears to be more frequent in Spanish cities when water services are privatized and companies introduce or raise these service fees (Saurí et al, 1998).

After almost a decade of struggling, the end result of the negotiation process between the Federation of Neighbourhood Community Groups and the regional government was an important change in water planning and management in Catalonia. A new Water Taxation Law was passed in 1999 and the Catalonian Water Agency was created. For the first time, executive powers to act on the whole hydrological cycle were concentrated in a single agency located within the regional Department of the Environment. As to the law, its most relevant feature was the reorganization and simplification of water taxes into a single tax called *Cànon de l'Aigua* (Water Tax). An agreement was reached with the Federation of Neighbourhood Community Groups by which this new tax was to be applied according to a progressive tariff structure starting from a minimum of 100 litres per person per day.

After the year 2000, water prices and taxes in many Spanish cities were expected to rise as a consequence of applying the full-cost recovery principle emanating from the 2000 Water Framework Directive (EU, 2000). However, despite calls from the Spanish Ministry of the Environment, city councils and regional governments (those with executive powers to do so) are reluctant to raise prices and, especially, taxes because they fear the potential for social opposition. Thus, a recent poll showed that about half of the people interviewed argued that water was a public good and that it should, therefore, be provided at no cost to consumers. In fact, this was until very recently the situation in many Spanish municipalities where water was never paid for. Nevertheless, this situation may be changing in the areas of strongest urban growth where city councils are not only expanding urbanization at the fastest rates in Europe, but also privatizing services such as water.

THE NEXT CHALLENGE: THE USE OF TREATED WASTEWATER

In 1991 the European Commission approved a new directive on wastewater. An assessment of the degree of compliance by member states undertaken in 2001 showed depressing results in this respect. Only Finland, Sweden and Denmark had responded to their commitments and treated their urban wastewaters with advanced methods. Spain, on the contrary, was far from fulfilling its obligations. In this country, where wastewater treatment is in the hands of regional governments (*comunidades autónomas*), in 2003 only 55 per cent of urban effluents were treated, and about 60 per cent of wastewater treated was concentrated in Madrid, Barcelona and Valencia. In 2006, Spain treated 3.375 cubic hectometres (hm^3) of wastewater, of which about 450hm^3 were reused (14 per cent of the total). According to the National Plan for Water Quality (2007–2015), by 2015 some 80 per cent of treated wastewater should be reused. In early 2007, and despite 3.8 billion Euros of European Union funds (some 50 per cent of all investment in wastewater plants), there were still more than 800 urban centres of more than 2000 people (6.5 million people equivalent in total) that did not treat, or treated insufficiently, their wastewaters, including provincial capitals such as Burgos and Ibiza (*El País*, 2007). Table 12.1 indicates the state of wastewater treatment plants in Spain at the end of 2006.

Table 12.1 *State of wastewater treatment in Spain (2006)*

State of conformity	*Pollution load (in people equivalents)*	*Percentage of pollution load*
Acceptable	56.608.111	77
Under construction	10.140.864	14
Non-acceptable	6.516.753	9
Total	73.265.728	100

Source: European Commission (2007)

To a large extent, this is related to the paucity of investments when compared to the needs generated by urban growth. In this sense, it is worth pointing out that in Spain, urban development is a local affair, while wastewater treatment plans are the responsibility of (and must be paid for) regional governments. Therefore, urban planning in the country does not consider the wastewater implications of new developments, and very few city councils are willing to charge the estimated cost of 80 Euro cents or more per cubic metre of wastewater to urban dwellers. This situation could well worsen because of the more stringent requirements emanating from Brussels regarding wastewater treatment. Since December 2005, the European Union has declared 'sensible areas' where high concentrations of organic waste need to be eliminated by 2012. In Spain, this will affect more than 25 million people (over half the country's population) and there is little doubt that meeting this target may be unfeasible even despite a new government plan to boost wastewater treatment amounting to about 17,500 million Euros that was recently approved.

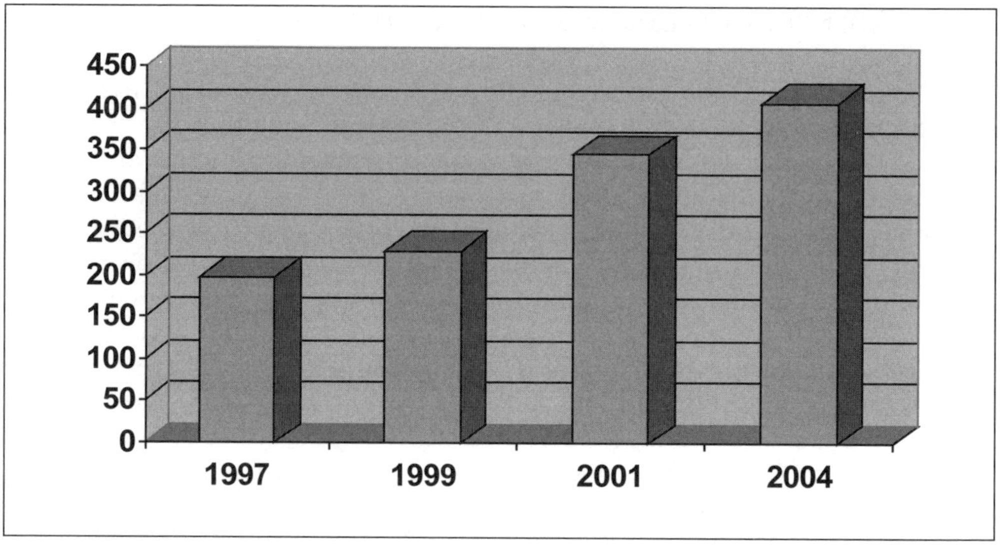

Figure 12.1 *Reused wastewater in Spain (cubic hectometres) (1997 to 2004)*

Source: Rico-Amorós et al (1998) and Iglesias (2005)

Taking into account that a significant proportion of Spanish urban wastewater is insufficiently treated or simply released untreated, it is no surprise that water reuse remains largely underdeveloped (see Figure 12.1). In part, water reuse may be hampered by the lack of interest in this issue implicit in the European norms. The 1991 Wastewater Directive does not mention water reuse beyond its environmental applications. However, treated wastewater could be a critical resource for water-stressed areas such as the Mediterranean countries. In Spain, only 14 per cent of treated urban wastewater (about 400hm^3 per year) is reused in activities such as supply for wetlands and riparian ecosystems, agricultural irrigation and, to a lesser extent, leisure activities such as the irrigation of golf courses or the watering of public parks. There are also significant differences among the *comunidades autónomas* in terms of the volumes of water reused. On the one hand, in some areas such as Valencia, Murcia, and the Canary and Balearic Islands, water reuse is certainly gaining momentum in the sphere of water planning and management. In 2005, Valencia (who leads Spain in this endeavour) reused some 226hm^3 per year of the 506hm^3 of wastewater generated. In 2008, it is foreseen that approximately 60 per cent of the treated wastewater will be subject to some kind of reuse, especially for orchard irrigation, groundwater recharge and coastal wetland conservation. At the other extreme, some regions barely make use of their treated wastewater, including Catalonia, which has serious water supply problems.

Two factors may explain the relative paucity in water reuse in Spain. First, there are health concerns about the quality of the treated wastewater, especially in areas such as Catalonia, where pollution loads may be high and contaminants persistent and difficult to remove. In this respect, the Spanish government issued in December 2007 an Executive Order (Real Decreto 1620/2007 of 7 December) regulating the uses of treated wastewater. First, reused water is banned for all uses involving drinking, personal hygiene and even recreational activities (e.g. swimming pools). Equally, the order forbids the use of treated wastewater in refrigeration towers (a key use for industrial activities) since it may contain bacteria and other pathogens that may cause respiratory diseases. Second, the price of this resource is still relatively high, about 0.30 Euros per cubic metre for agricultural irrigation and about 1 Euro per cubic metre for urban uses, compared to conventional supply sources, although it is still more competitive than desalted water. Another important issue is whether reused water is going to be considered 'private', as the Conservative party (*Partido Popular*) proposed in 1997 when it held the majority of the Spanish Parliament, or public, as the present socialist government prefers. Should reused water become private, it would become a new and welcomed source of income for the private water companies, which today are facing an interesting dilemma: while they must advocate the need for efficiency and promote savings in water consumption, they may also suffer from decreasing water sales as a consequence of these conservation campaigns. In public hands, reused water would probably go mostly to environmental or agricultural uses rather than to urban uses.

Beyond the actual impact of the Executive Order (highly criticized by the private water sector because of a supposed over-concern with health issues), the reuse of treated wastewater will probably become one of the key features of future Spanish water policy simply because, price and energy wise, as mentioned earlier, treated wastewater appears more attractive than desalted water. In this respect, an interesting debate may arise in the years to come in Spanish cities. Under the provision of actions to implement Local Agenda 21s, some city councils are enacting municipal ordinances that require the use of (minimally) treated wastewater for certain household necessities such as toilet flushing or garden irrigation. Basically, the ordinances make mandatory the installation of simple wastewater systems for individual buildings so that new pipes carrying water from baths or showers would feed toilets and gardens. While this is seen as an optimal solution in terms of costs and decentralization in decision-making, other voices point towards the logistical difficulties that may incur, such as the need for a community of neighbours in a building to name a person responsible for the management and maintenance of the system, the extra time required for making the system work, etc. To overcome these difficulties, the private water companies and some water experts are advocating a centralized system of water reuse where all aspects will be taken care of by the water companies – of course, at a cost to be paid by the final consumers. In times where the income of many water companies is dwindling (because of the decline in domestic water consumption), the alternative of supplying treated wastewater for non-potable uses may become an interesting business opportunity. Together with desalinization, the use of treated wastewater appears as one area of future expansion for private companies should more conventional sources such as dams or inter-basin transfers fail for political or environmental concerns.

Conclusions

In this chapter we have attempted to lay out the main characteristics of urban water supply and sanitation in Spain. After a brief historical account, we have focused our attention in existing trends towards privatization, especially important during the last decade along the so-called Arco Mediterráneo, or the coastal areas of eastern Spain, also known for their strong rates of urban and tourist growth. Generally speaking, the progressive privatization of urban water in Spain has not yielded any relevant social conflicts despite strong increases in price in several cases. However, many municipalities, small and large, pay very little for water and, according to some recent polls, many Spaniards do not think that a basic good such as water should be paid for. Conflicts over water have taken place mainly because of public policies, and we have presented one example in this respect with direct implications for urban life. The 'Barcelona water war' of the 1990s was a popular revolt against what was perceived as an unfair taxation system designed mostly to pay for wastewater treatment plants. Finally, we have mentioned the issue of wastewater and water reuse as another pressing area of future concern for two main reasons: first, wastewater is going to increase in certain areas because of the fast pace of urbanization; and, second, the growing costs of wastewater treatment will be passed on to consumers as a result of the application of the full-cost recovery principle introduced by the EU Water Framework Directive. This could create social conflict in certain urban areas. Meanwhile, water reuse, which could be an essential piece in the newly emerging hydro social cycle in cities, still remains almost invisible to the eyes of Spanish policy planners, although not to the eyes of certain private water companies.

References

del Romero, L. (2006) 'La privatització de la gestió de l'aigua i govern del territori a l'arc mediterrani espanyol', *Documents d'Anàlisi Geogràfica*, vol 48, pp35–59

Domene, E. and D. Saurí (2006) 'Urbanization and water consumption: Influencing factors in the Metropolitan Region of Barcelona', *Urban Studies*, vol 43, no 9, pp1605–1623

El País (2007) '800 Spanish urban areas do not comply with the Wastewater Treatment Directive', *El País*, Madrid, 19 February, p36

European Commission (2007) 'Towards sustainable water management in the European Union. First stage in the implementation of the Water Framework Directive 2000/60/EC', European Commission, Brussels

EU (European Union) (2000) *Directive of the European Parliament and of the Council 2000/60/EC Establishing a Framework for Community Action in the Field of Water Policy*, European Parliament and Council, Luxembourg, http://europa.eu.int/comm/environment/water/water–framework/index_en.html, accessed May 2008

EUWATER (2005) *European Declaration for a New Water Culture*, New Water Culture Foundation, Saragossa, Spain, www.unizar.es/fnca/euwater/docu/europeandeclaration.pdf, accessed June 2008

Gandy, M. (2004) 'Water, modernity and emancipatory urbanism', in L. Lees (ed) *The Emancipatory City? Paradoxes and Possibilities*, Sage, London, pp175–191

Gómez-Ordoñez, J. L. (1987) 'García Faria i el seu projecte de sanejament', in VVAA (ed) *El Naixement de la Infraestructura Sanitaria a la Ciutat de Barcelona, Ajuntament de Barcelona, Sèrie Salut Pública*, no 6, pp21–28

Harvey, D. (2003) *The New Imperialism*, Blackwell, Oxford

Iglesias, R. (2005) 'Escenarios existents y propuestas para el avance de la regeneración y reutilización de aguas en España', Paper presented to the Technical Meeting on the Role of Treated Wastewater in Integrated Water Resource Management. LLoret de Mar, Spain (October)

Masjuán, E., E. Domene, H. March and D. Saurí (2006) 'Water uses and water conflicts in the Metropolitan region of Barcelona (1900–2000)', Paper presented to the Fourth International Round-Table on Urban Environmental History – 19th and 20th Century, Paris, 16–18 November

Rico-Amorós, A., V. Paños-Callado, J. Olcina-Cantos and C. Baños-Castiñeira (1998) *Depuración, Desalación y Reutilización de Aguas en España*, Oikos Tau, Barcelona

Saurí, D. (2003) 'Lights and shadows of urban water demand management: The case of the Metropolitan Region of Barcelona', *European Planning Studies*, vol 11, no 3, pp233–247

Saurí, D., A. Durà-Guimerà and F. Muñoz (1998) 'Sostenibilidad y conflictos distribucionales: El caso de las tasas sobre el agua en el area metropolitana de Barcelona', Paper presented at the Congreso Ibérico sobre Panificación y Gestion de Aguas, Saragossa, Spain, 14–19 September

Swyngedouw, E. (2004) *Social Power and the Urbanization of Water*, Oxford University Press, Oxford

UN (United Nations) (2000) *Millennium Declaration*, UN, New York, www.un.org/millennium/summit.htm, accessed May 2008

UN (2002) *Key Commitments, Targets and Timetables from the Johannesburg Plan of Implementation*, World Summit on Sustainable Development, Johannesburg, www.johannesburgsummit.org/html/documents/summit_docs/2009_keyoutcomes_commitments.doc, accessed May 2008

13

Decentralized Services: The Nordic Experience

Pekka Pietilä, Maria J. Gunnarsdóttir, Peder Hjorth and Susanne Balslev Nielsen

Introduction

This chapter addresses the main features of water and sanitation services typical to the Nordic countries: Denmark, Finland, Iceland, Norway and Sweden. Despite variations in size and natural landscape, all have decentralized public administration and local authorities that play a central role in the provision of essential services, including water and sanitation. Changes in the Nordic countries are typically not radical, but occur gradually. This can be partly explained by the multi-party democratic tradition. Some administrative reforms have taken place; but the central role of local authorities has remained, or even strengthened. Water and sanitation services are considered so essential that they are to be kept under close public control while the decentralized administration structure places the responsibility for them at the local level. An external factor that has contributed to the development of the decentralized water service structure is the relatively low population density, with the exception of Denmark. But even though the water and sanitation services are, in principle, public, the private sector is not excluded – the reverse is actually true. Municipal water undertakings commonly outsource non-core activities to such an extent that the majority of operational activities are carried out by private companies.

TRADITION OF DECENTRALIZED PUBLIC ADMINISTRATION

In the Nordic countries, municipalities have traditionally been strong self-governance units at the expense of regional-level administration. The present municipal system in Norway was established as early as 1837 by the Local Government Act. In Finland, local-level administration was formalized by legislation some 30 years later. The importance of local-level administration is clearly illustrated by the share of public-sector employees working for it (see Figure 13.1). A central pillar of municipalities' self-governance is their taxation power (Henning, 2001). Local-level (municipal) administration has been arranged in a fairly uniform way – the same legislation applies and administrative structures are similar in rural municipalities and large cities. However, there are some differences in the responsibilities and relations between public administrative bodies. In Finland, the role of municipalities is relatively stronger than, for instance, in Sweden or Denmark because in Finland the regional/provincial-level administration is rather weak (Mennola, 1999, p386). So far, Finland has not introduced elected regional-level councils like Sweden, Norway and Denmark (Council of Europe, 2002). Thus, Finnish public administration consists, to a large extent, of only two tiers: state and local (municipal). This is the case in Denmark as well after a structural reform introduced in January 2007, which reduced

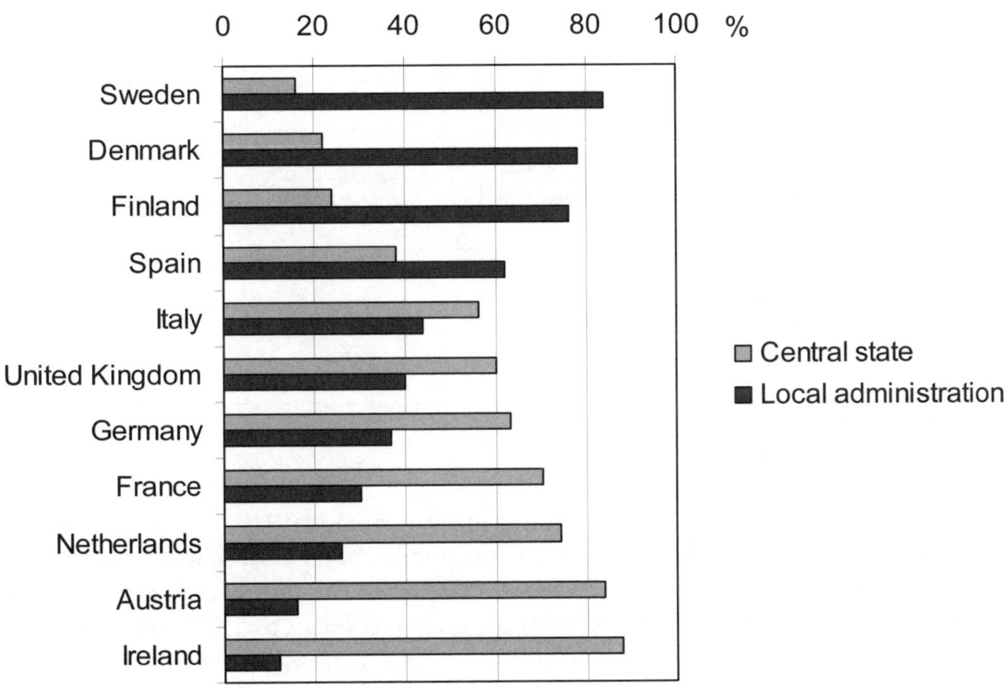

Figure 13.1 *Public-sector employees in some European countries*

Source: Dexia (2002)

the role of regional councils and made the municipalities and the state the only water authorities. Iceland also has only two levels of administration – municipal and central government – and municipalities have a strong say in managing their own affairs, and the tendency is to transfer more tasks from the central government to the municipalities.

The tradition of decentralized public administration is also a tradition of investments in public administration to ensure the democracy and 'law and order' of societies in areas of public interest. However, the organization of public administration has changed according to historical perspective. In Sweden, the number of municipalities reduced dramatically from about 2400 in 1950 to less than 300 in the 1970s. The reduction was mainly due to two nationwide municipal reforms, one in 1952 and the other in 1971. In Finland, the changes have not been as radical; the number of municipalities has reduced only from 547 in 1950 gradually to 416 in 2007. During recent years, there have been slightly more municipal mergers than previously, boosted by the financial incentives of the central government. However, a recent study established that mergers of municipalities have not brought the financial savings commonly believed (Moisio and Uusitalo, 2003). In Iceland, the number of municipalities has also reduced dramatically, especially since 1990, when the merging of municipalities started. In 1950, the number of municipalities was 229; in 1990, it was 204; and in 2006, it was down to 79. Mostly smaller municipalities are merged to enable them to better carry out their constitutional tasks; but this is done with a democratic process through general election in respective municipalities.

Table 13.1 *The Nordic countries (data as of 1 January 2007), with the exception of the right-hand column*

	Area (km^2)	*Population (millions)*	*Number of municipalities*	*Number of municipalities in 1950*
Denmark	43,000	5.4	98	1302
Finland	338,000	5.3	416	547
Iceland	103,000	0.3	79	229
Norway	324,000	4.7	431	~800
Sweden	450,000	9.1	290	~2400

Source: Authors' elaboration from different sources

In Denmark, a radical municipal reform was implemented in 1970, where the previous 1386 municipalities were reduced to 275, and the number of regional councils was reduced to 14. Another major reform took place in 2007, when the number of municipalities was further reduced to 98, and the number of regional councils to 5. The municipalities are now larger and have more responsibilities, while regions lost their influence on water management to the municipalities and the state. Despite the reduced number of municipalities, water service provision has remained rather decentralized. In 2005, there were 2623 public water suppliers, 155 of them owned by municipalities and 2468 private cooperatives. Municipal water suppliers are typically large and supply 60 per cent of the total amount of water in Denmark, whereas private cooperatives, even though numerous, supply only 40 per cent. Today, all building owners in urban areas have to connect to the local water supply system. In rural areas, there are still individual private wells in use, a total of 71,000 in 2003. The number of cooperatives and private wells is slowly decreasing due to groundwater pollution, which has caused the closure of some wells. Stricter administrative requirements have also caused some cooperatives to close down or merge with larger undertakings (DANVA, 2007).

In Norway, during the 1960s, the number of municipalities was reduced from some 800 to less than 500 through central-government forced mergers; but since then changes have been slow. Many Norwegian municipalities are small – over half have less than 5000 inhabitants, and only about ten have more than 50,000. Recently, some politicians have advocated a major reduction in the number of municipalities; but so far government policy has been that municipalities should choose for themselves whether to merge with others (ESA, 2007).

The number of municipalities in the Nordic countries has decreased over the last decades, but not at the same rate, and the change has happened during different periods in the various countries. Sweden and Denmark, and to some extent also Iceland, have undergone radical municipal reforms forced by the central government, which reduced the number of municipalities dramatically. In Norway and Finland, the reduction in the number of municipalities has been more gradual and not as dramatic. Traditionally, municipalities have played an important role in public administration in the Nordic countries, and the tendency has been to increase rather than decrease that role. The reduction in the number of municipalities does not indicate that less emphasis is put on local-level administration; it is the other way around: larger municipalities have even been given tasks that earlier were the responsibility of regional-level administration. However, at least in Finland and Sweden, the fact that municipalities have not been given the necessary financial resources to take on the new responsibilities is a growing problem.

Development of water and sanitation services

In the Nordic countries, the development of water and wastewater services has been closely linked to municipalities, whereas people in the vast sparsely populated rural areas have had to rely on their own initiative to improve their water and sanitation arrangements.

During the early 19th century, serious health problems caused major concern in large cities in England. In 1842, Edwin Chadwick produced a report on the living conditions of working-class people and stressed that clean water and proper sanitation are the keys to eliminating health problems (Goddard and Sheail, 2001). Chadwick's report describing the inadequate environmental sanitation in London also aroused concern in Sweden; as a result, the first urban water utilities were established in Stockholm and Karlskrona in 1861. In 1875, Sweden got its first Public Health Act. In 1919, the act was updated, with an emphasis on access to safe water and extension of service to areas previously uncovered.

In perspective, the long-term development of water and sanitation services in Sweden, as well as in the other Nordic countries, can be divided into the following key phases (Katko and Stenroos, 2005):

- some early private concessions simultaneously with city-managed undertakings, 1850–1880;
- establishment and expansion of water and sewerage services as municipal departments and works, 1890–1930;
- extension of systems, including water pollution control, 1950–1980;
- increasing autonomy, inter-municipal cooperation and outsourcing of non-core operations, 1990–2000.

On another count, the use of water sources – surface or groundwater – vary enormously between the Nordic countries, as can be seen in Figure 13.2. This is, to a great extent, due to varying physical conditions, but is also related to the policies of each country. During the last decades, the general policy has been to increase the use of groundwater at the expense of surface water. The reasons are mainly health related. On the one hand,

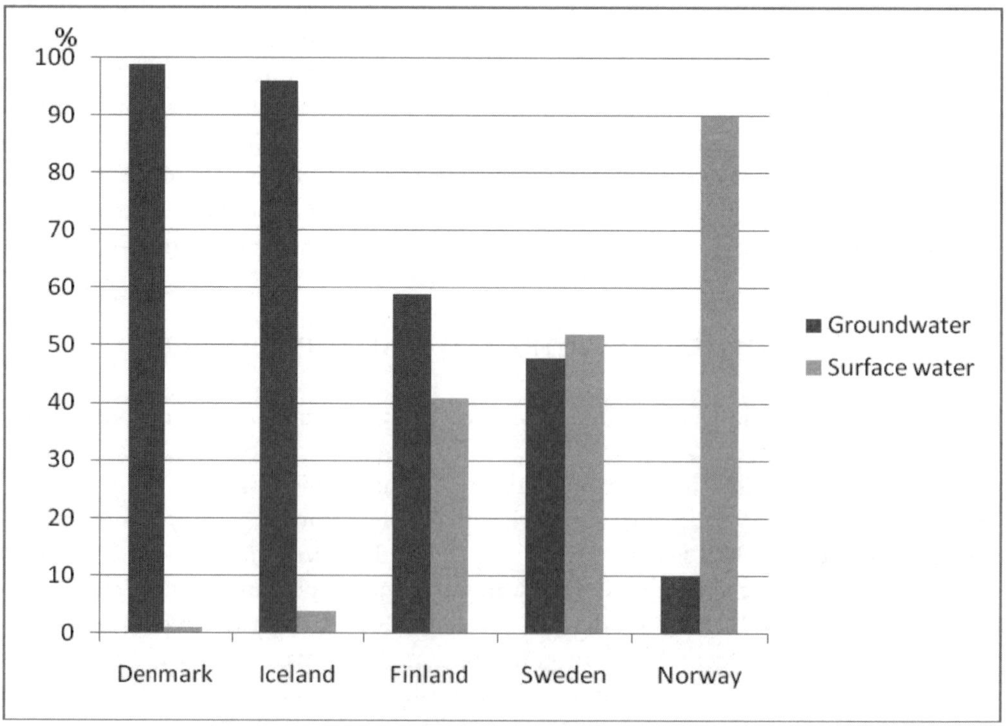

Figure 13.2 *Water sources in the Nordic countries*

Source: IWA (2006)

groundwater is regarded as a safer water source, less vulnerable to contamination. On the other hand, groundwater is generally of better quality than surface water as raw water; hence, fewer chemicals are needed in the purification process.

Wastewater treatment has been taken seriously in the Nordic countries since the 1960s. In Denmark, half of wastewaters were still discharged into watercourses, lakes and the sea without treatment in 1970. The rest largely underwent only simple mechanical treatment. Today, all industries and 90 per cent of all Danish houses are connected to advanced wastewater treatment plants with effective organic matter, phosphorus and nitrogen removal. The wastewater of the remaining 10 per cent of the houses is typically treated in septic tanks or individual micro-wastewater treatment plants.

In Sweden, release of nutrients increased drastically until the 1960s due to the construction of sewerage networks. Thus, during the intensive urbanization period in 1960 to 1980, the national government gave economic support to municipalities in need of extending their water and sewer networks and for the construction of wastewater treatment plants. During the 1970s, state support was limited to wastewater treatment plant construction only; but it was available to all municipalities. Since 1980, this kind of state support to water and wastewater facilities has ceased (Eriksson, 1994).

When Finland and Sweden joined the European Union (EU) in 1995, the countries had to adopt several EU directives, including the one on wastewater treatment. Consequently, the EU's Urban Waste Water Treatment Directive (91/271/EEG) has been incorporated within these countries' regulatory systems, partly in the legal framework and partly in rules issued by public authorities. Very strict requirements are imposed on all urban wastewater treatment plants for the removal of organic matter and phosphorus, while the requirements for nitrogen removal apply only to plants that serve more than 10,000 people (or equivalent) and are situated within catchment areas that drain into nitrogen-sensitive marine waters.

Iceland and Norway do not belong to the EU but are members of the European Economic Area (EEA) and, as such, have harmonized their national legislation with EU directives. Norway is a mountainous, long and narrow country with an extremely fractured coast cut by numerous narrow watercourses (fjords). The total length of the coastline is more than 20,000km – stretched out in a straight line it would extend more than halfway round the globe. Due to the fractured terrain, the country has as many as 2500 wastewater treatment plants serving at least 50 people. During the intensive wastewater treatment plant construction period from the 1970s until 2000, the state provided construction grants that covered a little over 10 per cent of the costs. The state grants system has now been discontinued.

Iceland, in turn, is a sparsely populated island with a long coast and a strong ocean current. Most people live in the coastal towns and villages. Around 90 per cent of the population is connected to a centralized sewerage system and the rest rely on septic tanks. Another 10,000 holiday homes, approximately, and a good number of tourist resorts outside centralized sewerage systems also rely on septic tanks. Iceland's wastewater treatment legislation is adapted to the local conditions; but the implementation process has been rather slow, with only 69 per cent of the population connected to adequate wastewater treatment systems.

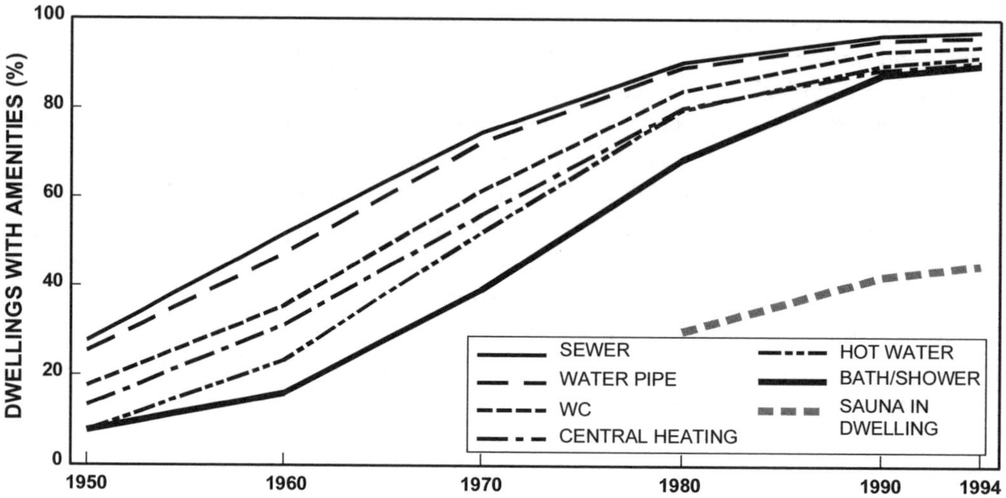

Figure 13.3 *Water-related amenities of Finnish dwellings from 1950 to 1994*

Source: Katko (1997, p249)

The development of centralized water supply, sewerage and wastewater treatment has been remarkably fast in Finland. As late as 1950, only 25 per cent of the Finnish population received piped water into their homes – by 1980, this figure had jumped to 90 per cent (see Figure 13.3). The development took place later than in Sweden; but one should keep in mind the hard times that Finland experienced from 1939 until 1944 when the country was at war with the Soviet Union, except for a good year of interim peace.

The development of municipal water supply systems in Iceland started at the beginning of the 20th century in major towns – often as an answer to recurring incidents of typhoid fever. In 1907, legislation on water supply for Reykjavik, and in 1912 for water supply to all authorized trading places, was enacted. The legislation gave the municipalities the monopoly to supply water and the right to use the water resources within the community. The municipalities were also granted permission to collect fees from house owners for providing them with water; but house owners had to lay the service pipes. This arrangement was in force until 1992, when new legislation on water supply obliged the water company to also lay the service pipes. In 1926, the Water Act was passed to govern all water use, including hydropower, water supply, sewerage and hot water for heating.

In smaller communities, progress was slow, and as late as 1947, only 27 per cent of the population received piped water from a centralized water supply. Elsewhere, people had individual connections to nearby wells. A housing survey carried out in 1940 in Reykjavik revealed that 99 per cent of the population had piped water to their houses and 80 per cent had a flush toilet (Björnsson, 1979). In 1947, legislation obliging the government to subsidize municipalities to build water supply and sewerage systems was passed, followed by further legislation in 1995 providing government support for building wastewater treatment plants. In 1992, legislation primarily concerning the management of municipal water supply came into force; it was reformed in 2004. Municipalities are obligated to supply water to their densely populated areas.

Unlike in several other countries, the Finnish state has not subsidized the construction or improvement of larger wastewater treatment plants. The responsibility for financing has rested entirely on the municipalities and their water utilities and companies. State subsidies are, however, available for the improvement of water services in rural areas and to support cooperation between rural municipalities in the water sector. Finnish rural areas have a long tradition of private consumer-managed water systems that operate on a small scale on a non-profit basis (Katko and Nygård, 2000). Municipally run water distribution and sewerage networks were built only in population centres, while those living in rural areas had to come up with their own water supply and sewerage systems. If good-quality water was not available close enough to the farm or house, people joined together to solve their water problem. A very common practice has been to establish a water cooperative. Cooperatives have also been established for other purposes, such as grain milling. Even though Finnish municipalities have been responsible for a number of basic services for their citizens since the late 19th century, in rural areas, many communal projects have been implemented jointly by the people with or without financial support from the state or municipality. Furthermore, during the 1960s, Finland passed water legislation directed at reducing water pollution, which included the creation of environment protection authorities. The development of this sector in the country has been driven by national norms and standards, where the focus has been more on effluent quality with less consideration for the state of the recipient bodies.

In the case of Sweden, the development of the wastewater sector has been heavily influenced by the creation of the Swedish Environmental Protection Agency (SEPA) in 1967. Given that the county administrations are, in fact, regional divisions of the central government, they have been eager to implement the policies of SEPA. Similar to the Finnish case, the focus here has also been on the quality of the water in the outlet pipe without much consideration for the state of the recipient water body. Consequently, the norms and standards advocated by SEPA have been adequate in many locations, but not everywhere. More attention should be paid to the impact of effluents on the water bodies: the southern part of Sweden suffers from eutrophication problems, while the northern part has problems with oligotrophic waters causing severe acidification.

Nevertheless, despite these shortcomings, there have been overall improvements in wastewater management, which can be partly attributed to the strong tradition of 'law and order' in the Nordic countries. Laws and policies set by the central government are obeyed, which has resulted, for instance, in significant improvements in water pollution control during the last decades – the watercourse flowing through the capital of Sweden, Stockholm, is swimmable again. The water pollution control measures were largely implemented before EU directives came into force, spurred by local interests, which is why compliance did not require massive investments. In Figure 13.4 we can see the consequence of tighter pollution control requirements as expressed by the number of municipal wastewater treatment plants built in Finland.

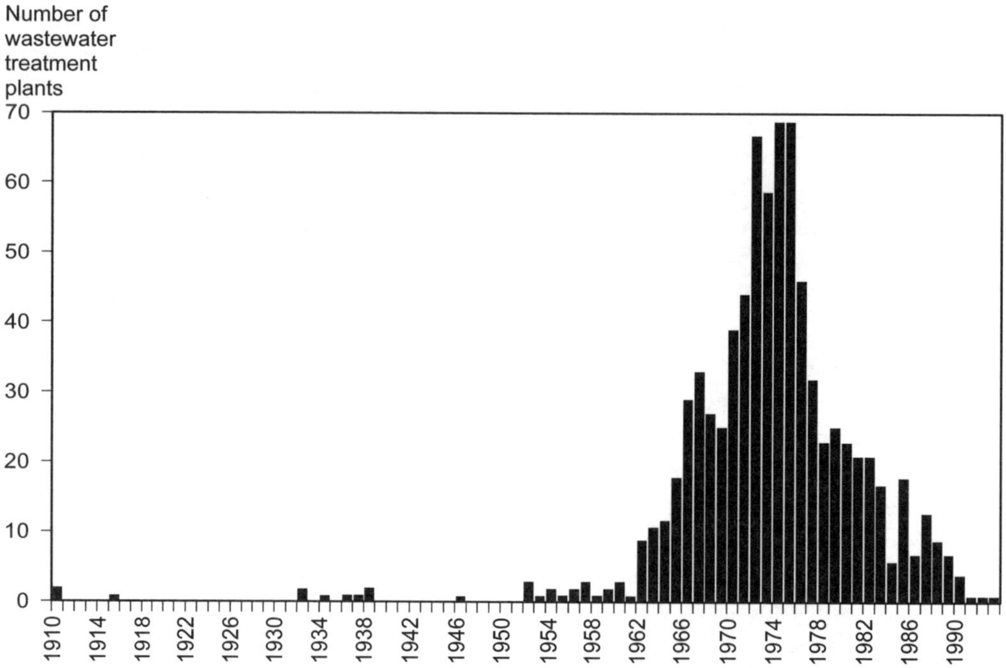

Figure 13.4 *Construction of urban wastewater treatment plants in Finland during 1910 to 1994*

Source: Lehtonen (1994, p64)

Treatment of wastewater from point sources – population centres and industries – is already quite comprehensive in the Nordic countries; but there is still a substantial challenge ahead in rural areas to reduce diffuse pollution from agriculture and households. In Denmark, wastewater is discharged into nature from approximately 350,000 houses and farmhouses, holiday cottages and allotment cottages scattered around villages. They account for 10 per cent of all sewage. The sewage from holiday and allotment cottages has until recently, been discharged untreated. The amount of sewage discharged by individual dwellings is small and has previously not been considered a problem. Most often, it was completely overshadowed by the pollution from dairies, villages and farm manure heaps. However, pollution from these sources has now been curtailed.

In Sweden, about 0.5 million dwellings, and roughly the same number of holiday homes, are not connected to centralized sewerage. Around half of these households just have a septic tank for sludge separation as a treatment method, which does not satisfy the present wastewater treatment requirements.

In Finland, the situation is fairly similar: about 2 million people live outside centralized sewer and wastewater treatment systems. Additionally, there are close to 0.5 million holiday residences – only few of them connected to centralized sewerage and treatment systems – mainly those by lakes. A new act was passed in 2003 that requires that every single household, even in rural areas, must have its wastewater properly treated by 2014. This can be accomplished either by leading wastewater to a centralized sewer network or by building an individual wastewater treatment system.

In this regard, the EU Water Framework Directive passed in the year 2000 (EU, 2000) prescribes a river basin approach that considers the ecological status of the watercourses. It also suggests that current problems could be overcome by means of a more participatory process of water management that involves citizens, users and other relevant actors. Both aspects present new challenges: the river basin approach requires new administrative models of water management, while public participation requires the introduction of organizational arrangements to foster new participatory practices in the water sector. However, this is not a straightforward process, as illustrated by the situation in Sweden, where five river basin districts have been identified and their respective competent authorities have already been created. However, there are worrying signs that the organizational approach adopted will be predominantly top down, reducing the autonomy of local government in issues concerning water and wastewater management.

MUNICIPALITIES AND THE PRIVATE SECTOR

Water supply and distribution, sewerage and wastewater treatment have been, and still are, statutory responsibilities of municipalities in the Nordic countries. This does not, however, imply that municipalities have to become involved in all aspects of these services – they are just responsible for ensuring that the services are provided. Private-sector involvement in water services has a long tradition in the Nordic countries, although perhaps not of the type that it is often promoted in current policy initiatives (e.g. private finance projects, long-term concessions, etc.). Municipal water undertakings outsource non-core activities and utilize the services of private companies based on competitive bidding. When contracts are retendered, typically every three years, there is real competition between the service or goods providers, and private-sector competition is used to benefit the water undertaking. The debate on private-sector participation and especially on the introduction of privatization and private operation emerged towards the late 1990s in the Nordic countries as well; but it never gained much wind and only a few cases have occurred. One case took place in 2002, when the Swedish city of Norrtälje (population 16,500 in 2007) awarded a ten-year concession to the French private company Veolia, and in the same year the Finnish municipality of Haapavesi (population 7600 in 2007), together with a local dairy, awarded a 12-year contract to a private company to rehabilitate and operate their common wastewater treatment plant (Pietilä, 2006). Since then, at least in Finland, the discussion on expanding private-sector participation in these ways has largely faded out – private companies have not actively pushed for the option, nor have municipalities shown much interest in it. The municipalities see that they can best benefit from private-sector competition by outsourcing non-core activities, rather than the entire operation.

The concrete details of the legislation governing the organization of water services in Nordic countries vary to some extent: in Denmark, Finland and Sweden, current legislation does not forbid private ownership and/or operation of water and wastewater facilities, but Danish and Swedish legislation clearly states that it is illegal for public water services to turn a profit, while in Finland the legislation allows a reasonable rate of return on capital investments. Sweden had a privately owned major water undertaking in the city of Norrköping (population 90,000 in 2007) for a while. In 1997, the municipality established a multi-utility company by merging the municipal water and wastewater utility with the energy company, the waste company, and the municipal street and park department. In 2000, the company was sold to Sydkraft, an energy utility owned by the German company EON Energy AG. Thus, for a while, water services were provided by a private multi-utility company; but in 2005 the municipality bought the water company back to 100 per cent municipal ownership (Norrköping, 2006).

In Iceland, municipal water supply legislation was renewed in 2004. Currently, municipalities are allowed to outsource water and wastewater services; but private ownership is limited to a maximum of 49 per cent. Yet, municipalities have the responsibility to secure that water is provided to the population in densely populated areas. The legislation stipulates that water utilities can only collect fees to cover operating costs, costs of capital and investment costs.

Even though municipalities are, in principle, responsible for providing water and wastewater services to their citizens, this does not mean that every citizen is literally entitled to these services from a municipal system. The Nordic countries have a relatively low population density compared with many other regions in the world. According to statistics covering 192 countries, Sweden is 154th, Finland 161st, Norway 165th and Iceland 189th, with population densities of 20, 16, 14 and 3 individuals per square kilometre, respectively. In this regard, Denmark is an exception, with a population density of 126 individuals per square kilometre – it ranks 61st (*WorldAtlas*, 2008).

In the Nordic countries, there are large sparsely populated rural areas between their population centres. Often, these areas are not covered by municipal water and wastewater services, leaving the owner of a property responsible for his own water supply and wastewater services. In many cases, people use their own wells; but where water sources have not been readily available, people have often joined forces together and established a common undertaking for their water supply. The Nordic countries also have, in addition to municipal water utilities, a large number of small water undertakings. Both in Denmark and Finland, the total number of public water undertakings is well over 2000, and the great majority of them are consumer-managed cooperatives. In Finland, the cooperative movement has a strong tradition and is still an important factor in economic and business life – one of the largest banking groups with a market share of about 30 per cent is a co-operative, as is also the case of a large consumer goods retail chain with roughly the same share of the market.

The decentralized water supply in Denmark is a result of the geology of the country, which gives easy access to groundwater resources throughout the country. Today, the decentralized supply system is appreciated because it causes relatively little stress on local environments and rather effectively prevents quality problems in water supply. Once a decentralized system is established, there are few economic incentives to centralize it since new infrastructure is expensive. In 2007, the national government initiated a water reform to increase the effectiveness of the Danish water sector. The aim of the reform, on which there is political agreement, is to optimize the water sector for the benefit of the users and the environment, to provide economic tools for meeting the needs of investments and implementation of new technology, and to separate the roles of the authorities from the responsibilities of operation (Ministry of Environment, 2007). The reform allows the reorganization of the water and sanitation sector and the creation of new types of utilities, as shown in Table 13.2.

MUNICIPAL COOPERATION

Cooperation between municipalities is increasingly common in water and wastewater service provision in the Nordic countries. Water sources are not evenly distributed, and often water from one municipality, with excess supply, is sold to a neighbouring

Table 13.2 *Organizational forms of the water and sanitation sector in Denmark*

	Form of organization				
	Municipality	*Cooperation between municipalities*	*Public or private company*	*Self-owned institution (fund)*	*Consumer cooperation*
Services	Water and wastewater	Water and wastewater	Water and wastewater	Water only	Water and wastewater
Ownership	Municipality	Two or more municipalities	Shareholders (municipality or private)	Institution (fund)	Members of co-operative (consumers)
Break-even principle	Yes	Yes	Yes	Yes	Yes
Separation between authority roles and operation	Mix is possible	Mix is possible	Full separation; mix is possible if municipal company	Full separation	Full separation
Tariffs	Municipality decides	Depends upon delegation of decisions	Municipality approves (both municipal and private)	Institution suggests and municipality approves	Co-operation suggests and municipality approves
Investments	Part of the municipality's investment programme	According to rules, typically set by the board	Investments set by the board	Investments set by the board or written into the operation agreement with the municipality	Investments set by the board or written into the operation agreement with the municipality

Source: DANVA (2007)

municipality, normally based on long-term agreements. The networks of neighbouring municipalities are often interconnected to guarantee the services in emergencies and exceptional situations.

There are several examples of bulk water supply companies and wastewater treatment plants serving the needs of several municipalities. In Finland, there are more than 20 bulk water supply companies owned by municipalities. The largest one is the Helsinki Metropolitan Area Water Company, which supplies water to about 1 million people (20 per cent of the population) in eight municipalities. This company is a pure bulk water supplier, and each of the municipalities has its own water utility, which takes care of water treatment and water distribution within its own area (Pietilä, 2006).

Sweden got its first bulk water supply company, the Käppala Water Federation, in 1957. It was established to ensure that Stockholm would be supplied adequate amounts of drinking water. In 1966, the Sydvatten Company was formed to ensure the water supply of 13 municipalities in southern Sweden (Sydvatten, 2007). Sweden also has 2000 publicly owned sewage treatment plants that treat about 1500 million cubic metres of wastewater annually. Yet, the six largest plants, listed in Table 13.3, take care of 30 per cent of the wastewater flow.

The requirements for the standards of water and wastewater services are getting tighter. Water users are increasingly more concerned about the quality of their drinking water, and water undertakings are looking for better raw-water sources, typically groundwater, even though the quality of existing water sources exceeds the requirements of drinking water standards. Better quality water sources can often be found only at large distances, and tapping these resources requires cooperation between neighbouring municipalities and water undertakings.

Table 13.3 *The largest municipal wastewater treatment plants in Sweden (2003)*

City	*Population served*	*Wastewater treated (million m³/year)*
Stockholm	970,000[a]	135
Gothenburg		
Ryaverket wastewater treatment plant	480,000	120[b]
Käppala wastewater treatment plant	400,000[c]	53
Malmö + four municipalities		
Sjölunda wastewater treatment plant	550,000[d]	50
Himmerfjärd wastewater treatment plant	250,000[e]	40
Uppsala		
Kungängsverket wastewater treatment plant	150,000[f]	19
Linköping	120,000	15

Notes:
[a] Three wastewater treatment plants in Stockholm also serve neighbouring municipalities.
[b] Includes 25 million cubic metres per year from six other municipalities.
[c] Wastewater load of 510,000 p.e. (person equivalent); serves 11 municipalities north of Stockholm.
[d] Treats a wastewater load of 550,000 p.e., including that of 270,000 people in Malmö.
[e] Treats an extra industrial wastewater load of 35,000 p.e.; serves several municipalities south of Stockholm.
[f] Uppsala and two other municipalities.
Source: Pietilä (2005, p187)

A noticeable indicator of the excellent quality of drinking water in Nordic countries is the fact that bottled water has gained only a limited market in the region compared with the situation in many other industrialized countries. In Italy, bottled water consumption has reached 190 litres per capita per year, while in Germany, France and Spain, the figure is over 130 litres per year. In the Nordic countries, consumption has hardly reached 20 litres per capita per year, and growth has more or less stalled during recent years. In Sweden, the use of bottled water decreased in 2007 for the first time since 1996. The low use of bottled water can be partially attributed to a number of studies that have shown that, on average, the quality of piped water is better than that of bottled water. A second reason for the relative unpopularity of bottled water is environmental – the wasting of natural resources in bottling it and the carbon dioxide emissions from its transportation. Tellingly, several public organizations have decided not to serve bottled water as refreshment in meetings and other events (Martinsson, 2008).

Despite this long tradition of sound water management, pollution of watercourses is a serious concern in the Nordic countries, which is why the requirements for wastewater treatment are often stricter than those of the EU Water Framework Directive adopted in the year 2000. In the face of mounting technical challenges and rising costs, many small wastewater undertakings have decided to close down their treatment facilities and channel their wastewater to a larger treatment plant in a neighbouring municipality. The trend towards larger wastewater treatment plants seems to continue.

Conclusions and perspectives

There seems to be a general tendency towards a more holistic approach in water and sanitation management in Europe, including the Nordic countries. The 2000 EU Water Framework Directive calls for integrated water resources management on the basis of river basins. In the Nordic countries, one step towards integrated water resources management had already been implemented during the 1970s, when thousands of administratively and operationally separate water and wastewater services were merged and their number significantly reduced.

The Nordic countries' water sector is *not* entirely dualistic, as it has been sometimes argued, where the central government would set the policies and regulations and the local authorities provide the services to the citizens. In fact, the strong local-level administration that has historically characterized these countries has been significantly strengthened as a result of the recent reforms, and municipalities have gained even more power at the expense of, in particular, the regional, but also the state, administration. For instance, in the field of environmental protection, including the issuing of wastewater discharge permits, municipalities have now stronger authority than earlier. One of the future challenges facing Nordic countries will be increased flooding due to heavier and more irregular rainfall. Decentralized and local solutions will likely play a key role in minimizing the seriousness of the consequences of climate change in the region.

In the Nordic countries, municipalities are clearly responsible for water and sanitation services. Water supply, sewerage and wastewater treatment systems have been built and financed by them: the development of water supply systems was typically funded through user charges, while sewerage and wastewater treatment were originally funded through municipal taxation and, more recently, increasingly through wastewater fees. The state used to partially subsidize the construction of wastewater treatment plants in some countries; but this has no longer been the case during the last decade.

The provision of water and sanitation services in Nordic countries involves a wide variety of institutional arrangements and a number of stakeholders. Moreover, their provision is not the preserve of the public sector; in fact, publicly owned utilities buy a variety of goods and services from the private sector and reap the benefits of competition. Very clearly, one-size-fits-all approaches are not suitable for the Nordic countries, even though central government authorities have recently entertained policies that seem to lead in that direction.

REFERENCES

Björnsson, L. (1979) *History of Municipality in Iceland*, Almenna bókafélagið, Reykjavik, Iceland

Council of Europe (2002) 'Outlines, syntheses and overviews of six models of regional self-government', Paper presented to the Conference of European Ministers Responsible for Local and Regional Government, 13th Session, Helsinki, 27–28 June 2002

DANVA (Danish Water and Waste Water Association) (2007) *Matrice over forskellige organisationsformer I vandsektoren og deres egne egenskaber* [*Characteristics of Different Organizational Forms in the Water and Sanitation Sector*], Danish Water and Waste Water Association, Skanderborg, Denmark, www.danva.dk/graphics/DANVA/VidenogInnovation/ALL/15.01.07%20Skema%20over%20organisationsformer.xls, accessed 30 November 2007

Dexia (2002) *Local Government Employees in the Fifteen Countries of the European Union*, Dexia Editions, Paris, France

Eriksson, Ö. (1994) 'Trends of organizational development of water supply in Sweden', in J. Hukka, P. Juhola, T. Katko, H. Morange and P. Pietilä (eds) *Sound Institutional Strategies for Water and Sanitation Services*, Seminar proceedings, Institute of Water and Environmental Engineering, Publication B 59, Tampere University of Technology, Tampere, Finland, pp18–23

ESA (European Space Agency) (2007) 'Norway', http://earth.esa.int/applications/dm/archdm/disman/db/synthesis_reports/SRNorway.html, accessed 26 November 2007

EU (European Union) (2000) *Directive 2000/60/EC of the European Parliament and of the Council Establishing a Framework for the Community Action in the Field of Water Policy*, EU Water Framework Directive (WFD), EU, Brussels

Goddard, N. and J. Sheail (2001) 'Victorian sanitary reform: Where were the innovators?', in C. Bernhardt (ed) *Environmental Problems in European Cities in the 19th and 20th Century*, Waxmann, Münster, Germany

Henning, R. (2001) *Regional Governance in the Nordic Capital Areas*, Nordregio Working Paper 2001:8, Nordregio, Nordic Centre for Spatial Development, Stockholm, Sweden

IWA (International Water Association) (2006) *International Statistics for Water Services*, IWA World Water Congress, Beijing, September 2006

Katko, T. (1997) *Water: Evolution of Water Supply and Sanitation in Finland from the Mid-1800s to 2000*, Finnish Water and Waste Water Works Association, Helsinki, Finland

Katko, T. and H. Nygård (2000) 'Views of research on the evolution of water, wastewater and solid waste services', *Tekniikan vaiheita,* vol 18, no 4, pp14–19

Katko, T. and M. Stenroos (2005) 'Sweden', in P. S. Juuti and T. S. Katko (eds) *Water, Time and European Cities*, European Commission-funded Watertime project, University of Greenwich, Greenwich, UK, pp197–203

Lehtonen, J. (1994) *Jäteveden käsittelyn kehitys Suomessa [Development of Wastewater Treatment in Finland]*, Institute of Water and Environmental Engineering, Publication B 58, Tampere University of Technology, Tampere, Finland

Martinsson, T. (2008) 'Ruotsi kulauttaa nyt hanasta' ['The Swedes drink tap water'], *Aamulehti*, vol 126, no 48, p12

Mennola, E. (1999) 'The institutionalization of regions in Finland – a historical perspective', *Finnish Local Government Studies*, vol 27, no 3, pp385–395

Ministry of Environment (2007) *Aftale for en mere effektiv vandsektor 1. februar 2007 [Agreement to make water sector more efficient, 1 February 2007]*, www.DANVA.dk, accessed 30 November 2007

Moisio, A. and R. Uusitalo (2003) *Kuntien yhdistymisen vaikutukset kuntien menoihin [Consolidation of Municipalities versus Municipal Expenses]*, Kuntaosaston julkaisu 4/2003, Ministry of the Interior, Helsinki, Finland

Norrköping (2006) 'Norrköping vatten i korthet' ['Norrköping's water in brief'], Norrköping, Sweden, www.norrköping.se/miljö-natur/vatten/om, accessed 25 November 2006

Pietilä, P. (2005) 'Sweden', in A. Danilenko and A. Child (eds) *Water Market Europe: Opportunities in EU Accession, the Framework Directive and the CIS*, Global Water Intelligence, Oxford, UK, Chapter 18, pp184–190

Pietilä, P. (2006) *Role of Municipalities in Water Services*, Publication 617, Tampere University of Technology, Tampere, Finland

Sydvatten (2007) *Årsredovisning 2007 [Annual Report 2007]*, Sydvatten, Malmö, Sweden, www.sydvatten.se/, accessed 26 May 2008

WorldAtlas (2008) 'Countries of the world by highest population density', www.worldatlas.com/aatlas/population/ctydensityh.htm, accessed 14 April 2008

14

The Development of Water Services in Europe: From Diversity to Convergence?

Bernard Barraqué

Introduction

In this chapter, we wish to provide an analysis of the variety of formulas for water services management existing in Europe, and try to explain why different countries chose their own paths and the way in which they did. We also attempt to show what these experiences reveal in terms of common perspectives, and possible challenges and even threats. Lastly, we propose to draw some insights on the situation of water services in developing countries.

The idea of domestic water supply was first implemented by private companies in England and France; but they were unable to generalize the service once the invention of bacteriology in the late 19th century turned it from being a luxury into an essential service. In the most advanced countries, but also in areas such as Nordic Europe, it was the municipalities that took over the responsibility for domestic water supply services and eventually managed to deliver treated water to all. Once the services were generalized and water industries became mature, the long-term reproduction of the heavy infrastructure called for new solutions away from pure direct and local procurement: concentration of local water services in joint boards or centralization at regional level, integration with other utilities and funding from customer bills, among others. In this process, the well-known and longstanding public versus private debate was taking place simultaneously with, and intimately related to, two less well publicized controversies: centralization versus decentralization, and unbundling versus integration. Nevertheless, beyond the differences between individual countries, as discussed later, there emerges a common European model that is fundamentally different from, to link it to an ongoing debate, the policies advocated by economists and implemented in England and Wales since the 1980s: regionalization plus privatization and regulation by an independent administrative authority.

In relation to the challenges faced, this chapter emphasizes the potential crisis due to rising water prices, which leads to opening the traditionally closed water and sanitation services (WSS) community to users, taking into account the social dimension of these services in order to meet the goals of sustainability. In this regard, the quest to reduce the ever-growing costs of WSS has prompted a search for territorial solutions instead of purely technological fixes. This, in turn, implies establishing cooperation between institutions at various territorial levels, which poses a significant challenge for the democratic governance of water and WSS.

MUNICIPAL PROCUREMENT OF WATER SERVICES AS THE INITIAL SUSTAINABLE SOLUTION

Towns and rural communes in Europe have a longstanding experience in bringing potable water from nearby sources to local public fountains, and during early times some private landlords or religious buildings would also get an in-house connection. In the 19th century, and more precisely once the Koch and Pasteur discoveries were popularized, public water supply (PWS) systems with pressurized water conducted through pipes were developed. However, cities often faced two competing alternatives: pumping potentially unsafe water from a nearby river or drawing water from sources located at long distances from the urban areas. The building of the PWS implied substantial and risky investments. With the Industrial Revolution and under the pressures of rapid urbanization, large cities in particular would have to get water from longer distances. This normally implied the intervention of central governments at two levels: financing the aqueducts (or subsidizing them), and granting a water right or a concession on distant water sources to the cities. This type of intervention often resulted in a *de facto* centralization of water supply services. From another perspective, the chief motivation for building PWS in this period was to clean the city, and as a result few people imagined having to pay for the services in those early days. Among other challenges facing early water utilities, they were obliged to respect the right to common use of water resources by peasant communities for their domestic needs. Lastly, the idea that the service should be free was supported by the relatively low operational costs of early urban water systems. The bulk of the investments needed were related to the building of infrastructure, which was paid for with public money; there was no tradition to depreciate the investments.

Yet another important development in the 19th century was the creation of private water companies after the assumption that water could be delivered in private residences, including condominiums, in exchange for an acceptable fee. In many cities of the time, local authorities considered that private water provision was a luxury and they did not want to become directly involved. They granted concessions to the private companies, which often had to deliver water at cheap or zero prices to fountains and other public outlets, while most of their revenues would come from rich subscribers to the 'private' water service (Pezon, 2000). These private companies were left by themselves to produce and distribute water, having to solve several technological problems (leak control, metering, etc.). These initial ventures usually lacked both enough capital and political support to be able to generalize the service. The initial model of the concession, in turn, was hampered by distrust between companies and the population. Conflicts developed when municipalities became convinced that water supply was not a luxury, but a fundamental public health issue. If operators took water from nearby rivers, it was normally of bad quality; but they had neither the financial nor the legal possibilities to tap water from distant sources – hence, central government intervention was required to solve these problems. In most countries, during the second half of the 19th century, municipalities terminated existing private contracts or refused to extend them, and went on to universalize the services themselves.

Glasgow was one of the first European cities to reach this temporary equilibrium: in the middle of the 19th century, it took over the water services from a deficient private water company, provided the investments needed to connect the whole population, and at the same time, thanks to government grants, built an aqueduct to tap clean water from Loch Katrine, some 55km away. It was a momentous occasion:

> *The official opening by Queen Victoria on an appropriately wet autumn day in 1859 was an event of enormous significance for Glasgow … Loch Katrine was unquestionably the prime municipal showpiece for the city, combining the wonders of Victorian technology with the nurturing quality of pure Highland water.* (Maver, 2000, pp90–91)

However, the invention of filtration systems gave a new impetus to the use of surface and nearby water sources, which, in turn, helped to reduce the dependence upon central government funding and water rights licences. The country that played the leading role in these developments was, indeed, England: despite the relative water scarcity compared to population density in that country, water supply and sewerage were generalized earlier there than on the continent. The engineering skills available in the country allowed for innovation in infrastructure development in an overall context of administrative decentralization and the rise of what should be termed 19th-century British municipalism (rather than municipal socialism, as depicted in France, or water and gas socialism, as the tradition is usually derided in the UK). Indeed, it can be argued that 'welfare policy' and central government involvement in the economy were both anticipated by at least two generations of municipal welfare policy in England, followed by other European countries.

With the emergence of a middle class and of a qualified working class also came the savings banks, which would eventually provide loans at cheap rates for developing local welfare facilities. Municipal bonds as funding instruments for essential services were found attractive by the public, and on top of this the government would subsidize projects. This is why, in most cases, the early private concessions were not renewed or were even terminated to be replaced by municipal direct management. But, overall, in this still initial phase of development and extension of PWS systems, the prevailing idea was that water services should be covered by local taxation (e.g. by rates proportional to the renting value of the properties). This proved to have important redistributive effects and played a significant role in the acceptance of domestic water supply systems. For instance, in Montreal, Quebec, a private company had started to develop a water system as early as 1798, but met severe profitability problems. The municipality terminated the contract soon after being authorized to incorporate by the British Crown; but it was soon to discover that revenues from water bills were far from covering the costs involved in expanding the service. Around 1850, the Crown authorized the municipality to make the connection to the water supply system compulsory for all citizens, and to charge them for the service through local taxes. Fifteen years later, every house in the city was connected (Fougères, 2004).

Following Britain, European cities resorted to various forms of filtration first, and later to purification systems, which helped to turn the overall *problematique* of water quantity into, at least partly, one of water quality. However, the supply-driven approach remained dominant in the New World, and was extended to the rest of the world after World War II due to a combination of factors, including low-cost funding for essential infrastructure offered by the international financial institutions (IFIs) and various forms of financial support offered by national governments (whether Keynesian or socialist) through direct intervention in infrastructure provision. A noticeable case is the involvement of the US federal government in building large hydraulic works, which offered a long-lasting example to other countries. Interestingly enough, this supply-driven approach also remained dominant in the Mediterranean countries of Europe during the period of authoritarian governments, which increased the importance of central and regional authorities, reduced the role of municipalism, and gave low priority to domestic water services compared to large-scale infrastructure such as multipurpose water transfers.[1] The large hydraulic projects of the 1950s and 1960s were increasingly devoted not to cities, but to irrigated agriculture, which was then associated with development. Today, many governments in developing countries still base their water policy on large water transfers so as to indirectly subsidize the production of irrigated cash crops for the world market. As a result of these developments, water has often been delivered to local utilities at very low prices, which has encouraged wasteful use. Thus, in many countries, the level of unaccounted-for water is high, which creates a negative cycle compounded by underfunding of operations and maintenance. As could be expected, by the end of the 20th century, it became more and more difficult to develop large hydraulic infrastructure works along the lines of supply-side water policy approaches. Interestingly, one of the outcomes of the *de facto* alliance between environmental protection movements and liberals advocating full-cost pricing in recent times is the possibility of checking the unsustainability of these past policies. In this regard, California, which indeed became the largest artificial river basin in the world thanks to huge state and federal investments in large-scale water infrastructures, is a top candidate for change in the current policy environment.

WATER TREATMENT AND CUSTOMER BILLING

Back in the 1880s, once Pasteur and Koch found that water contamination was a major cause of diseases, it became evident that even distant and pure water would eventually have to be treated. This is how sanitary engineering was developed to complement civil (hydraulic) engineering and solve the quality crisis of water services. Typical of the times, a 30-year-long debate started in 1890 in Paris City Council concerning a project to build an aqueduct from Geneva Lake, which would 'solve all the city's water quantity problems forever'. The project was discarded after World War I for geostrategic reasons; but, indeed, it had been made superfluous after the inauguration, in 1902, of the slow filtration plant in Ivry, which operated with Seine River water just upstream of Paris. The Ivry plant is still in use today, and its construction was legitimized by a typhoid epidemic in 1899 that was caused by contamination of the Loing springs, which at the time were the furthest and supposedly cleanest water sources serving the city.

In the temperate climate of Northern Europe, the main problem was not resource availability, but rather water quality. Because direct medical action to treat waterborne diseases was not easy, towards the end of the 19th century it was decided that water should be filtered; filtration was complemented with disinfection technologies (chlorination, ozone or activated carbon beds) around World War I. These advances supported the development of water supply at the local level: taking surface water just upstream, cities would induce economies of scale and would allow substantial savings in the investment for water-carrying infrastructure. However, treating water would lead to a serious rise in operation costs, which are usually more visible to the public and to the city councils. This, in turn, supported metering and billing. Even publicly provided supplies would then be at least partly covered by water bills. It was certainly also an important change for the public, and charges were initially limited to cover operation and maintenance. Slowly, however, delivery of pressurized water to homes changed status from being a luxury good to a commercial service (and not to an administrative service), and made water billing a normal practice. In Europe, only the UK and the Republic of Ireland have kept the ancient charging system based on rateable housing values (but so have many municipalities in the US, such as New York, and in Canada).

Covering an ever-larger fraction of the costs by billing increased the self-financing capacity of water services, which improved their sustainability. Turning the citizen into a water consumer was found to be very attractive, and in many continental countries after World War II, authorities included sewerage charges in the drinking water bills, taking advantage of the need to treat wastewater before discharging it back into the environment (which was also a visible cost). This was done despite the compulsory character of sewer connections, which would normally imply coverage of the costs by local taxes. Sewerage then also became a commercial service. In many countries, local sewers remain in the hands of communes and are funded by local taxes, while interceptors and wastewater works are run by a publicly owned company, and the service is billed together with the water supply. Eventually, increased self-financing capacity reinforced the legitimacy of local authorities as service providers, or at least organizers. However, in order to achieve economies of scale, it was often necessary to develop the joint efforts of neighbouring communes. Usually, central governments allowed and supported the creation of joint boards of municipalities in order to bring together the institutional, technical and management dimensions. In turn, these innovations in both water supply and wastewater services provision supported the development of new territories and of new relationships between politics and expertise, and between elected representatives and engineers. It also supported the development of WSS as a policy sector separate from water resources management.

VARIOUS ADAPTATIONS OF THE MUNICIPAL MODEL IN EUROPE

In France, as in other Northern European countries, important efforts have been made on city sewerage since the 1950s and on sewage treatment works from the 1970s. However, by then, PWS itself had become a mature business, which, in particular, meant that utilities increasingly had to face the challenge of renewing ageing infrastructure without any subsidies.[2] This is the fundamental reason why European municipalism in WSS had to be adapted to respond to the new challenges arising in the late 20th century.

One of the crucial changes introduced was the return of the services to some form of private legal status: traditional public accounting rules were embedded in a financial culture that was reluctant to incorporate the depreciation of assets and to make provision for infrastructure renewal, while private accounting had always done it. However, there were obviously other possibilities, such as the concentration of undertakings at supra-local level, the development of cross-subsidization via ear marked funds, or the integration of water and other utilities in the same company. What are the results today of these complex arrangements?

First, the direct management formula (direct labour) still remains the top preference today in terms of the number of WSS utilities organized on this principle. It mainly concerns small undertakings, where technical issues tend to be less complex. In some countries such as Denmark, or in north-western Spain and northern Portugal, one can even still find PWS operated at a neighbourhood level by communities of residents themselves. But the growing technicality and complexity of regulations reduces the scope of these units. Conversely, some large WSS undertakings have also kept the direct management formula, even in France where several large cities have maintained *régies* (direct labour) to operate PWS in their jurisdictions. These utilities are often administratively separated from the general accounting system of the city, and in this case they are compelled to balance costs with recipes. Additionally, they now apply some form of asset depreciation and make provisions for infrastructure renewal, which makes them more sustainable and able to compete with alternative private formulas (e.g. concessions or service contracts).

The first step in the reforms introduced to the municipal model is often to organize joint boards of neighbouring municipalities to facilitate a fair distribution of the investment burdens. This is quite frequent in Germany, The Netherlands, Belgium, Italy, Switzerland and, in particular, France, where very small communes have long continued to resist government proposals for mergers. This type of arrangement was also quite frequent in England and Wales before the 1974 regionalization.

Often, the reorganization of the municipal model at the supra-local level can be combined with the adoption of, and industrial and commercial status for, WSS services. Many cities in continental Europe created public or mixed economy WSS companies in order to adopt private accounting practices, and occasionally to facilitate access to the banking system. These changes have also taken place simultaneously with the slow replacement of pricing systems based on property valuation by consumption charges. For instance, a 1903 law in Italy forbade the concession to private (and foreign) companies, but supported the creation of the *azienda municipalizzata* or *azienda speciale* (specialized public company) formula, with separate accounting from the main budget. In The Netherlands, a progressive concentration of PWS utilities leads to the formula of private companies owned by the concerned municipalities, although the provinces frequently also have shares in these utilities. While there are almost 500 communes in the country, the number of *waterleidingbedrijven* (water supply companies) has been reduced to just ten.

In turn, the process of concentration can lead to centralization: the government decides that utilities are reorganized at county or regional level. The best example was the creation of the Regional Water Authorities (RWAs) in the UK in 1974: all of the utilities except 29 small water providers that had remained privately owned (later called 'water only companies') were merged into ten river basin institutions in charge of water planning, abstraction and discharge licences, water supply, and sewage collection and treatment. A similar process is currently happening in Italy: the 1994 Galli Law is promoting the merger of small utilities into optimal territorial units (ATOs), although these are usually not organized at catchment scale (as planned initially), but more often at the level of the 100 *province* (counties). Once created, the new authority must choose between two management formulas: a public company or a delegation to a private operator after a tendering process. Direct labour is phased out.

Another clear outcome of centralization is opening the way to full privatization of WSS services, as occurred in England and Wales in 1989 after 15 years of regionalization: RWAs were constrained by fiscal austerity imposed by the government, and they were also under criticism for mixing the functions of operator and regulator. Hence, the privatization of the water industry and the development of national-level administrative authorities outside the central government or QUANGOs (quasi non-governmental organizations), such as the Office of Water Services (OFWAT) for social and economic dimensions or the National Rivers Authority for the environment (now part of the Environment Agency). Nevertheless, parts of the regulatory activities are carried out by central government bodies, such as the Drinking Water Inspectorate for public health. However, today, as in the past, this full privatization model implemented in England and Wales remains fragile: the private sector turns out to be embarrassed to run infrastructure that is very costly but depreciates very slowly.[3] In Wales, the private water company took advantage of the creation of the regional parliament in 2001 to sell them their assets. Italy has also adopted the pricing formula and the type of economic regulatory body used in England and Wales; but implementation will be even more difficult in this country since local authorities retain membership of the ATOs and have responsibility for authorizing investments and tariffs. The national regulator in this case does not face private companies, but political entities. The very high increase in prices, which is likely to result from the implementation of the new system in Italy, has stirred a strong anti-privatization and pro-public management movement.

Portugal has recently adopted an intermediary solution, with the partial centralization of WSS utilities: after the country's return to democratic government in 1974, the 300 communes recovered the powers that they had lost during the long dictatorship. WSS were clearly identified as a local government duty, except for the water utility of Lisbon, which was a public company owned by the state. But in a growing number of cases, communes could not face the difficult but long-needed modernization. Since Portugal was eligible for European subsidies for this matter, the government created Aguas de Portugal (AdP), a national WSS company, to channel the European and national funds towards the communes. In turn, it was planned that communes would eventually create a mixed company where the national company would have 50 per cent of the shares, and municipalities would share the rest. AdP is in charge of producing potable water and delivering it to communes, as well as providing sewage treatment. Local water and sewer systems remain in the hands of each commune. This leads to a differentiation of roles that the Portuguese term *agua alta* (high water) and *agua baixa* (low water).

This Portuguese reorganization is consistent with a Mediterranean specificity: due to seasonal rainfall variation, local water supply is frequently threatened by scarcity. This often led to the creation of regional bulk water storage and transfer infrastructures, resulting in a dual-level PWS system – for instance, Compagnie du Bas-Rhône-Languedoc (1956), Aguas Ter Llobregat in Barcelona (transfers started in 1966) or Canal de Provence in Marseilles (set up in 1967). In Madrid, the utility Canal Isabel II, created during the 19th century, finally took over all water services in the autonomous community. A similar process has taken place with Aquedotto Pugliese in southern Italy.

In Germany, as well as in some large cities elsewhere in continental Europe, on top of concentration and 'formal privatization', in the early 20th century, various utilities were merged to form the *Stadtwerk*. Originally the system was implemented for water, gas, electricity, public transportation and district heating; but, more recently, some of these utilities also run sewage collection and treatment, solid waste, cable TV and other services. This organizational choice was consistent with the long trust that German citizens placed in their local governments and with their pragmatism: a private (PLC) and transversal but fully publicly owned company would be more creditworthy for the banking system, and would pay business taxes to the main city budget. Eastern Germany under Soviet rule took a different path and WSS services were reorganized at the regional level. On this topic, the World Bank regretted the process of re-municipalization of these services carried out by the German government after reunification, but I have shown elsewhere why the German WSS model is, indeed, interesting (Barraqué, 1998).

The last management formula to describe is delegation to a private company, which is quite frequent in France: the infrastructure remains in public hands, but operations are carried out by a private company, under either management or lease contracts. One of the reasons for this particular development was the historical confrontation between communes and the central government, and the tight limits placed by the latter on local budgets and on the possibility of creating mixed-economy companies. Centralization also reduced possibilities for public undertakings to obtain cheap loans from savings banks.

Communes then preferred to find or keep a private partner, particularly when they decided to create joint boards: delegation reduced local political rivalries. Private water companies, in turn, concentrated their efforts on developing financial business strategies, and also diversified into other public services. This unique situation gave French private multiservice companies the capacity to invest in other countries and to become global players in the private water market, as we well know today. The 'French model' is influential in Spain and in Eastern Europe. But in the French homeland, paradoxically, one of the outcomes of the decentralization process of the 1980s was the growing role of the *départements* (counties) and their elected boards, which increasingly tend to get involved in WSS services in support of local authorities. This has eventually led to a better balance of power with the private companies. Moreover, recent laws have obliged delegating authorities to organize tenderings at the end of all contracts, and a yearly report on activities must be made public: the private WSS companies report to the delegating authorities, and the latter have to report to the public. It is then more difficult for companies to take advantage of the weakness of small communes to make more profits.

THE KEY WORDS: CROSS-SUBSIDIES AND MULTILEVEL GOVERNANCE

What can we gather from this presentation of the various models of WSS services in Europe? It seems that the initial model of municipal involvement in the provision of these services had to be adapted to face the long-term challenges posed by running a costly capital-intensive infrastructure at a low price for water users. The various models examined show that water suppliers or authorities have developed a number of cost-averaging or cross-subsidy mechanisms to limit the impact of 'heavy and lumpy' (though long-lasting) investments upon water prices. In some cases, this has been achieved through the spatial integration of utilities, and there is, indeed, a concentration process going on all over Europe, which tends to dispossess local authorities of their former responsibilities for water services, at least partly. Another common strategy is cost averaging across time: earmarked funds, modernization of public accounting to allow for depreciation, water banks, etc. For instance, the French *Agences de l'Eau* are not only in charge of funding the new policies of integrated river basin management (IRBM), but also of providing subsidies to local authorities for improving WSS through the redistribution of the water levies that they collect. In a way, this system allows the French to keep a very decentralized WSS provision structure. The third possibility, frequent in Germany, is to integrate several local utilities, but going beyond water and wastewater management to also include other services, such as gas, electricity, district heating or communications.

From another angle, social forms of averaging out heavy capital costs are well known, even if they are not always presented as such. For instance, paying for wastewater via local taxes based on the rental value of the properties makes this service more expensive for those who have a large house, and who are usually wealthier. Under municipalism, water suppliers did not pay much attention to the detailed breakdown of potable water uses or the distributive effects of tariffs because they wanted the best quality in unlimited quantities to serve all purposes (in a *commonwealth* vision). In contrast, modern water suppliers try to encourage more efficient water use; but they are still reluctant to really consider the distributive effects of pricing mechanisms: they just want water services to be covered by consumption charges rather than by taxes on properties, and they claim that it is more equitable that way. However, the English and French examples show that excessive consumer orientation of water services is dangerous, even for the companies themselves: users will not readily accept that a good service is costly and they will have an even harder time accepting the fact that if they conserve water, their unit price (per cubic metre) will probably go up![4] Another crucial outcome of the ongoing transformations is that the current opening of the traditionally closed PWS policy community to a whole range of newcomers, particularly the general public, makes water engineers feel awkward and insecure. But is there any alternative to this opening if what is at stake is the public's general confidence in the service and in those who provide it?

Overall, however, the complex arrangements found in Europe, thanks to the cooperation between different levels of government and sometimes with the private sector, allowed the utilities, or 'urban water', to be kept as a separate issue from water resources allocation. Most people in Europe are connected to WSS collective systems, and those who are not live in low-density areas, where the improvement of decentralized technologies allows them to get a good service too.[5]

NEW CHALLENGES FOR WATER AND SANITATION SERVICES (WSS): IS FULL-COST RECOVERY POSSIBLE, AND, IF YES, HOW?

The sustainability of the European models might be challenged in the near future: how will it be possible to maintain good service quality in the long run (even if everybody is connected to the network) if the full cost-recovery model is imposed?

For instance, adopting depreciation and making provisions for infrastructure renewal mean a long-term rise in bills. As a result, an increasing number of large users today (industry, services) either quit PWS, change their processes to become more water efficient or work hard to reduce leaks. This largely explains the recent stagnation and even decrease in water volumes sold (Barbier, 2000). In some countries, even domestic consumers have reduced their demand for PWS, either by changing fixtures and domestic appliances, adopting water-saving garden designs, or switching to rainfall storage or other alternative sources of water for non-drinking uses. Yet, this apparently positive demand reduction ultimately worsens the already fragile financial balance of networked WSS!

In addition, water suppliers in Europe are evaluating the fact that consistent compliance with the stricter drinking water standards adopted by the European Union at reasonable costs will be a much harder objective to achieve in coming years. Public policies on drinking water standards tend to privilege a traditional 'no-risk' strategy (Lave, 1981), which often fails to take the actual costs involved into account. Moreover, the multiplication of standards is slowly making the situation unnecessarily complex and even riskier: chlorination by-products give cancer (Okun, 1996). There are many other problems to consider, including the contradictions between public perceptions and the actual long-term evolution: for instance, media reports on a growing proportion of people receiving non-compliant water services may give rise to a very negative public perception of PWS, while, in fact, water treatment systems are consistently improving in the long run.

All of these factors lead part of the European public to question PWS, particularly once they have abandoned the 'out of sight, out of mind' approach to these services that characterized the municipalist era. However, there are important differences between countries and regions. For instance, criticism of WSS pricing has been stronger in France and England and Wales, where the infrastructure reached maturity long ago, than in Southern Europe, where the infrastructure is still being developed and water prices continue to be far from matching the costs (which, in turn, means government subsidies).[6] The example of France and England and Wales shows that the involvement of large private WSS companies generates suspicion among the public, but also that *rationalization* (both in economic and environmental terms) cannot really start until *rationing* is over. In this regard, to lower the risk of failure and the high prices, water suppliers, along with local, national and European authorities, have turned to new strategies: water resources protection and demand management. These strategies are part of a new approach to water management developed by engineers: environmental engineering. However, the implementation of these strategies is blurring the traditional frontier between water services and water resources; in turn, it also casts light on problems faced, particularly in developing countries: when a significant proportion of the population is not connected to WSS services, it is more difficult to imagine that there is a difference between access to services and access to resources, or else, between water rights and the right to water.

Having originated from the field of sanitary engineering in the US, environmental engineering aims at protecting not only populations from negative environmental factors, but also global and local environments from potentially dangerous human activities. Of course, knowledge of natural processes has been an important consideration in its development; but environmental engineering programmes have also focused on urban problems and technical systems (Barraqué, 1993). WSS services have also become an object of attention for environmental engineers; but their approach to these services has been more global in scope than has been the case of traditional sanitary and civil engineering. Indeed, the characteristic of the traditional WSS policy community is to focus on technology and supply-side solutions, while our new environmental approaches recognize the need to also consider demand in the analysis.

Until recently, urban systems usually called 'basic infrastructure' were hidden from people, despite the fact that they provided people with more freedom as time- and space-saving devices. In addition, from another angle, it was not the (usually sunk) infrastructure, but the plot of built or buildable land, which provided political legitimacy. Thus, city aldermen and elected representatives became competent on issues such as valorization/de-valorization of urban land; but they did not use to know much about the infrastructure systems. Conversely, early sanitary engineers were convinced that public health was too important an issue to be negotiated with either landowners or their tenants, and they preferred to impose connection to infrastructure systems located in the public space. The choice of the *tout-à-l'égout* public policy in Paris is a typical example (Dupuy and Knaebel, 1982). In turn, people became ignorant of the importance of urban infrastructure systems: 'out of sight, out of mind' and NIMBYism (the not-in-my-backyard approach) tended to characterize the public's attitude, which is at odds with a conservation ethic (Melosi, 2000). WSS operators also had no interaction with the public or with demand-side problems: they just had to match the demand with the more or less invisible infrastructure, and that was it. Now when urban services are finally meeting diseconomies of scale, the demand-side approach is at last considered as a potential tool for rationalization.

But what is 'demand side'? While economists equate supply and demand with market prices, environmental engineering considers that demand and supply are not independent as in an ideal transparent market, but are, rather, interrelated. Thanks to the concept of environment, we have learned that sometimes causes act on consequences not linearly, but exponentially, that there are feedback loops, and that supply and demand are not really independent, but interacting. This is what we call the 'network effects'. In other words, in Europe, WSS have become a particular kind of club good: almost everybody is a member of the club, and collectively they pay little for the infrastructure they use. However, if some members quit the club, then the others necessarily pay more! In turn, this stirs up the old question: is water a club good or a public good to be provided to all for social and sanitary reasons?

This leads us to redefine the very notion of WSS 'operator'. If supply and demand interact permanently, then the traditional separation between supply side and engineers, and demand side and elected representatives, is questioned. A direct contact of operators with the public becomes necessary, especially to get away from the tradition of coarse linear and over-optimistic demand projections. Furthermore, it might be useful to get WSS users to become more efficient and thus help to alleviate the negative network effects, although if a traditional operator promotes such changes, it may be resented by the elected authority as undermining its power.

It seems that the ongoing changes will require local authorities to move away from traditionally patronizing attitudes towards their 'client-citizens' to become mediators between the conflicting interests of the various groups of users. This is contractual policy, and political leaders should not be afraid of organizing consultative commissions where various types of users are qualitatively represented. The new WSS operator should not be left alone to make decisions about technology or organizational changes. For instance, too often, local authorities decided to privatize some urban infrastructure, hoping to get rid of a problem that was becoming too difficult or too costly on the back of private companies, which would then take responsibility for rising prices. This is partly what happened in Germany in 1999 with the privatization of the Berliner Wasserbetrieb, Berlin's WSS utility, and also in France in the Grenoble corruption scandal involving a private concession established in 1989 and finally cancelled by the municipality in March 2000. In the end, this type of privatization often fails because rather than less, what was needed was more direct involvement of the local authority in defining and controlling the WSS.

Another important factor of change is that citizens will have to accept the outcomes of modern urban society: they are not little landowners anymore, they are mainly wage earners, and users of all sorts of public utilities and facilities, which will become unsustainable if they do not participate in the definition and provision of the services. Urbanites will need to learn how to conserve drinking water to help diminish the need for future investments. In California, for instance, several utilities subsidize people who adopt low-water consumption fixtures. Water conserved is then made available for extra PWS users, and this water is far cheaper than that tapped through long-distance transfers or even than bought from farmers (Dickinson, 2000). However, if too many users quit the networked PWS or dramatically reduce their consumption, then the utility will be in trouble because it usually has to reimburse long-term debts: an important reduction in volumes billed will oblige the operator or the local authority to raise the unit prices. This may be only a short-term negative outcome; but it has to be carefully explained to consumers to avoid building distrust.

In the end, it can be argued that under the new conditions, the operators of grid-based urban service systems are, in fact, a complex mix of people and institutions who can only master the networks' effects if they interact. Sharing information then becomes crucial in order to achieve success when the provision of WSS requires the interaction of a growing number of institutions and other actors. This is a clear case for multilevel governance.

What does this mean for developing countries?

It is only after a long historical development that, in Europe, WSS services became differentiated from water resources allocation, with separate budgets and legal status. In the initial phases of WSS development, funding through local taxes and subsidies became the rule. Billing and consumerism came later. Indeed, the success of European WSS is largely due to the strong legitimacy of local councils that granted them a good capacity to recover the taxes, or the rates, and later the bills in order to improve the self-financing capacity of WSS.

Conversely, in developing countries, the development of basic infrastructure by national states since World War II was supported by IFIs, which largely overlooked the role of local authorities. In some of these countries, it is possible to identify a typical Anglo-Saxon distrust of local authorities, who are deemed to be irredeemably hostage to corrupt politics. The continued weakness of local public bodies in most developing countries has, in turn, helped to support the idea that if governments did not succeed in organizing the provision of safe WSS, then the only viable alternative was privatization. However, the lessons from the privatization policies of the 1990s in developing countries suggest that, in most cases, the decision to privatize did not respect one of the rules that proved to be crucial in the evolution of European WSS: no rationalization before rationing is over. For instance, privatization of WSS in developing countries sometimes caused people who were not even connected to the network, or who were connected but received poor-quality service, to be charged as if the service were normal. Yet, in Europe, charging people before WSS reached minimum quality standards would not be an option. In turn, good-quality WSS could only be developed once a large majority of people were connected and paid for the services, and once individual solutions, such as drilling local wells or water distribution by street vendors, were phased out.

Going back to the European experience, we can hypothesize what was at the heart of municipalism: the European city of the 20th century was the prime locus for the emergence of a middle-class vision where citizens should have equal access to public services. This was a 'commonwealth' ideal that partly substituted a prior communitarian vision according to which separate communities living in different neighbourhoods should obtain differentiated (yet equitable) access to resources. But what if, in large developing country cities, this commonwealth vision does not prevail? There is no technical way to resolve the *de facto* socio-spatial fragmentation resulting from lack of solidarity and rapid rural–urban migration. On top of this, the sanitary situation has changed: during the 19th century, if there was a waterborne disease epidemic in a city, even the rich who had good water services would eventually die. This gave them a strong incentive to fund the extension of WSS services to poorer neighbourhoods. Today, individual protection against these epidemics is available (for the rich) thanks to medical progress. Why would urban elites then care for the poor immigrants?[7]

This leads us to a concluding remark in terms of territorial scales: in any given country, the optimal size of WSS organization depends not only upon physical-natural dimensions such as rainfall, catchment sizes and population densities, but also upon the degree of trust that society can place in its institutions at various levels (see Maria, 2007). In Europe today, the optimal level of organization is clearly moving towards supra-local, as well as infra-regional multilevel arrangements. In developing countries, it may well be that the municipal level is already too large. One could then imagine public WSS improving, but doing so at 'various speeds': networked domestic supply for the middle-class downtown, but other forms of provision in the peripheral urban areas. Wherever a community (e.g. an ethnic group) has a strong degree of legitimacy, why not entrust them with the distribution of bulk water produced at the municipal level? For many neighbourhoods, it may turn to be more urgent to moralize the profession of water vendors than to propose connections to a hypothetical network (Zérah, 1997). This is an equitable, if not equal, solution; but it may turn out to be the sole possibility until a common feeling of citizenship emerges at the municipal and metropolitan level. There is some room for the private sector in this development; but not if private actors do not understand that one cannot bill customers before they are accustomed to being served with a continuous and reliable public service.[8]

NOTES

1 Some examples of this are the Acquedotto Pugliese in southern Italy, mostly built under Mussolini; the systematic damming of rivers for irrigation and hydroelectricity purposes in Franco's Spain from 1940 to 1975 (although the approach was not entirely abandoned after Franco's death); etc. (Swyngedouw, 2007).

2 Economists arguing in favour of full or at least fair-cost pricing put pressure on politicians to phase subsidies out.

3 For a general analysis of WSS privatization in England and Wales, see Bakker (2004).

4 This is what happened in the city of Paris in January 2007: the socialist mayor had said, upon being elected, that he would lower water bills; but one year before the next election he actually had to raise bills because water consumption had structurally decreased, and the law compels WSS utilities to match expenses (mostly fixed costs) with revenues (water bills).

5 However, in some countries, the levels of people unconnected to PWS remain above 20 per cent, as happens in areas such as western Spain and northern Portugal, southern Italy, Eastern Europe and large parts of Ireland. This, by the way, offers grounds for a comparative reflection with developing countries.
6 Water prices, including sewage collection and treatment, are above 4 Euros per cubic metre in Germany, The Netherlands and Denmark, around 3 Euros in France and the UK, but around 1.5 Euros in Spain and under 1 Euro in Italy.
7 Note that this reasoning is fragile in the face of new diseases such as the Dengue epidemics in Rio de Janeiro (2008), which are largely due to the deficiencies of the wastewater systems in the poorer neighbourhoods.
8 There is a dramatic case in South Africa, where the generalization of billing with free volumes after the end of apartheid resulted in a worse situation for some very poor people, compared with free standpipes in townships before: there even was a cholera outbreak. See Jeter (2002).

References

Bakker, K. (2004) *An Uncooperative Commodity: Privatising Water in England and Wales*, Oxford University Press, Oxford

Barbier, J. M. (ed) (2000) 'Dossier: Evolution des consommations d'eau', in *Techniques, Sciences et Méthodes – Génie Urbain – Génie Rural*, no 2, pp14–65

Barraqué, B. (ed) (1993) *La Ville et le Génie de l'Environnement*, Presses de l'École Nationale des Ponts et Chaussées, Paris

Barraqué, B. (1998) 'Europäisches Antwort auf John Briscoes Bewertung der Deutschen Wasserwirtschaft', in *GWF Wasser-Abwasser*, vol 139, no 6, pp360–366

Dickinson, M.-A. (2000) 'Water conservation in the United States: A decade of progress' in A. Estevan and V. Viñuales (eds) *La Eficiencia del Agua en las Ciudades*, Bakeaz and Fundación Ecología y Desarrollo, Bilbao and Saragossa, Spain

Dupuy, G. and G. Knaebel (1982) *Assainir la VilleHier et Aujourd'hui*, Dunod, Paris

Fougères, D. (2004) *Approvisionnement en Eeau à Montreal: Du Privé au Public 1796–1865*, Septentrion, Sillery

Jeter, J. (2002) 'South Africa's driest season', in *Mother Jones*, November–December, www.motherjones.com/news/feature/2002/11/ma_145_01.html, accessed July 2008

Lave, L. (1981) *The Strategy of Social Regulation*, Brookings Institution, Washington, DC

Maria, A. (2007) *Quels modèles techniques et institutionnels assureront l'accès du plus grand nombre aux services d'eau et d'assainissement dans les villes indiennes*, PhD thesis, University of Paris, Dauphine

Maver, I. (2000) *Glasgow, Town and City Histories*, Edinburgh University Press, Edinburgh

Melosi, M. V. (2000) *The Sanitary City, Urban Infrastructure in America from Colonial Times to the Present*, Johns Hopkins University Press, Baltimore

Okun, D. A. (1996) 'From cholera to cancer to cryptosporidiosis', *Journal of Environmental Engineering*, vol 122, no 6, pp453–458

Pezon, C. (2000) *Le Service d'Eau Potable en France de 1850 à 1995*, Presses du CEREM, Paris

Swyngedouw, E. (2007) 'Techno natural revolutions – the scalar politics of Franco's hydro-social dream for Spain, 1939–1975', *Transactions, Institute of British Geographers*, vol 32, no 1, pp9–28

Zérah, M. H. (1997) 'Inconstances de la distribution d'eau dans les villes du Tiers Monde: Le cas de Delhi', in *Flux, Cahiers Scientifiques Internationaux*, no 30, October–December, pp5–15

15

'From East to Western Sea': Canada – A Country of National Abundance and Local Shortages

Mark W. Rosenberg

Introduction

The words of Canada's national anthem, 'O Canada', come from a poem by R. Stanley Weir. In the second stanza, he wrote:

> O Canada! Where pines and maples grow.
> Great prairies spread and lordly rivers flow.
> How dear to us thy broad domain,
> From East to Western Sea,
> Thou land of hope for all who toil!
> Thou True North, strong and free! (Government of Canada, 2008)

On the Arms of Canada, the motto reads: '*A mare usque ad mare*', or 'From sea to sea'. There is, arguably, no country in the world where water is so ingrained in the psyches, symbols, lore and everyday activities of people's lives as in Canada.

With a population of approximately 32.5 million in 2007 and a land mass of almost 10 million square kilometres, Canada is a country of abundant freshwater in contrast to most countries around the world. There are approximately 2 million freshwater lakes covering approximately 7.6 per cent of the land mass. Another 2 per cent of the land mass is covered by glaciers and ice fields. The Great Lakes (Superior, Michigan, Huron, Erie and Ontario), Lake Winnipeg, Great Bear Lake and Great Slave Lake are among the largest freshwater lakes in the world. The Mackenzie, Columbia, Fraser, Saskatchewan and St Lawrence rivers are among the longest and largest freshwater rivers in the world. In addition, there are giant freshwater aquifers in various parts of the country that are sources of freshwater (Environment Canada, 2008a). As a result, Canadians generally take for granted that their local governments will provide safe, inexpensive (i.e. by global standards), abundant clean water for consumption and sanitation (see Shrubsole and Draper, 2007).

Bakker (2007, p16) argues 'that Canadian water management and governance are at a crossroads'. Her expert view stems from a reading of the complex interplay of three levels of government (the federal, provincial and local), where the political discourse of water management and governance has grown more intense in recent years, the desire to experiment with the introduction of various forms of private sector and not-for-profit partnerships has increased, and the changing public perception resulting from the effects of high-profile failures of government to ensure the quality of the water supply and sanitation.

In the following sections, first an overview of the provision of water and sanitation services is provided. Then two recent events, the death of 7 people and the illness of 2300 others in Walkerton, Ontario, in 2000, and the evacuation of the Kashechewan First Nations in 2005 because of a failed water system, are used to raise the spectre of a very different image of the water supply and sanitation systems that Canadians still, for the most part, take for granted. The argument presented is that Walkerton and the Kashechewan First Nations are emblematic of more general issues of ageing infrastructure, declining institutional responsibility and governance facing Canadians that challenge the taken-for-granted notion of a safe, inexpensive, abundant water supply and sanitation, and provide more general lessons that apply to other countries where water is far less abundant than it is in Canada.

An overview of Canada's water and sanitation systems[1]

More than 80 per cent of Canada's population lives in urban areas (Statistics Canada, 2007, p20) and as such they are likely to receive their water from a municipal water system and are dependent upon a municipal sanitation system for the removal of wastewater. The source of water in most municipal water systems (88.4 per cent) is from surface flows. There is, however, a high degree of variability even at the provincial/territorial scale. For example, in the small province of Prince Edward Island, no municipal water comes from surface flow sources. In contrast, in the far north in two of the territories, 100 per cent of the municipal water comes from surface flow sources.

Municipal water systems cover 90.6 per cent of their populations.[2] Water use is divided among residential users (56.4 per cent), industrial users (30.6 per cent) and, remarkably, 13 per cent is accounted for in system losses.

By international standards, Canadians are heavy users of water: 329 litres per capita per day (lcd). There are, however, significant differences in use between people living in large municipalities as measured by population size (population > 500,000 use 291lcd) compared to people living in small municipalities (populations 2000 to 5000 use 497lcd).

Wastewater treatment also varies geographically and by the population size of municipalities. About 88.7 per cent of municipal populations are connected to sanitary sewer systems, while 85.9 per cent have some form of sewage treatment. In provinces such as Newfoundland and Nova Scotia, the percentage of the municipal population who are connected to some form of sewage treatment facility falls to 17.2 and 29.5 per cent, respectively. In municipalities of less than 1000 people, only 33.2 per cent of the population is connected to some form of sewage treatment facility compared to 97.6 per cent of the people living in municipalities with populations greater than 500,000. Of the 22 million Canadians served by sanitary sewers and living in municipalities, 23 per cent are served by primary treatment levels, 46.6 per cent by secondary treatment levels, 21 per cent by tertiary levels, 6 per cent by waste stabilization ponds (WSPs), 0.2 per cent by preliminary treatment levels, and 3.2 per cent are not served by wastewater treatment facilities at all.

While the data above provide evidence of extensive access to modern water and sanitation systems, there are regional differences and differences related to the population size of municipalities. Particularly in municipalities that serve smaller populations, where the hinterlands of the municipalities are likely to be rural and in rural areas, access to modern water and sanitation systems becomes less likely. In the next section, two case studies are reviewed that illustrate the challenges faced by local governments in providing modern water and sanitation systems.

TWO CASE STUDIES

The story of Walkerton, Ontario[3]

Walkerton, Ontario, is located in south-western Ontario. As the result of municipal restructuring, it is now part of Brockton Municipality. The municipality had a population of 9658 and a population density of 17.1 people per square kilometre in 2006 (Statistics Canada, 2006a). Walkerton is a quintessential Ontario small town, providing services for a mainly rural population. Like many small towns, water, the sanitation system, electricity and natural gas are provided by a Public Utilities Commission (PUC) that reports to the municipal government. The PUC is a small group of workers mainly drawn from the local population, in most cases with little or no professional training. In the case of Walkerton, one person, Stan Koebel, was mainly responsible for the town water supply. Part of his responsibilities was to take water samples, send them to a private laboratory that is contracted with the Ontario provincial government to do water testing, and return the information to the PUC for *appropriate* action to be taken. As it turned out, this was a fatal flaw in the system.

On 18 May 2000, Koebel received a fax from the private laboratory informing him that the water samples sent on 15 May 2000 from Walkerton were contaminated with *E. coli*. He did what he had done for years, which was to add more chlorine to the water supply and keep the fax a secret. On 19 May 2000, the Regional Medical Health Office (RMHO) learned that the local hospital that serves the Walkerton population had been treating patients since 17 May 2000 who presented themselves to the hospital with bloody diarrhoea, vomiting, cramps and fever. Koebel assured the RMHO that the water was safe and the RMHO launched an investigation looking for other possible sources of contamination. As more people continued to arrive at the local hospital with symptoms of *E. coli* poisoning, Koebel continued to assure the RMHO that the water supply was safe. On 21 May 2000, the RMHO warned people to stop drinking water from the local water supply and launched its own investigation into the water supply of Walkerton. Using its publicly owned laboratories, the RMHO confirmed that the water supply was contaminated with *E. coli*; they also found out about the fax sent on 18 May 2000, and that equipment used to inject chlorine at one of the wells used to supply water for the Walkerton water and sanitation system had not been working for some time. By this time a full-scale public health emergency was in progress and by the time the emergency was over, 7 people were dead and approximately 2300 people were ill from the water contamination.

What happened in Walkerton was arguably the worst case of water contamination in modern post-World War II history in Canada. As a result, the provincial government asked Judge Dennis O'Connor to conduct a full inquiry into what went wrong and to recommend changes to how water is managed in Ontario in order to guarantee safe water and sanitation systems. In January 2002, Judge O'Connor delivered Part I of his report. Box 15. 1 summarizes selected key points from his report.

What the key points highlight is that, first, farming practices were not to blame for the initial water contamination. Second, the operators failed in a number of critical areas to carry out their duties appropriately; they had knowingly failed in their responsibilities over a long period of time (i.e. the contamination was not the result of one-time breakdown in practices); and they lacked the training and expertise to ensure that an outbreak did not occur and how to address the problem appropriately should an outbreak occur. Third, beginning in 1996, the provincial government made a series of decisions to privatize the water-testing laboratories of the Ministry of the Environment (MOE), but failed to create an effective system of communication among the PUCs, the private laboratories, the MOE, and the provincial medical officer of health. Furthermore, the MOE, as the result of budget cuts, did not have the resources to ensure effective inspections of local PUC practices.

Summing up, the events before and after what took place at Walkerton raised doubts in the minds of many Canadians as to whether the water they drink and the sanitation systems that they used are safe. Especially in small towns and rural areas, the expertise of those responsible for the water and sanitation systems is now a focus of concern. Beyond the local issues, the privatization of provincial laboratories raises more general questions about the neoliberal agenda of many governments, local, provincial and even federal, as to the claims that the private sector can deliver public goods at lower costs and, in this

Box 15.1 Selected key findings from the report of Judge Dennis O'Connor on the Walkerton, Ontario, case

- The contaminants, largely *E. coli* O157:H7 and *Campylobacter jejuni*, entered the Walkerton system through Well 5 on or shortly after 12 May 2000.
- The primary, if not the only, source of the contamination was manure that had been spread on a farm near Well 5. The owner of this farm followed proper practices and should not be faulted.
- The outbreak would have been prevented by the use of continuous chlorine residual and turbidity monitors at Well 5.
- The failure to use continuous monitors at Well 5 resulted from shortcomings in the approvals and inspections programmes of the Ministry of the Environment (MOE).
- The Walkerton Public Utilities Commission (PUC) operators lacked the training and expertise necessary to identify either the vulnerability of Well 5 to surface contamination or the resulting need for continuous chlorine residual and turbidity monitors.
- The scope of the outbreak would very likely have been substantially reduced if the Walkerton PUC operators had measured chlorine residuals at Well 5 daily, as they should have, during the critical period when contamination was entering the system.
- For years, the PUC operators engaged in a host of improper operating practices, including failing to use adequate doses of chlorine, failing to monitor chlorine residuals daily, making false entries about residuals in daily operating records, and misstating the locations at which microbiological samples were taken. The operators knew that these practices were unacceptable and contrary to MOE guidelines and directives.
- The MOE's inspections programme should have detected the Walkerton PUC's improper treatment and monitoring practices and ensured that those practices were corrected.
- The PUC commissioners were not aware of the improper treatment and monitoring practices of the PUC operators. However, those who were commissioners in 1998 failed to properly respond to an MOE inspection report that set out significant concerns about water quality and that identified several operating deficiencies at the PUC.
- On Friday, 19 May 2000, and on the days following, the PUC's general manager concealed from the Bruce-Grey-Owen Sound Health Unit and others the adverse test results from water samples taken on 15 May and the fact that Well 7 had operated without a chlorinator during that week and earlier that month. Had he disclosed either of these facts, the health unit would have issued a boil water advisory on 19 May, and 300 to 400 illnesses would have been avoided.
- In responding to the outbreak, the health unit acted diligently and should not be faulted for failing to issue the boil water advisory before Sunday, 21 May. However, some residents of Walkerton did not become aware of the boil water advisory on 21 May. The advisory should have been more broadly disseminated.
- The provincial government's budget reductions led to the discontinuation of government laboratory testing services for municipalities in 1996. In implementing this decision, the government should have enacted a regulation mandating that testing laboratories immediately and directly notify both the MOE and the medical officer of health of adverse results. Had the government done this, the boil water advisory would have been issued by 19 May at the latest, thereby preventing hundreds of illnesses.
- The provincial government's budget reductions made it less likely that the MOE would have identified both the need for continuous monitors at Well 5 and the improper operating practices of the Walkerton PUC.

Source: MAG (2002)

case, safely, and whether governments can provide the safeguards to ensure that the private sector meets its commitments to the public.

The story of Kashechewan[4]

In Canada, 1,172,790 people, or 3.8 per cent of the population, self-identified themselves as part of the Aboriginal population in the 2006 Census (Statistics Canada, 2006b). The majority of the Aboriginal population live in southern Canada on or off reserves. The remainder live on reserves in semi-remote or remote locations in the northern parts of provinces and the far north. Ethnically and politically, the Aboriginal population is divided into three groups: the First Nations (North American Indians) who mainly live in the provinces; the Métis, who are mainly the descendants of the mixing of French explorers and settlers with First Nations' members in what are now the western provinces of Canada; and the Inuit, whose bands mainly live in the far north of Canada in what are now the three territories (Yukon, Northwest and Nunavut).

The people of Kashechewan are part of the Fort Albany and Kashechewan First Nations. The Kashechewan reserve is among the most isolated in Canada. It is situated on the Albany River just inland from James Bay, between 52 and 53 degrees latitude north and between 81 and 82 degrees longitude west. It is between 400km and 480km from the nearest service centre and has no year-round road access. In 2007, the on-reserve population of the Fort Albany and Kashechewan First Nations was 2395. It is governed by an elected chief and tribal band council, as is common among First Nations in Canada (Statistics Canada, 2006a).

In 2005, two major evacuations took place because of flooding resulting from ice dams on the Albany River downriver from the reserve. Most of the population had to be flown south over 400km to Timmins, Ontario. At the time of the second evacuation in October 2005, high *E. coli* levels were found in the reserve's drinking water and chlorine levels had to be increased to 'shock' levels. It turned out that the process of increasing chlorine levels to shock levels had been a long-time practice and that due to excessive chlorine use, people were suffering from chronic skin conditions (e.g. scabies and impetigo) and other health problems.

The two evacuations, the poor health of many of the residents of the reserve and the cost of the evacuation (estimated to be about 16 million Canadian dollars spent by the provincial government of Ontario and the federal government of Canada), brought to the fore a number of issues. The water system in use at the time was built in 1995 to replace an old one, which presumably was even less adequate. When it was built, the water intake pipe was placed downstream from the reserve sewage lagoon. Some members of the Kashechewan reserve believed that the water system was too small to serve the approximately 1900 people living on the reserve at the time.

In November 2006, three solutions were proposed to address the problems of the failed water and sanitation system and the problems of periodic flooding of the reserve: move the reserve to Timmins, Ontario; move the reserve 30km to higher ground; or rebuild the community with an improved water system and dykes to stop the flooding. The tribal band council chose the third option. At the beginning of 2008, the rebuilding had still not been completed.

While no deaths occurred as a result of the water problems in Kashechewan, the health problems resulting from over-chlorinated water, poor sanitation and, ultimately, the use of contaminated water shocked Canadians. The plight of the people of Kashechewan brought attention to the over 100 other reserves across Canada that have similar water problems. It was also another high-profile example of how Aboriginal Peoples are treated by the federal and provincial governments (both accused the other of negligence).

EXPLAINING TWO STORIES

In research, policy and planning, and the popular press, there is always the possibility and temptation to see the stories of Walkerton and Kashechewan as unique events explained, in the former case, by the failure of individuals and, in the latter case, as the vagaries of living in an isolated place under harsh conditions. There are, however, both empirical evidence and theoretical reasons to argue that the two stories are exemplars of more general problems related to the supply of clean water and safe sanitation systems in Canada.

Water availability and quality

In their own analysis, Environment Canada (2007) acknowledges that water availability and quality continue to be issues of significance in the delivery of water and the treatment of waste across Canada. In their survey of municipal water systems, Environment Canada reported that 72 municipalities out of 510 reported experiencing water shortages in 2004, 107 out of 507 reported water quality problems between 2001 and 2004, and 44 out of 507 reported one or more 'boil water days' between 2001 and 2004.

The issue of water quality is even more acute among First Nations and Inuit communities: 'In 1999, 65 First Nations and Inuit communities were under a boil water advisory for varying lengths of time – an average of 183 days of boil water advisories per affected community' because their water supplies were contaminated (Health Canada, 2000, p66). This is all the more remarkable because 98 per cent of all First Nations people living on reserves are connected to a water supply service and 95 per cent have sewage disposal services (Indian and Northern Affairs Canada, 2005, p16).

Canada's record on the treatment of wastewater also needs to improve. Only approximately 57 per cent of Canadians are served by wastewater treatment plants, compared with 74 per cent of Americans, 87 per cent of Germans, and 99 per cent of Swedes (Environment Canada, 2008b).

Issues of improving the infrastructure for water in Canada remains an ongoing issue (see below). The events at Walkerton and Kashechewan fit with the broader problems that municipalities and Aboriginal communities continue to have in supplying water and appropriate waste treatment facilities; but the two case studies are also exemplars of three more general arguments about ageing infrastructure, declining institutional responsibility and issues of governance.

Ageing infrastructure

In a study for the Federation of Canadian Municipalities (FCM), Mirza (2007) provides a picture of the problems of the ageing infrastructure of Canada's water and sanitation system. Looking at all of Canada's infrastructure stock – transportation (roads, bridges, curbs, sidewalks), water infrastructure (distribution, supply and treatment), wastewater systems (sanitary and storm sewers and related treatment facilities), transit systems (facilities, equipment and rolling stock) and other public infrastructure (including cultural, social, community and recreational facilities, and waste management facilities) – 28 per cent is between 80 and 100 years old, 31 per cent is between 40 and 80 years old and only 41 per cent is between 0 and 40 years old (Mirza, 2007, p10).

In analysis that focuses on sewer systems and wastewater treatment facilities, Gaudreault and Lemire (2006, p9) suggest that 58 per cent of local government wastewater treatment facilities had reached the end of their service life in 2003, and at the provincial and federal levels, 76 and 71 per cent, respectively, had reached the end of their service lives. Their analysis of sewer system infrastructure is more positive because new sewer systems have been installed as new roads have been built, particularly in the fastest growing urban areas of Canada. In 2003, 47 per cent of local government-owned sewer systems had reached the end of their service lives in comparison to provincial and federal sewer systems, where 71 and 68 per cent had reached the end of their service lives (Gaudreault and Lemire, 2006, p8).

Taking into account the ageing of the infrastructure and that water and wastewater infrastructure made up about 30 per cent of Canada's infrastructure stock in 2000, Mirza (2007, pp7, 12) reports that it would cost US$31 billion to meet existing needs, while future needs are estimated at US$56.6 billion. Other analysts have carried out similar exercises, and although the dollar amounts that result are sometimes higher or lower than those reported by Mirza because of the various ways in which water and wastewater infrastructure can be defined, the time periods used, and whether measurements are in current or constant dollars, there is a consensus that the local governments cannot keep up or can barely keep up with their existing infrastructure needs. There is also consensus that, in the future, substantial new investment will be required to replace ageing infrastructure and to meet the demands for new developments (Harchaoui et al, 2003; Mirza, 2007; Roy, 2007). The other fact that no one denies is that the responsibility for water and wastewater infrastructure has fallen increasingly to local government authorities who have the least fiscal ability of all levels of government to fund infrastructure investment (see below).

Declining institutional responsibility

Legal responsibility for water is divided among three levels of government: the federal government; the provinces and territories; and local governmental authorities. The federal government only has shared jurisdiction with the provinces and territories where water is international (e.g. the Great Lakes and the coastal waters of Canada). The provincial and territorial governments have jurisdiction over all inland lakes and rivers. It is the local governmental authorities, mainly through their water departments or public utility commissions, which are responsible for the day-to-day delivery of safe, clean water;

operating the sewer and wastewater system; and ensuring that the water quality meets the standards set by the federal, provincial and territorial governments. In fact, local government authorities own more than 70 per cent of the water supply facilities and more than 80 per cent of the sewer systems infrastructure in Canada (Gaudreault and Lemire, 2006, p9).

De Loë and Kreutzwiser (2007) describe how water management in Canada has become increasingly complex since the 1980s, while policy interest in water by the federal and provincial governments has declined. The complexity of public management is illustrated in the two case studies just by considering the governments and the departments involved. In the case of Walkerton, you have the municipal government and the PUC. At the provincial level, you have the Ministry of Health and Long-Term Care through the Office of the Provincial Medical Officer becoming involved once people began presenting themselves at the local hospital in significant numbers, as well as the Ministry of the Environment, which was involved in the monitoring. In the case of Kashechewan, the tribal band council, Indian and Northern Affairs Canada (the federal government ministry responsible for the federal role on reserves) and Health Canada (the federal government ministry responsible for First Nations health) were the main governmental actors prior to the evacuations. With the evacuations, the Ontario government and the local municipal government of Timmins, Ontario, were drawn into providing healthcare, temporary housing and schooling for the children, while most of the Kashechewan people stayed in Timmins for months until it was safe to return.

In both the cases, the higher levels of government were aware of the problems the two communities were having with their water supplies well before the events that brought them to the attention of the Canadian public. In the case of Walkerton, Judge O'Connor noted in 1998 that an MOE report identified problems in how the water system was being managed, but neither the PUC commissioners nor the MOE ever followed up to ensure that the problems were rectified. In the case of Kashechewan, the problems of the water system can be traced back to its 1995 installation, and ongoing problems from that time forward, which were ultimately the responsibility of Indian and Northern Affairs Canada and Health Canada.

ISSUES OF GOVERNANCE

Much has been written about the privatization of public utilities responsible for water internationally (e.g. see Bakker, 2004). In Ontario, the supply of water and responsibility for sewer and sanitation systems, however, remain mainly the responsibility of local government authorities either through water departments or public utility commissions with respect to ownership. Like other jurisdictions around the world, there is some experimentation taking place with how water and sewer and sanitation systems are managed (see Bakker and Cameron, 2005).

What did change in Ontario in the 1990s was the approach of the provincial government to all aspects of water. In 1994, the Progressive Conservative Party of Ontario under the leadership of Mike Harris took control of the government and instituted what they called the 'Common Sense Revolution' (Progressive Conservative Party of Ontario, 1994). Their neo-conservative agenda was to cut taxes and deregulate, on the one hand, and, on the other, to create a smaller role for the provincial government through downloading responsibilities to local government authorities or through privatization (for a critical general analysis, see Ralph et al, 1997). Crucial to the analysis in this chapter is that those parts of the government responsible for the environment were among the most severely restructured in terms of privatization and deregulation (Winfield and Jenish, 1998).

What one sees in the Walkerton example is how the neo-conservative agenda contributed to the breakdown in governance that ultimately resulted in what took place. Judge O'Connor, in his findings, pointed directly to the lack of resources that the Ministry of the Environment had to enforce its regulations. He also pointed to the failure in the legislation that privatized the water testing laboratories of the Ministry of the Environment, leading to a situation where the private laboratory was only obligated to communicate with the local water department or PUC.

The Kashechewan example is also about a breakdown in governance. What happened in Kashechewan is the result of the flawed system of reserves that has its roots in the 19th century and remains unworkable today (see Dussault and Erasmus, 1996) because of the dynamics of a distant central government (in this case, the federal government through Indian and Northern Affairs Canada in Ottawa) that continues to control the day-to-day lives of the people of Kashechewan thousands of kilometres away. In a summary document, the Institute on Governance (1997) noted a key finding of the Royal Commission on Aboriginal Peoples related to healthcare was that only through self-government would Aboriginal Peoples have the ability to improve the natural environment (i.e. water being a key issue) and improve their health.

What also comes out of the examples of Walkerton and Kashechewan is a governance issue that is common in many other countries, as well, and that is how the legal and regulatory responsibilities for water and sanitation and sewage systems are divided up among various levels of government and among various departments within the same government. In the case of Walkerton, while the Ministry of the Environment was responsible for monitoring and regulation, it was the Ministry of Health and Long-Term Care that was responsible for ordering people to stop drinking the water and the health implications that ensued for those who had become sick. In the case of Kashechewan, you have the federal government involved through Indian and Northern Affairs Canada and Health Canada and the provincial government involved through the Ministry of the Environment and Ministry of Health and Long-Term Care (see the above section for the examples of the levels of government involved).

Conclusions

The stories of Walkerton and Kashechewan have been used to motivate a broader picture of the challenges faced by Canadians, Aboriginal Peoples and the various levels of government to deliver clean water for drinking and sanitation across Canada. While the working assumption has always been that there is an abundance of clean water in Canada, Canadians and Aboriginal Peoples remain at risk for waterborne diseases, and there remains a significant part of the Canadian and Aboriginal population who cannot count on the water quality that enters their homes and/or are not connected to a wastewater treatment plant.

To explain this fractured picture at a systemic level, the ageing of the infrastructure, declining institutional responsibility and issues of governance have been discussed. In the delivery of water and the development of sanitation systems, the specific issues that underlie these three themes might be unique to Canada; but the themes themselves are being played out in developed and less-developed countries around the world. In Canada, the Aboriginal Peoples are paying a particularly high price for the policy failures in the delivery of clean water and the development of sanitation systems. This, too, is unique in the details, but has its parallels in many other countries where indigenous populations and other demographic groups marginalized either by economics, politics, race or ethnicity have inadequate access to clean water and sanitation systems. The challenge in Canada and, indeed, in many other countries is to make clean water and sanitation systems accessible for all.

Notes

1 The statistics in this section all come from Environment Canada (2007).
2 The remainder of the population who live in municipalities are not connected to the municipal system and are likely to get their water from private wells on their property. This normally occurs in municipalities where only part of the municipality is urban and the remainder is rural.
3 For background material, this section relies heavily on CBC (2004).
4 For background material, this section relies heavily on CBC (2006).

References

Bakker, K. (2004) *An Uncooperative Commodity: Privatizing Water in England and Wales*, Oxford University Press, Oxford

Bakker, K. (2007) 'Introduction', in K. Bakker (ed) *Eau Canada: The Future of Canada's Water*, University of British Columbia Press, Vancouver, pp1–20

Bakker, K. and D. Cameron (2005) 'Changing patterns of water governance: Liberalization and de-regulation in Ontario, Canada', *Water Policy*, vol 7, no 5, pp485–508

CBC (Canadian Broadcasting Corporation) (2004) 'Canada's worst-ever *E. coli* contamination', www.cbc.ca/news/background/walkerton, accessed 4 January 2008

CBC (2006) 'Kashechewan: Water crisis in Northern Ontario', www.cbc.ca/news/background/aboriginals/kashechewan.html, accessed 4 January 2008

de Loë, R. and R. Kreutzwiser (2007) 'Challenging the status quo: The evolution of water governance in Canada,' in K. Bakker (ed) *Eau Canada: The Future of Canada's Water*, University of British Columbia Press, Vancouver, pp85–103

Dussault, R. and G. Erasmus (1996) *Report of the Royal Commission on Aboriginal Peoples*, Royal Commission on Aboriginal Peoples, Ottawa

Environment Canada (2007) 'Municipal water use, 2004 statistics', www.ec.gc.ca/water/en/info/pubs/sss/e_mun2004.htm, accessed 27 May 2008

Environment Canada (2008a) 'Freshwater facts', www.ec.gc.ca/water/en/info/facts/e_contnt.htm, accessed 4 January 2008

Environment Canada (2008b) 'A sampling of water quality facts', www.ec.gc.ca/water/en/manage/qual/e_facts.htm, accessed 4 January 2008

Gaudreault, V. and P. Lemire (2006) *The Age of Public Infrastructure in Canada*, Statistics Canada, Ottawa

Government of Canada (2008) 'National anthem of Canada', www.canadianheritage.gc.ca/progs/cpsc-ccsp/sc-cs/anthem_e.cfm, accessed 4 January 2008

Harchaoui, T. M., F. Tarkhani and W. Warren (2003) *Public Infrastructure in Canada: Where Do We Stand?*, Statistics Canada, Ottawa

Health Canada (2000) *A Statistical Profile on the Health of First Nations in Canada*, Health Canada, Ottawa

Indian and Northern Affairs Canada (2005) *Comparison of Socio-Economic Conditions, 1996 and 2001: Registered Indians, Registered Indians Living On Reserve, and the Total Population of Canada*, Indian and Northern Affairs Canada, Ottawa

Institute on Governance (1997) *Summary of the Final Report of the Royal Commission on Aboriginal Peoples: Implications for Canada's Health Care System*, Institute on Governance, Ottawa

MAG (Ministry of the Attorney General) (2002) *Report of the Walkerton Commission of Inquiry*, Government of Ontario, Ontario, www.attorneygeneral.jus.gov.on.ca/english/about/pubs/walkerton/part1/Default.asp, accessed June 2008

Mirza, S. (2007) *Danger Ahead: The Coming Collapse of Canada's Municipal Infrastructure*, Federation of Canadian Municipalities, Ottawa

Progressive Conservative Party of Ontario (1994) *Ontario PC: The Common Sense Revolution*, Progressive Conservative Party of Ontario, Toronto, Ontario

Ralph, D. S., A. Regimbald and B. St-Amand (1997) *Open for Business, Closed to People: Mike Harris's Ontario*, Fernwood Publishing, Halifax, Nova Scotia

Roy, F. (2007) 'From roads to rinks: Government spending on infrastructure in Canada, 1961 to 2005', *Canadian Economic Observer*, September, pp3.1–3.22

Shrubsole, D. and D. Draper (2007) 'On guard for thee? Water (ab)uses and management in Canada', in K. Bakker (ed) *Eau Canada: The Future of Canada's Water*, University of British Columbia Press, Vancouver, British Columbia, pp37–54

Statistics Canada (2006a) '2006 community profiles', www12.statcan.ca/english/census06/data/profiles/community/Index.cfm?Lang=E, accessed 4 January 2008

Statistics Canada (2006b) 'Aboriginal peoples in Canada in 2006: Inuit, Métis and First Nations', 2006 Census, www12.statcan.ca/english/census06/analysis/aboriginal/surpass.cfm, accessed 17 January 2008

Statistics Canada (2007) *Portrait of the Canadian Population in 2006, 2006 Census: Population and Dwelling Counts, 2006 Census*, Minister of Industry, Ottawa, Ontario

Winfield, M. S. and S. G. Jenish (1998) 'Ontario's environment and the "common-sense" revolution', *Studies in Political Economy*, no 57, pp129–147

16

The US Experience on Water Supply and Sanitation: The Interaction between Public Policy and Management

Venkatesh Uddameri and Vijay P. Singh

Introduction

The ascendancy of the US as a major agricultural and industrial nation in the modern world has always been guided by the ability to supply adequate amounts of water for human consumption and growth. The nation also has a long tradition of recognizing the importance of proper sanitation and suitable waste management practices to ensure sustainability of its water supplies. While incidences of improper management of water due to overexploitation or loss in the quality due to pollutant discharges can be found in US water history, they have provided valuable lessons for conservation. Compared to many populous nations in the world, the urban centres of the US provide continuous water supply and high-quality sanitation services to even its poorest residents. However, the rise of major cities such as Los Angles and San Antonio has come at the cost of endangering sensitive ecosystems such as Mono Lake, Owens Lake and the Edwards aquifer. The US has learned many lessons from these failures and responded by developing a comprehensive socio-political and technological framework that continues to evolve even today. Health risks arising from improper waste disposal practices, such as those documented at Love Canal, New York, and Woburn, Massachusetts (Harr, 1996), have provided the impetus for developing comprehensive environmental legislation. While the US has made great strides to provide access to adequate amounts of high-quality water to over 95 per cent of its population, it will need hundreds of billions of dollars to rehabilitate its ageing infrastructure and reach out to economically distressed communities (Ward, 1999; Melosi,

2004). This chapter aims to provide a brief overview of the US experience on water supply and sanitation, with particular reference to the interaction between technical and social sciences.

US WATER SUPPLY AND AVAILABILITY

The three major sources of water in the US include inland surface water resources, such as lakes, reservoirs, streams and rivers; groundwater in aquifers; as well as saline water from oceans and seas and deep underground systems. There are nearly 250,000 rivers in the US spanning a length of over 3.5 million miles (5.6 million kilometres). The US is home to some major river systems, including the Mississippi, Missouri, Columbia, Ohio and Colorado rivers. Natural lakes are also a major source of water, especially in the northern sections of the country. The five great lakes – namely Superior, Michigan, Huron, Erie and Ontario – cover over 94,500 square miles (244,660km^2). In addition, the US has an extensive network of man-made reservoirs created by over 75,000 dams (FEMA, 2005). The average annual rainfall is approximately 35 inches (89cm) per year and ranges from a low of about 7 inches (18cm) per year in the desertic south-west to about 60 inches (152cm) per year in the humid east. Furthermore, the rainfall also exhibits considerable temporal variability as well, necessitating the dense network of dams to store water during periods of high water availability for use during dry periods.

The nation is also endowed with significant groundwater resources. There are 66 principal aquifer systems and hundreds of other minor aquifer formations (Maupin and Barber, 2005). Groundwater resources are extensively used in the south-western sections of the nation. In addition to these freshwater sources, the US also has a long coastline along the Atlantic, Gulf of Mexico, Pacific and Arctic oceans. The natural coastline of the US is nearly 12,400 statute miles (19,950km) with a shoreline of over 88,600 statute miles (142,550km). Coastal waters are mostly used to meet thermoelectric power and industrial demands. There is also a renewed interest in using coastal waters to meet drinking water demands, especially in the gulf states of Florida and Texas.

The US Geological Survey (USGS) estimated that the total fresh and saline withdrawals during 1995 were approximately 402,000 million gallons (1.5 billion litres) per day for all off-stream water-use categories (Solley et al, 1998). Nearly 85 per cent of this water could be classified as freshwater and the remaining 15 per cent classified as saline. Thermoelectric power and agricultural irrigation were the two largest water-use categories, which accounted for nearly 80 per cent of the total water use (see Figure 16.1). The largest water withdrawals were noted to occur in the states of California and Texas. The US per capita water use is one of the highest in the world and was about 1500 gallons (5678 litres) per day (Solley et al, 1998). Nearly 25 per cent of the total water that is withdrawn is used consumptively. As depicted in Figure 16.2, nearly 66 per cent of the water withdrawal was supplied by fresh surface water sources and nearly 20 per cent was obtained from groundwater resources.

In addition to anthropogenic demands, the ecological importance of water is increasingly being recognized in the US. Lakes, rivers and outflows from aquifers

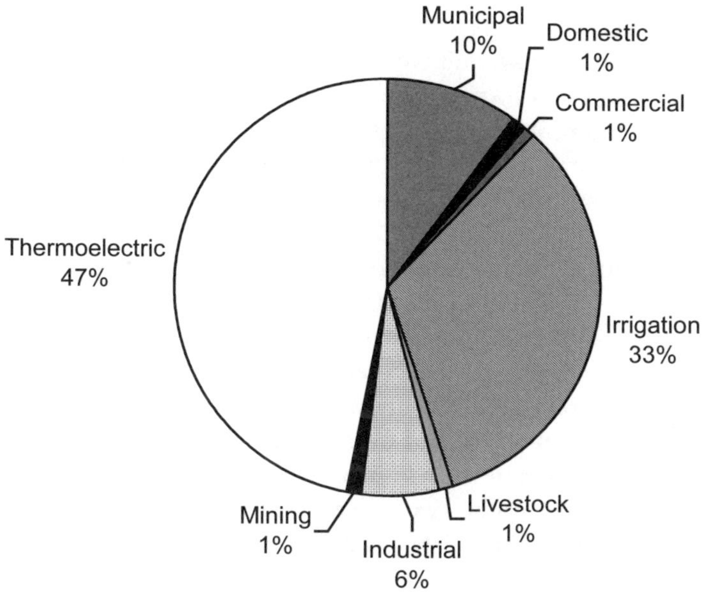

Figure 16.1 *Relative withdrawals for different water-use categories (US)*

Source: based on data from Solley et al (1998)

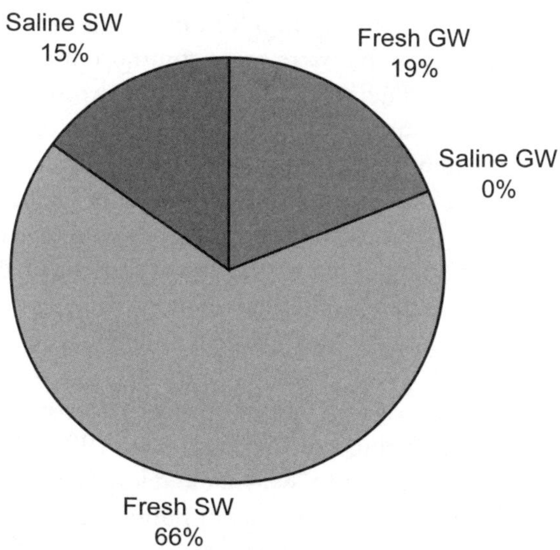

Figure 16.2 *Relative fractions of fresh and saline groundwater and surface water withdrawn in 1995 (US)*

Source: based on data from Solley et al (2005)

(springs and other discharges) help to meet the water needs of terrestrial wildlife and aquatic species. Discharges to coastal bays and estuaries regulate salinity, temperature and other biogeochemical conditions and help to sustain estuarine ecosystems. The natural variability in stream flows has significant ecological implications. High flows (floods) help to transport sediments and facilitate the movement of some species to their spawning grounds. On the other hand, low flows reduce the number of larger species and facilitate the growth of smaller species (Postel and Ritcher, 2003). The concept of environmental flows aims to preserve a certain threshold flow in the river for aquatic species, and to preserve or restore the natural variability of the stream. There is a growing interest in understanding freshwater needs of bays and estuaries (e.g. Pulich et al, 2002). While surface water resources and groundwater reservoirs represent two different pools of water storage, they are intimately interconnected. Groundwater discharges (base flows) sustain flows in rivers during dry summer months, and overexploitation of aquifers, especially near rivers and lakes, can curtail flows in them. The interactions between surface water and groundwater resources are being studied extensively and attempts are under way to conjunctively manage these resources (Sophocleous, 2002).

Institutional frameworks for water supply

During the early days of colonization, the populations of American cities were small, and naturally occurring springs, lakes and shallow-dug wells were used to meet water supply needs. As communities expanded, small wooden conduits were used to transport water to meet hinterland demands. Significant increases in population post-American independence led to the construction of large reservoirs. In addition to serving potable water needs, water was also essential to fight fires that were more common due to wood burning and the lack of proper safety devices. The availability of water throughout the city became necessary to combat fires. This requirement, coupled with the luxury of having water on site, led to the laying out of extensive pipe networks. Technological advancements, such as the development of low-cost clay pipes and pumps, greatly facilitated the movement of water to significant distances away from the source and led to the expansion of cities. Further advancements in technologies led to the establishment of dense water-supply networks. Increased population also brought forth contamination of water resources and necessitated frameworks and policies for the treatment and management of water supplies (WFA, 1984).

The formalization of institutions and frameworks for water supply occurred during the early years of the nation (i.e. during late 1800 and early 1900). Before the mid-19th century, water supply was mainly a private endeavour and less than half of the major water works were publicly owned. However, the number of public waterworks multiplied at an accelerated rate from 45 in 1830 to 9850 in 1924 (Jacobson et al, 1985). Historically, the regulation and management of water was largely a local issue. Many city leaders believed that operation of water systems would be profitable for city government. Furthermore, such a move also helped to address the issues of maintaining adequate sanitation and to reliably deliver high-quality water. The consolidation of water distribution operations brought about economies of scale and helped to provide water and sanitation services to all citizens. The US has over 600,000 water companies and nearly 40 per cent of the water utilities are publicly owned, while nearly 60 per cent are operated with private investment. However, most large water-supply corporations are publicly owned and only 15 per cent of the total water supplied comes from the private sector. A vast majority of the private water companies can be classified as ancillary and constitute small supplies such as schools, hospitals and homeowner associations (IATP, 2007). While the 1990s saw some efforts towards privatization, the current trend appears to be towards re-municipalization and private disinvestment (IATP, 2007).

Water allocation laws were developed by different states. In particular, the eastern states with abundant water supplies largely followed the riparian doctrine for water allocation. According to this doctrine, mills and mill dams belonged to the riparian landowner one either side of the stream and could be transferred when the property was sold. Excess water could not be diverted from the stream and had to be returned unimpaired in both quality and quantity. Navigable rivers were considered to be public domain and any water use could not render the river non-navigable or obstructed in any form. The development of mill dams, however, obstructed fishing paths and reduced fishing areas. The protection of fishing rights was an early major issue, and conflicts between mill owners and fisherman often arose and were settled in the court of law.

In contrast to the humid east, the water allocation in the arid and semi-arid west followed the doctrine of reasonable use during the period of 1800 to 1847, when the population in the western US was fairly limited. Community-based irrigation projects were common during the early days of western settlement. However, during the period of 1848 to 1899, the doctrine of prior appropriation developed primarily to meet the demands of the California gold rush. This doctrine permits the diversion of water from the stream for use in a non-riparian area. The diverted water has to be used beneficially and the ability to divert water is not automatic, but is based on a priority system that respects the concept of 'first in time, first in right'. The first person to appropriate water and place it for beneficial use is termed the senior appropriator, and all others who come later are referred to as junior appropriators. The junior appropriators cannot access water until the senior rights are fully exerted, even if they (junior right holders) are upstream of the river.

The state of California developed a water allocation system that is a mixture of both riparian doctrine and prior appropriation system. This water law is referred to as the California doctrine and allows users in the humid parts of the state to follow the riparian doctrine, while those in the water-scarce regions are to follow the prior appropriation doctrine. This mixed approach has been adopted by several other states, including Hawaii, Kansas, Nebraska, North Dakota, Oregon, Oklahoma, South Dakota and Washington, all of which lie west of the 100 meridian and have both humid and dry regions (Cech, 2005). In addition to individual water rights, there are several interstate compacts that have been developed to share rivers that cross multiple states. In addition, transboundary compacts between the US and Canada, as well as between the US and Mexico, exist to share surface waters that cross between these nations.

Groundwater resources in most of the US are privately owned and their regulation varies from state to state. For example, groundwater is considered a property of the state in New Mexico, but deemed private property in the neighbouring state of Texas. The prior appropriation type doctrine (i.e. first in time, first in use) is typical when states have jurisdiction over groundwater. The rule of capture doctrine is common in the western US states and states where each landowner has the right to produce as much groundwater as beneficially needed, even if it precludes their neighbours' ability to extract groundwater. This idea, also dubbed as the law of the largest pump, is noted to be inefficient and is slowly making way to decentralized democratic management of groundwater using elected boards (groundwater conservation districts).

While water supply and management is largely a state issue in the US, there are several federal agencies that have a direct or indirect involvement with water issues. The most notable water-related federal agencies include the US Army Corps of Engineers (USACE); the US Bureau of Reclamation (USBR); the US Geological Survey (USGS); the US Fish and Wildlife Service (USFWS); the Bureau of Land Management (BLM); the National Park Service (NPS); the National Marine Fisheries Service (NMFS); the National Weather Service (NWS); the Federal Emergency Management Agency (FEMA); and the US Environmental Protection Agency (USEPA). These agencies work closely with state and local agencies and provide valuable assistance with regard to construction and maintenance of water-related infrastructure, as well as management of water resources on federal lands. The USEPA has direct and indirect regulatory oversight over water quality issues. The federal agencies, such as the USACE and USBR, played a major role during the late 19th to mid-20th century in developing the over 75,000 dams that exist in the country today. Agencies, such as the USGS, help to develop scientific tools for water resources management. The Natural Resources Conservation Service deals with soil erosion and agricultural water quality issues. The National Marine Fisheries and National Fish and Wildlife Service help to protect aquatic species in coastal and inland waters, respectively. Their role in water resources management is increasing significantly as the environmental and ecological needs of water are becoming clearer and important. The USEPA has developed comprehensive regulatory programmes to protect the quality of US waters as required by various legislative mandates. The impairment of water bodies (both surface water and groundwater) due to improper waste disposal and accidental spills presents a grave threat to water availability and supplies. The USEPA estimates that more than 21,700 surface water bodies are impaired and cannot meet their intended use. In the same vein, there are over 20,000 known abandoned or uncontrolled hazardous waste sites that have led to groundwater contamination. Other localized groundwater contaminant plumes also exist due to leaking septic tanks, underground storage tanks and other improper disposal of wastes. Proper sanitation and waste disposal continues to be a challenge in the US and is the next topic of discussion.

SANITATION AND WATER QUALITY ISSUES IN THE US

The quality of the water defines its intended use and, as such, is a critical driver of water availability. Water quality can limit the amount of water that is available for a specific purpose (such as drinking water or industrial use), even when sufficient quantities are available. Clearly, maintaining good water quality is even more critical in areas with limited supplies. The quality of water is affected by both naturally occurring impurities, as well as anthropogenic contamination. Natural impurities include suspended and dissolved solids and a variety of cations, anions and minerals that naturally occur in the soil and sediments, and dissolve into the water. Anthropogenic pollutants include both organic and inorganic constituents and can stem from municipal, agricultural and industrial sources. Improving water quality requires treatment of both natural and anthropogenic constituents. Natural waters typically contain a variety of impurities that are physical, chemical and biological in nature. Polluted water can cause a variety of diseases and ailments and increase the risk of fatality.

The primary impetus for water treatment in the US stemmed from the high death rate due to waterborne diseases, such as cholera and typhoid. Hazards arising from industrial wastes were largely ignored in the early part of the 20th century as they did not normally contain disease germs (Tarr et al, 1984). During the early 1900s, water and wastewater treatment was not actively pursued and untreated sewerage, as well as industrial wastewater, was directly discharged into streams and rivers. Many communities withdrew drinking water from places very close to wastewater outfalls and used it directly without adequate treatment (Veslind and DeStefano, 2006). Establishment of federal drinking water standards was considered unlikely as it was not part of the US constitution. The US Public Health Service (USPHS) under the US Department of Treasury set the nation's first bacteriological standard of 2 coliforms per 100 millilitres of water in 1914. It was initially not intended to be a nationwide standard, but designed to be implemented in those communities where trains stopped to obtain water. The goal was to protect the safety of interstate travellers, and trains would not stop if the water did not meet the standards. The economic disincentives of trains not stopping coupled with the social factors of governmental prestige and public opinion led to wide acceptance of this standard and the development of water treatment plants in many communities, even those not served by trains, and paved the way for federal regulation of water quality.

In 1942, the USPHS set the first comprehensive drinking water standards. In 1948, the federal pollution control act was passed to reduce pollution entering lakes and streams. These legislations had minimal effects and were largely ignored, and there were no enforcement mechanisms. The Safe Drinking Water Act (SDWA) was passed in 1974 and is regarded as the most comprehensive legislation aimed to protect potable water (i.e. water used for drinking, cooking and cleaning). The SDWA regulates several organic (e.g. benzene and DDT) and inorganic chemicals (e.g. lead, arsenic and chromium), as well as bacteriological pollutants. The SDWA requires that public water systems (i.e. which provide water through 15 service connections or 25 individuals) must monitor for and comply with certain contaminant levels. The primary SDWA standards are mandatory and enforceable and cover those constituents that pose the highest health risks. The standards (also referred to as maximum contaminant levels, or MCLs) are set such that they would not pose significant health risks over the lifetime of an individual. The secondary standards are not enforceable but are aimed to make the water more palatable by reducing colour, odour and corrosion characteristics. The Safe Drinking Water Act not only charged the USEPA with the development of appropriate standards, but also provided them with enforcement authority. The SDWA was amended in 1996 and includes the requirement that water providers inform customers of the quality of the drinking water. In addition, the providers must inform the public within 24 hours if drinking water becomes contaminated with constituents (such as bacteria) that could cause immediate illness.

While the SDWA aims at regulating contaminants at the receptor, the Clean Water Act (CWA) aims to control pollutants at the source. The CWA stems from the Federal Water Pollution Control Act Amendments of 1972 and was amended in 1977, with further amendments in 1981 and 1987. The CWA statute employs a variety of regulatory and non-regulatory tools to sharply reduce direct pollutant discharges into waterways, to finance municipal wastewater treatment facilities, and to manage polluted runoff. These tools are employed to achieve the broader goal of restoring and maintaining the chemical, physical and biological integrity of the nation's waters so that they can support 'the protection and propagation of fish, shellfish, and wildlife and recreation in and on the water'. The early focus of the Clean Water Act was the reduction of point-source discharges from municipal and industrial wastewater treatment plants. More recently, the focus has shifted to non-point sources that arise from storm-water runoff from urban areas and agricultural fields. The management and implementation aspects associated with the plan have therefore moved from focusing on individual dischargers to more holistic watershed-scale assessments. The CWA aims to delineate total maximum daily load of the pollutant (TMDL) that can be assimilated by a receiving body and allocate waste loadings among different dischargers within the watershed. The CWA does not concern itself with groundwater resources or water quantity issues, and exclusively focuses on surface water bodies.

The land disposal of solid and hazardous waste through sanitary landfills became common in the US post-World War II. However, land disposal was *ad hoc* and carried out with little or no regard to future environmental impacts. Many landfills were unlined and caused pollutants to leach and contaminate groundwater. The Resource Conservation and Recovery Act (RCRA) was passed in 1976 and is the cornerstone of US efforts to protect groundwater resources by regulating solid and hazardous wastes. Although the RCRA was passed in 1976, it was not promulgated until 1980. The RCRA aimed to put a ban on 'midnight dumping' of hazardous waste into ditches, vacant lots and open spaces, a common practice that led to large-scale groundwater pollution. The RCRA is a cradle-to-grave concept that places the responsibility of waste disposal on the generator of the waste. The regulation of municipal solid waste is governed under the subtitle 'D' of the RCRA, while the disposal of hazardous waste is under the subtitle 'C'. The RCRA has been amended three times since its inception, most notable of which is the Hazardous and Solid Waste Amendments (HSWA) of 1984. Subsequent amendments to the RCRA were in 1992 and 1996.

As per the RCRA guidelines, entities involved with hazardous wastes are classified as generators, transporters, and treatment, storage and disposal (TSD) facilities. The generators are responsible for correctly classifying the wastes, the transporters are charged with properly transporting the waste to a TSD facility determined by the generator, and TSD facilities treat, store or dispose of wastes following strict RCRA guidelines. As the RCRA places the burden on generators, it emphasizes the minimization of hazardous waste generation. The HSWA amendment of the RCRA initiated a ban on land disposal of hazardous waste to minimize groundwater contamination. The land ban was, however, not absolute and certain chemicals could be placed in landfills if they were altered substantially to reduce their toxicity and mobility prior to any land disposal.

The HSWA also addressed the issue of leaking underground storage tanks. However, the HSWA was concerned with tanks containing hazardous wastes and not hazardous products such as gasoline *per se*. It was estimated that nearly 25 per cent of the 5 million underground storage tanks containing petroleum products were leaking but not regulated by the RCRA (Uddameri, 1998). The HSWA amendments of the RCRA placed the burden of cleaning up gasoline-contaminated soils on the state. Risk-based corrective action guidelines were developed in response to dealing with the large number of sites that were contaminated with petroleum products (ASTM, 1995). In addition to leaking fuel tanks, groundwater contamination can also arise from the improper placement of on-site wastewater treatment systems (e.g. septic tanks and pit-prives) and land application of fertilizers and pesticides. These latter pollution problems are more prevalent in rural areas of the US. Currently, there is limited regulatory oversight, if any, concerning these diffuse rural groundwater pollution problems.

Treatment technologies for water quality improvement

During the 18th and 19th centuries, domestic waste management was rudimentary and wastes were discharged directly into the yard, street gutter or an open channel that served as a sewer (Burian et al, 2000). The sanitation problems were not substantial as populations were low and waste loadings were minimal. However, as cities grew, so did the sanitation problems and nuisances. The majority of the residents accepted nuisances, such as odour, as part of urban life. However, increased frequencies of epidemics resulted in the search of better technologies for waste disposal. The early domestic waste treatment systems used dry sewage systems. Human excreta were collected in containers and disposed of. However, crews hired to remove wastes from homes and to use them on lands as fertilizers or dispose them in rivers did not perform adequately, leading to waste accumulation and improper disposal in rivers and streams, and contamination of drinking water sources.

The goal of protecting public health through proper management of human wastes led to the construction of water-carriage sewer systems during the late 19th and early 20th centuries. These centralized systems were thought to be cost effective and were preferred over decentralized alternatives. This idea continues until today in the US, and decentralized systems are not favoured not only due to health risks and water quality protection aspects, but also to the concern of curtailing urban sprawl. The early centralized wastewater treatment technologies in the US were based on European experiences. Most cities used the combined sewer system (CSS) to carry both human wastes and storm water following similar designs found in England and Germany (Burian et al, 2000). The separate sewer system (SSS) technology became viable with the development of small diameter clay pipes and was available in the mid- to later 19th century. However, a complete switch to this technology did not occur until the early 20th century, when cities realized that CSS systems simply transferred the nuisance from the city to the river and that wastewater treatment would be necessary to completely take care of the human waste disposal problem.

Most cities and towns in the US have centralized wastewater systems with thousands of kilometres of pipe networks buried underground that carry household wastes to centralized treatment facilities. Most municipal wastewater treatment plants utilize physical processes, such as settling, filtration and communition for primary treatment and activated sludge treatment as a secondary treatment. The effluent discharges from the treatment plants are regulated for oxygen demanding substances: biochemical oxygen demand (BOD) and total suspended solids (TSS). However, with increased discharges, wastewater treatment plants are increasingly being regulated for ammonia, nitrate and phosphorus, as well, requiring additional tertiary treatment. In addition to municipal wastewater treatment, industries are required to treat their wastes prior to discharge to a receiving body. Industrial wastewater treatment plants are often regulated for chemical oxygen demand (COD), total suspended solids and other specific contaminants, and employ a variety of physical, chemical and biological treatment processes. In some instances, treatment of agricultural water is also required. Aquaculture operators are regulated in some states and have to treat their water prior to discharge.

Alternative wastewater treatment systems such as facultative ponds and constructed wetlands are more common for treating agricultural and aqua-cultural wastes, and these technologies are also being used for the treatment of municipal and industrial wastes (WEF, 2001). Discharge of treated effluent often occurs to surface water bodies, but under rare occasions can occur to aquifers as well. In some instances, the treatment systems operate under zero-discharge permits (i.e. they cannot discharge the water and will have to evaporate the wastewater). The treatment of municipal and industrial wastewater often generates solid wastes (e.g. biosolids) that need to be disposed of as well. The disposal of such solids is usually subject to state regulations. In many instances, disposal of solids can be more complicated and add substantially to the wastewater treatment costs.

The treatment of wastewater is also a local issue in the US and there are currently over 16,000 wastewater treatment plants in the nation. Of these, only 2 per cent are privately owned and less than 10 per cent are privately operated (ITT, 2008). However, private companies are intimately involved with wastewater treatment operations and provide a variety of wastewater-related services, including construction and rehabilitation, chemicals and other supplies, as well as engineering and ancillary services. While the USEPA expects increased public–private partnerships in the years to come, it is likely that wastewater treatment will largely remain in the public sector given the consumer confidence in its abilities to provide appropriate disposal and treatment options.

New trends in water supply and water quality augmentation

Water demand continues to grow at an alarming rate in many parts of the US. The areas experiencing significant population growth unfortunately do not coincide with regions having abundant water supplies. Water deficits are projected in several areas along the US–Mexico border and south-western US over the next few decades. The sustainability of both existing surface water and groundwater sources is at peril. These deficits are further exacerbated by erratic climatic patterns that minimize the reliability of existing supplies. Otherwise humid areas, such as the state of Georgia, are experiencing water shortages in recent times. There is also a general aversion towards the construction of large-scale reservoirs, given their negative environmental impacts. New water policies have made it difficult to pursue large-scale inter-basin transfers of water from low-density areas to those with high demands.

Innovative policy instruments and technological solutions are required to address the future water supply challenges in the US. There is also a growing recognition not to rely too heavily on one source. Conjunctive water management of both groundwater and surface water is gaining ground and long-held legal obstacles for such use are being worked upon. There is a shift towards proper valuation of water and efforts are under way to move water from low-value usages to higher-valued functions. Newer sources of water arising from wastewater reuse, desalination and rainwater harvesting are aggressively being explored in many parts of the US. The quality of water is a major issue with these technologies. The city of Los Angles has initiated a 'toilet to tap' paradigm that aims to reuse some of the treated wastewater to meet domestic water demands. However, the ability to detect trace and emerging contaminants, such as pharmaceutical and other household products, and to treat them effectively needs to be further refined and is being studied. Artificial recharge of excess rainwater and wastewater is also being pursued in some places and the risks associated with the movement of pathogens in aquifers are being explored. Increased energy costs and disposal of concentrate still limit the economic viability of desalination technology, and attempts are being made to address this issue. Clearly, there is a shift towards integrated water management approaches that simultaneously consider quantity and quality aspects in a holistic manner.

Challenges facing water supply and water quality issues

The US has a wide network of water supply and sanitation infrastructure, although the coverage is still not universal. Many low-income communities, such as the *colonias* along the US–Mexico border, lack adequate water and sanitation facilities. There is a general reluctance among water and wastewater utilities to integrate them within their infrastructure despite financial incentives from the government (Ward, 1999). A similar situation exists in inner cities and low-income neighbourhoods of many cities in the US.

While the US was once a leader in building large water and wastewater treatment systems, the infrastructure in many locations has become antiquated and is in dire need of rehabilitation. The American Society of Civil Engineers (ASCE) estimated that the US faces a shortfall of US$11 billion annually to replace ageing water infrastructure and to meet the US Safe Drinking Water Act requirements; the Congressional Budget Office (CBO) estimates the shortfall to be between US$10 billion to US$20 billion annually. The ASCE also states that ageing wastewater management systems discharge billions of gallons of untreated sewage into US surface waters each year. The USEPA estimates that the nation must invest US$390 billion over the next 20 years to replace existing wastewater treatment systems and to build new ones in order to meet increasing demands. In 2001, the USEPA performed a gap analysis and concluded that US$151 billion would be needed over 20 years to repair, replace and upgrade the nation's 55,000 community drinking water systems in order to protect public health (ASCE, 2005). The CBO concluded in 2003 that current funding from all levels of government and current revenues generated from ratepayers will not be sufficient to meet the nation's future demand for water infrastructure, indicating the role of governmental investment in water and wastewater infrastructure.

Proper planning of short-term and long-term needs, as well as flexible disbursement of funds as loans, grants and other subsidies, are needed to rehabilitate water supply infrastructure requirements. There is a growing recognition of the importance of overall ecosystem health and a movement towards sharing aquatic resources, not only with other humans, but with wildlife and other living beings as well. This move will most likely call for more stringent cleanup of municipal and industrial wastewater. Stricter cleanup goals are likely as our scientific understanding related to the fate and transport of emerging contaminants (such as endocrine disruptors) increases. All of these factors are likely to increase the treatment costs in the future.

CONCLUSIONS

The primary goal of this chapter was to look at some of the factors that motivated the development of water supply and wastewater treatment in the US and to review major policies that were enacted to facilitate the development of water infrastructure. The situation of current water availability, trends in water and wastewater management, as well as challenges facing water infrastructure were also evaluated. Historically, the provision of adequate amounts of water has been an important factor for the growth of the nation. By the same token, there has also been a concerted effort to minimize the outbreak of water-related epidemics and to improve the quality of life of its citizens. The nation has made considerable investments in the past in developing a comprehensive water network and continually adapted to meet new water and wastewater challenges using technology and policy instruments in a synergistic manner. Population increases, uneven economic growth, and ageing infrastructure once again appear to challenge the need to develop creative solutions to help meet the basic human needs of adequate water and high-quality sanitation. Technological and policy efforts are under way to meet this grand sustainability challenge, despite certain economic shortcomings.

REFERENCES

ASCE (American Society of Civil Engineers) (2005) *Report Card on Infrastructure*, American Society of Civil Engineers, Washington, DC, www.asce.org/reportcard/2005, accessed May 2008

ASTM (American Society for Testing and Materials) (1995) *ASTM E1739-95(2002) Standard Guide for Risk-Based Corrective Action Applied at Petroleum Release Sites*, American Society for Testing and Materials, West Conshohocken, PA

Burian, S. J., S. J. Nix, R. E. Pitt and S. R. Durrans (2000) 'Urban wastewater management in the United States: Past, present, and future', *Journal of Urban Technology*, vol 7, no 3, pp33–62

Cech, T. V. (2005) *Principles of Water Resources – History, Development, Management and Policy*, John Wiley and Sons, New York

FEMA (Federal Emergency Management Authority) (2005) *Dam Safety and Security in the United States*, Federal Emergency Management Authority, Washington, DC, www.fema.gov/plan/prevent/damfailure/ndsp.shtm, accessed May 2008

Harr, J. (1996) *A Civil Action*, Knopf Publishing Group, New York

IATP (Institute for Agriculture and Trade Policy) (2007) *Privatizing US Water*, Institute for Agriculture and Trade Policy, Minneapolis, MN

ITT (International Telephone and Telegraph Corp) (2008) *Guidebook to Global Water Issues*, ITT, New York, www.itt.com/waterbook, accessed May 2008

Jacobson, C., S. Klepper and J. A. Tarr (1985) 'Water, electricity, and cable television: A study of contrasting historical patterns of ownership and regulation', *Technology and the Future of Our Cities*, vol 3, no 9

Maupin, M. A. and N. L. Barber (2005) *Estimated Water Withdrawals from Principal Aquifers in the United States*, United States Geological Survey, Reston, VA

Melosi, M. V. (2004) 'Full circle: Public goods versus privatization of water supplies in the United States', in *Proceedings of the Summer Academy 2004 Urban Infrastructure in Transition: What can we Learn from History*, International Summer Academy on Techology Studies, Graz, Austria, www.ifz.tugraz.at/index_en.php/filemanager/download/309/Melosi_SA%202004.pdf, accessed May 2008

Postel, S. and B. Ritcher (2003) *Rivers for Life: Managing Water for People and Nature*, Island Press, Washington, DC

Pulich, W., J. Tolan, W. Y. Lee and W. Alvis (2002) *Freshwater Inflow Recommendations for the Nueces Estuary*, Texas Parks and Wildlife Department, Austin, TX

Solley, W. B., R. R. Pierce and H. A. Perlman (1998) *Estimated Use of Water in the United States in the Year 1995*, United States Geological Survey, Reston, VA

Sophocleous, M. A. (2002) 'Interactions between groundwater and surface water: The state of the science', *Hydrogeology Journal*, vol 10, no 1, pp52–67

Tarr, J. A., J. McCurley, F. C. McMichael and T. Yosie (1984) 'Water and wastes: A retrospective assessment of wastewater technology in the United States, 1800–1932', *Technology and Culture*, vol 25, no 2, pp226–263

Uddameri, V. (1998) *Methodologies to Derive Soil Cleanup Goals at Contaminated Sites*, PhD thesis, University of Maine, ME

Veslind, P. A. and T. D. DiStefano (2006) *Controlling Environmental Pollution – An Introduction to the Technologies, History and Ethics*, DEStech Publications Inc, Lancaster, PA

Ward, P. M. (1999) *Colonias and Public Policy in Texas and Mexico: Urbanization by Stealth*, University of Texas Press, Austin, TX

WEF (Water Environment Federation) (2001) *Natural Systems for Wastewater Treatment*, Water Environment Federation, Alexandria, VA

WFA (Wallace Floyd and Associates) (1984) *A History of the Development of the Metropolitan District Commission Water Supply System*, Wallace Floyd and Associates, Boston, MA

17

Discrimination by Default: The Post-Colonial Heritage of Urban Water Provision in East Africa

David Nilsson and Arne Kaijser

Introduction

It is 12 November 2007 and, at last, the Wamalanda water supply in the Nyalenda informal settlement in Kisumu, Kenya, is about to be reopened. Nyalenda is said to host some 100,000 people, to whom essentially no basic infrastructure services such as water and sanitation have been availed over the years. The direct beneficiaries in the Wamalanda Development Group – a community-based organization – have been looking forward to the reinstatement of the supply for months now, after it was completely vandalized shortly after its construction in January 2007. This so-called 'delegated line' will at long last bring in water into the hitherto un-served area, and at a much lower price than what consumers have been paying to water vendors operating in the jurisdiction. For years, most of the vendors have obtained water illegally from the main pipeline of the public water utility Kiwasco, bypassing the fringe of Nyalenda. The only remaining step before the new system is fully operational is to cut the illegal connections – which is now thoroughly done on this November day.

Late the following night, the tranquillity of the neighbourhood is suddenly pierced by the shouts of a large group of young men and of the menacing noise of wilful destruction. For the second time in less than a year the water supply to this poor area is being destroyed. Methodically and purposefully, the vandals move from one connection point to the next, smashing up the infrastructure and stealing the meters, leaving little to be pondered over on our visit two days later at the scene of the crime. All we can do is to lament the sad state of affairs and to ask ourselves: why has this happened? For the leaders in the Wamalanda

Development Group, and the public asset owner, the Water Service Board, the answer is obvious. There is simply too much at stake for some people in the 'water cartels' selling water to the poor people in the slums (Ombogo and Morel, Wamalanda Development Group, pers comm, 2007). But why has the service provision in a country such as Kenya deteriorated to such a state, what are the causes and what is there to do about it? With challenges such as this, how will the situation ever improve for the more than 600 million poor people today living in towns in developing countries without adequate access to safe and affordable water (UN-Habitat, 2003)? And how will we understand these situations, as well as the processes leading up to them?

As argued in Chapter 1 of this book, technical systems such as water and sewage systems are formed by their contexts – by political and economical systems, cultural values, geographic and physical features of the environment, etc. But such technical systems also influence the societies surrounding them; there is, thus, a mutual shaping of technology and society. In this chapter we discuss water and sewage systems as 'socio-technical systems', including not only physical artefacts, but also the people (and organizations) who design, build and operate these artefacts, as well as their legal and economic frameworks. Socio-technical systems evolve over long periods of time, especially where large-scale infrastructure systems are involved, and these systems are often associated with inertia that makes them resistant to abrupt change (Summerton, 1994; Kaijser, 2003).

For the last decade, the water sectors in Kenya, Uganda and Tanzania have been undergoing reforms, led by the governments of each country but ushered in and supported by their international development partners (Bayliss, 2003; Ballance and Trémolet, 2005). These countries are, just like so many other developing countries, at the same time undergoing a rapid social, demographic and economic change, including a high rate of urbanization. The social context of technology for water supply provision is, hence, changing fast and urban growth puts increased pressure on the already weak infrastructure. Despite the recent reforms and increased investments in the sector, access to safe water supply in all three countries is still low (see Figure 17.1). In Uganda and Kenya, around one third of urban dwellers lack reasonable and affordable access to safe water. Those living in informal settlements, in particular, have to resort to buying water from vendors, water kiosks and neighbours, just like in Nyalenda (Collignon and Vézina, 2000; WUP, 2003, p10). If the ongoing reforms are to be of any benefit to the urban poor, then the historically shaped shortcomings of the existing water system need to be understood and, consequently, addressed in the reform.

We have three objectives in this chapter. The first is to outline and explain the historical process that has led to widespread discrimination of poor people when it comes to public water provision in urban settlements in East Africa, with a focus on Kenya and Uganda. The second is to analyse some of the shortcomings of the socio-technical systems for urban water supply when it comes to providing services to informal settlements. The third is to review what kinds of changes are needed for public water supply to become responsive to the needs of poor people. In order to accomplish this somewhat tall order, we need to start from the beginning. We need to go back more than 100 years, when urban water supply in its present large-scale format was introduced as a socio-technical system in East Africa.[1]

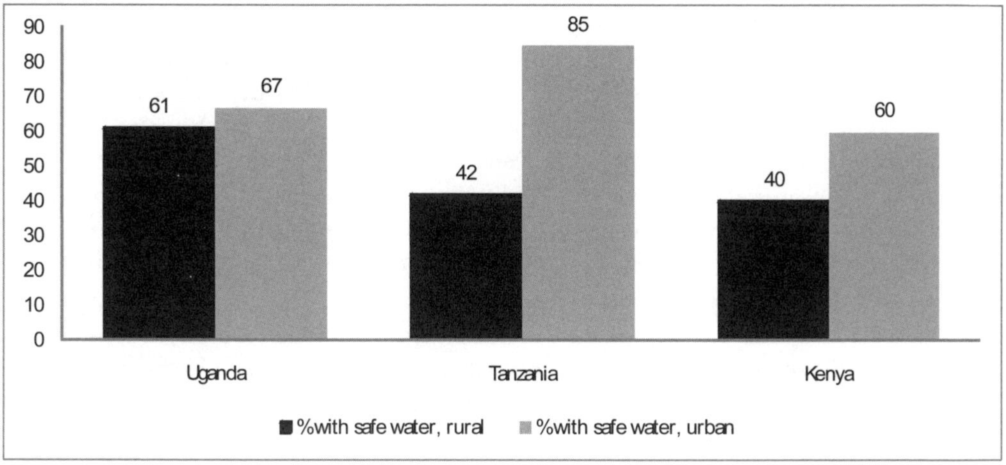

Figure 17.1 *Access to safe water supply as a percentage of the population: Kenya and Uganda (2006), Tanzania (2002)*

Source: World Bank (2006); Republic of Kenya (2007a)

THE INTRODUCTION OF PIPED WATER SUPPLY IN EAST AFRICA

In Britain's strive to win the race against the other European super-powers for the control of the headwaters of the Nile in Uganda, the 'Uganda railway' from the coast to Lake Victoria was completed in less than six years, a remarkable achievement at the time. The railway-head reached the lakeshore at today's Kisumu in 1901 after conquering numerous technical, physical and logistical challenges along the 1090km route from the coast (Miller, 1971, p315; Pakenham, 1991, p506). One of these challenges lay in water supplies. The steam engines were dependent upon water; hence, at regular intervals a station had to be put in place, each offering a reliable water supply. These stations gave rise to towns as people settled around the stations. What began as small humble railway stations little more than a century ago have today transformed into sizeable cities such as Nairobi, Nakuru and Kisumu (Obudho and Obudho, 1992).

The Uganda railway thus provided a vector for introducing European technology for urban water supply into the interior of East Africa. In the small urban centres that grew up, piped water supplies were constructed and operated by the Uganda Railway to serve both machines and man. Distribution systems were limited in capacity as well as geographical reach, and standpipes were commonly used (Bransby Williams, 1907; British East Africa Protectorate, 1915). Nevertheless, these small proto-systems provided a blueprint for subsequent expansion of water supply. From around World War I, the state – in the form of the colonial administration – gradually took over responsibility of water supplies in East Africa. Motives for an increased involvement by the state were mainly public health aspects and the strategic importance of water supplies for the social and economic development of the colonies.

The period from the mid-1920s to the mid-1930s saw an expansion phase for water supplies in East Africa. From the 1930s onwards, the colonial state assumed responsibility for urban water supply in both Uganda and Kenya. The first urban water supply in Uganda was opened in Jinja in 1928, followed by Kampala in 1930. Whereas operation and ownership of urban water supplies in Kenya were often delegated to local authorities, in Uganda they were solely run by the Public Works Department. After World War II, the British government boosted water supplies in both Uganda and Kenya through large investment programmes. As a result, urban water supplies in the two countries were relatively well developed at the onset of decolonization during the early 1960s. Large-scale piped systems adopted from Europe were in place, with individual household connections as the norm (Nilsson, 2006b; Appelblad and Nilsson, 2007; Nilsson and Nyangeri, 2008). The World Health Organization (WHO) even claimed that Kenya had near universal coverage of water supply in 1963, when it became independent (WHO, 1963).

However, the service provision systems in both Uganda and Kenya were *de facto* based on racial segregation. As an illustration, normal quantities allowed for domestic uses of water in Kenya were officially given as outlined in Table 17.1.

Table 17.1 *Excerpt from Kenya Official Gazette (supplement no 52, 2 July 1953): The 1951 Water Ordinance, the 1953 Water (General) Rules, form for application for water permit*

User	*Gallons of water allocated per head per day*
Non-natives	50
Africans	10
Large stock	10
Small stock	2

A similar racial division was applied in Uganda. Furthermore, design norms and technology in the colonial period were mainly chosen to fit the needs, preferences and income levels of the colonialists. However, it should be noted that the tariffs paid by these well-to-do households in the colonial period were high enough to actually make the water systems economically self-sustaining (Nilsson, 2006b).

Independent nations with colonial provision systems: A growing mismatch

Independence came to Uganda in 1962 and to Kenya the year after. The democratically elected governments inherited systems for urban water supply that built on modern European technology suitable for the small high-income segment of the population, but not for meeting the needs of the millions of Africans who would flock to the towns of East Africa within a decade or two (Nilsson, 2006a). Within a few years after independence, the new governments declared universal water supply as a key objective. The Government of Kenya stated its 'water for all' policy for the first time in *Development Plan 1970–1974* (Republic of Kenya, 1969).

Throughout the 1970s and 1980s, the Government of Kenya pursued an expansionist policy of urban water supplies, with steadily increasing budget allocations. However, the spiralling inflation that set in from the mid-1970s eroded much of the budget increases. The financial sustainability of these systems simultaneously started to crumble due to less focus on cost recovery by the authorities. Notably, the Kenyan government policy stated that the full cost for operation, maintenance and capital should be recovered from urban consumers; in practice, this was not followed. Water supplies were therefore running huge deficits by the end of the 1980s. Uganda saw a similar development, exacerbated by the political turmoil of the 1970s and early 1980s (Tumwine, 2002, p8; Appelblad and Nilsson, 2007; Nilsson and Nyangeri, 2008).

Due to – *inter alia* – financial shortcomings, an extremely rapid urbanization and lack of capacity, water services in urban centres throughout East Africa deteriorated between the 1960s and the 1990s (Thompson et al, 2000). The inability of the public water providers to expand services hit the poor worst and informal provision systems mushroomed to fill the gap. Many East African households became dependent upon buying water from neighbours, or from water vendors selling at prices up to ten times the price from public systems (Collignon and Vézina, 2000; UN-Habitat, 2003; Kjellén, 2006). While the urban poor in East Africa were left outside the public supply, the middle-income and high-end consumers with access to public water did not pay the full cost of service. Through such a practice the state, in fact, subsidized the rich while excluding the poor from a reasonable service access (Hukka et al, 1992; Kjellén, 2006, p91).

There are, without doubt, many causes for the failure of public water provision in East Africa, and the sector reforms launched in the 1990s are attempts to address these causes. However, in the debates surrounding these reforms, little attention has been given to the importance of history and the inertia of socio-technical systems. Given that the systems taken over from the colonial regime were designed for the European lifestyles of the high-income segment, independence would have required a drastic reformulation of these systems to better suit the situation of the average citizen. This was not done. In the post-colonial situation of Kenya during the 1960s and 1970s, not only was the infrastructure overtaken, but also institutions, organizational structure and even development ideals and norms. The socio-technical systems for urban water supply therefore were preserved well into the post-colonial period (Nilsson, 2006b). Infrastructure was expanded according to old norms and engineering practices, but within a new economic and social context. Even the geographic context had changed with the sprawling of informal settlement, often posing unclear land tenure situations. This growing mismatch between context and technology has not been given due attention, while the urban poor have suffered.

REFORMS OR REPRODUCTION?

During the 1990s, water sector reforms were initiated throughout East Africa. Cost recovery and private-sector participation have been key elements of these reforms, and donors have been egging on these policies. Whereas actual privatization of utilities has not proved very successful, commercialization of services through pricing based on cost recovery and a more business-like approach by suppliers have been undertaken more vigorously (Ballance and Trémolet, 2005; WASREB, 2007). Institutional reforms have been carried out, which in Kenya has involved a completely new legislation: the 2002 Water Act. The sector organization has also been overhauled, especially in Kenya, where completely new public authorities with responsibility for water supply have seen the light of day. While local authorities have been banned from delivering services, the responsibility has been vested in seven new water service boards (WSBs), operating on a regional scale. They, in their turn, contract out the operations of supplies to water service providers (WSPs), which in most cases are semi-public companies formed by the old municipal water departments. As the Kenyan sector emulates a market approach, a regulatory body has been instated through the Water Services Regulatory Board (WASREB), leaving only the oversight and policy-making in the sector to the Ministry of Water and Irrigation (Republic of Kenya, 2007b). The sector in Uganda has retained the parastatal National Water and Sewerage Corporation (NWSC), created during the 1970s, as the main water service provider in larger towns; but it has been revamped according to commercial principles. The process of privatizing the operations that was envisaged early on in the reform in Uganda has been halted during recent years. As mentioned, privatization of water services has not been implemented on any larger scale in any of the three countries, except for the (in)famous and short-lived lease of the water utility in Dar es Salaam to an international water operator (Ballance and Trémolet, 2005, p207; Kjellén, 2006, p205; Appelblad, 2009).

But what has really been changed by these reforms? The countries in East Africa have seen water sector reforms before in terms of reorganization and new legislation. Even private provision and cost recovery principles are nothing new to the region. It can, in fact, be argued that the key characteristics of the sector organization in Uganda and Kenya have been preserved since colonial times. Whereas Uganda has carried forward its tradition of one centralized public provider – first the Public Works Department and later the NWSC – Kenya has preserved its more fragmented and delegated sector structure, hosting a multiplicity of actors (Appelblad and Nilsson, 2007; Nilsson and Nyangeri, forthcoming).

When looking at social or economic change at large, incentives are crucial (North, 2005). However, from the perspective of the urban poor, what has actually changed in the sector? What new incentives will make providers more responsive to the needs of the poor than has been the case before? Although there has been great improvement in terms of cost recovery in Uganda and Kenya, any major improvement in terms of access to water supply for the urban poor is yet to materialize (NWSC, 2006; Republic of Kenya, 2007a). Cost recovery for the sector as a whole has sometimes been held out as a prerequisite for expansion of services to the poor, and for the possibility of cross-subsidies (WUP, 2003). This notwithstanding, what is it in the 'new' sector set-up that will provide incentives to expand services to low-income people in the informal settlements and not just perpetuate a discriminating service provision regime? In the next section we try to identify the main barriers for creating such incentives, and then discuss three different examples of how to overcome these barriers in practice.

The barriers for extending large-scale systems to the periphery

The dilemma of how to extend services to poor people in the peri-urban areas while still adhering to cost-recovery principles is not a new one, nor is it anything unique to East Africa. More than a century ago many cities and towns in Europe and North America were struggling with this problem. Processes of expanding large-scale networks often show a pattern of initial establishment in high-income areas and then a gradual expansion, first, to middle-income and, finally, to poorer areas. This is the case not only for water and sewerage systems, but also for gas and electricity provision. The common denominator in these expansions has been the ability to find technical and management solutions for linking the economies of scale in the large-scale production of clean water, gas or electricity with a suitable distribution network. The most profitable consumers of these services are large factories, offices, hotels, shops and well-to-do households in downtown locations. They consume large quantities in relation to the length of pipelines or wires needed to connect them. In contrast, poor households in the outskirts of a town or in the countryside consume small quantities in relation to the cost of connecting them (Kaijser, 1986; Tarr, 1988).

However, it is not only the tangible cost of building the additional network that is of importance. There are also more intangible 'transaction costs' at play. They are of two kinds. The first has to do with metering and billing. For the supply of water, gas and electricity, payment is commonly done in proportion to consumption, which calls for special devices for measuring consumption and for administration to collect payments from the users. For large consumers, this does not entail a substantial extra cost; but for small consumers it does. For poor households, just the cost of a meter may be prohibitive and the variation over time of income for a poor household does not go well with the post-paid billing system, adding to the transaction costs for collection of payments.

The second transaction cost has to do with property rights. Large-scale infrastructure is capital intensive, with a high proportion of sunk costs. This requires the asset holder to have a long time perspective in order to recuperate the costs; consequently, property rights and enforcement is a critical risk factor for the asset holder or the investor. In the setting of informal settlements, these implications are particularly important. Quite often, informal settlements have sprung up in the periphery of towns, sometimes outside their jurisdiction. The informal character of these settlements generally implies that ownership and use of land is weakly regulated and protected by law (UN-Habitat, 2003, p103). Both the insecurity of tenure and use of land, as well as the risk for illegal connections, may translate into transaction costs for the asset holder of large-scale infrastructure.

To summarize, the characteristics of large-scale systems have a tendency to create barriers for expansion into low-income and informal areas due to:

- high marginal infrastructure expansion cost; and
- the transaction costs associated with property rights and payments.

THREE MODELS FOR OVERCOMING THE BARRIERS

Let us now return to our question of incentives for extending water services to the urban poor. Under a cost-recovery paradigm, the barriers described above must be lowered to create incentives for the provider (public or private) to extend these services. In this section we illustrate how some of these barriers can be reduced by looking at three examples of provision of infrastructure services to 'uneconomical' households in the periphery. These examples, spanning from the energy sector in Sweden during the early 1900s to contemporary approaches for water supply in Kampala and Kisumu, may offer useful models for provision of water services in informal settlements.

Making large-scale systems work for the population of the periphery

Sweden is a country endowed with many waterfalls and many of these were exploited for power generation in the first decade of the 20th century. Special power companies were formed in many parts of the country, which built hydropower plants and regional power networks to enable the transmission of this cheap hydropower to distant towns and industries. Moreover, in 1909, the Swedish state established a parastatal called Vattenfall with the purpose of exploiting the hydropower in a number of large waterfalls that were owned by the state.

The population in rural areas did not get access to this new hydropower as the cost for distributing power to scattered households in the countryside was seen as prohibitive by the power companies and Vattenfall. However, during 1910 to 1920, farmer associations began lobbying for the extension of electricity distribution to the countryside. The parastatal Vattenfall felt a special obligation to try to serve the rural areas; but it also clearly realized that the costs would be very high if it would take on the task of distribution by itself. To overcome this dilemma, it proposed a new solution: the establishment of independent distribution cooperatives that would build and operate electricity networks on their own and buy their power from Vattenfall. This solution was adapted to a long tradition in the Swedish countryside of cooperatives of different kinds, which meant that there already existed a skill in organizing collective efforts of this kind.

The first distribution cooperative of this nature was established in 1915 in a rather rich rural area about 100km north-west of Stockholm. A local landlord was the driving force in the formation of this co-operative and he cooperated closely with the regional manager of Vattenfall. All farmers in the area were invited to become joint owners of the cooperative by buying shares in proportion to the acreage of their farms. These shares were not too costly; but each shareholder also had to grant a security for a loan that the cooperative received from the local bank for buying the electrical equipment. Vattenfall provided technical assistance for the design of the system, and the members of the cooperative did much of the actual work. They provided poles for the power lines and dug holes for these poles on their land. In this way, the cost of the network could be kept fairly low, and disputes about the rights of way for the power lines were avoided. Furthermore, the costs for billing and other administrative tasks were minimal as this was handled by the board of the cooperative. The board members received a rather low annual remuneration for their work. Finally, the cooperative could buy its power from Vattenfall at a low rate for bulk supply. All in all, this meant that the cooperative was able to provide its members with electricity at a reasonable and affordable cost (Modig, 1984).

From Vattenfall's point of view, this cooperative solution was in many ways ideal. It became a bulk supplier to the cooperatives and could sell large amounts of power without having to take any risks in building distribution networks. Furthermore, the transaction costs for selling the power were very low as each cooperative was a single customer. Finally, it gained political goodwill by supplying power to rural populations. Vattenfall thus strongly encouraged the formation of more co-operatives, not only by providing technical assistance, but also by drawing up standard by-laws for such co-operative associations in order to streamline the local institutional set-up (Kungl. Vattenfallsstyrelsen, 1932). Similar co-operatives were established in many other parts of Sweden: at first in the rich farmlands on the plains, and later in poorer areas (Olsson, 1984).

Hence, in this example, a large-scale system for production and bulk supply of the good was successfully combined with a local distribution network that was separated from the supplier, but regulated through a contract and a metered connection point. The economies of scale of the regional power supply were combined with flexible local solutions for end-consumer service provision. More importantly, transaction costs were substantially reduced since fee collection by the bulk supplier was limited to one metered connection per cooperative, while economic and physical risk associated with the local distribution network was fully born by the consumers themselves. The technical coupling of these two systems – one large scale and one small scale – was matched with two different sets of institutions: one parastatal governed by national laws and the cooperatives governed by local by-laws (see Figure 17.2). This two-layer institutional coupling is what Elinor Ostrom has referred to as 'nested institutional systems' (Ostrom, 1990, p90).

Leaping in time and space to today's Kampala, the parastatal for urban water services, the NWSC, is (since 2006) testing a system for pro-poor water services on a pilot scale in three low-income areas in the central division of Kampala. The idea here is also to utilize the economies of scale of the existing large-scale provision system, but in this case for water supply. What is currently being tried on a pilot scale is a system of pre-paid water supply. The main network is slightly expanded into informal areas, where water is dispensed through an automated system of vending machines where water is sold (see Figure 17.3). Although it is too early to assess the results of the pilot project, in theory it would be able to facilitate pro-poor service provision while still being able to recover its costs. The investment cost will be reasonable since the distribution network is kept small. In comparison with the provision through water kiosks and vending that have previously been the main modes of water provision in these informal areas, transaction costs may be lowered through the elimination of middlemen. Furthermore, the NWSC will have full control over the price for the end user, which may be otherwise hiked by the water kiosk operator or the vendor (for a more detailed discussion, see Appelblad and Nilsson, 2007).

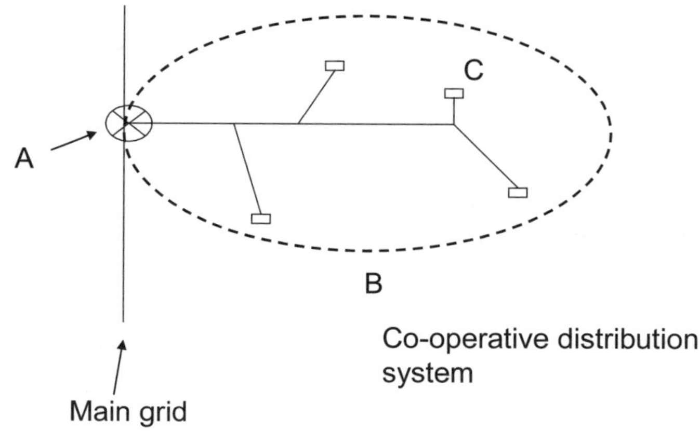

Figure 17.2 *The co-operative electricity distribution system in Sweden during the early 1900s*

Note: The bulk supplier supplied electricity in accordance with a contract at the metered connection point 'A'. The distribution cooperative 'B' constructed and operated the distribution network and took up fees from their members 'C' for electricity consumed.

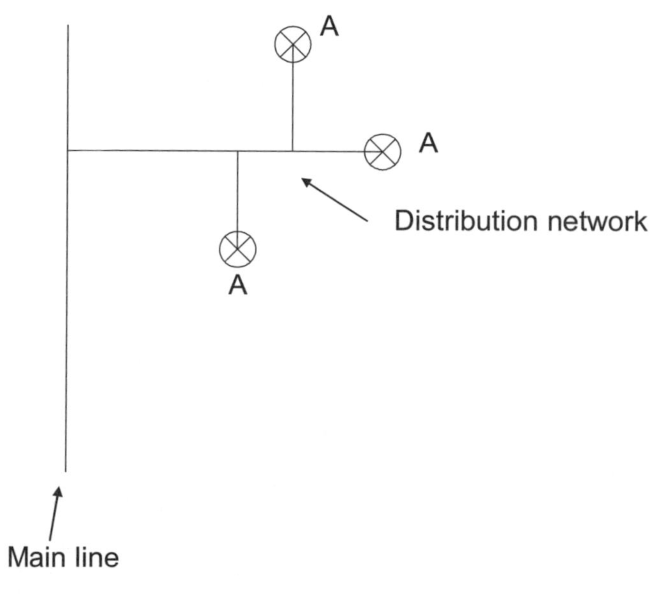

Figure 17.3 *The system of pre-paid water supply currently being tested in Kampala*

Note: The NWSC owns and operates the whole distribution network, which is marginally expanded into informal areas. Water is sold at regulated prices through automated vending machines at 'A'.

In our last example, we return to where our journey started: to the Nyalenda informal settlement in Kisumu. The unfortunately delegated Wamalanda line is just one of five lines that have been built through a pilot project trying out a new management model of water services in Nyalenda. The pilot project is a joint effort by the asset-holding institution (Lake Victoria South Water Services Board, or LVSWSB), the operator of the water utility (KIWASCO) and the World Bank's Water and Sanitation Programme. Instead of just bypassing Nyalenda, KIWASCO has now built five pipelines, each extending a few hundred metres into Nyalenda. Along each line, a number of 'meter chambers' have been built to which households will join their private connections. But what is so innovative about the Nyalenda model is that each of the five 'delegated lines' is operated by a private entity called the 'master operator', who can be a cooperative of users, a local community-based organization or an individual. KIWASCO and LVSWSB have built the delegated lines, but have outsourced their operation to the master operators, regulated by contracts. The master operator pays according to consumption at the connection point to the main pipeline and according to the agreed tariff. Each consumer builds his or her own connection and pays to the master operator according to consumption metered in the meter chamber on the delegated line (see Figure 17.4). Since every connection has to be done according to standards set by KIWASCO, not all households can afford a connection. However, connected households are allowed to sell to neighbours, but at an official and widely publicized pricelist (Ombogo and Morel, pers comm, 2007).

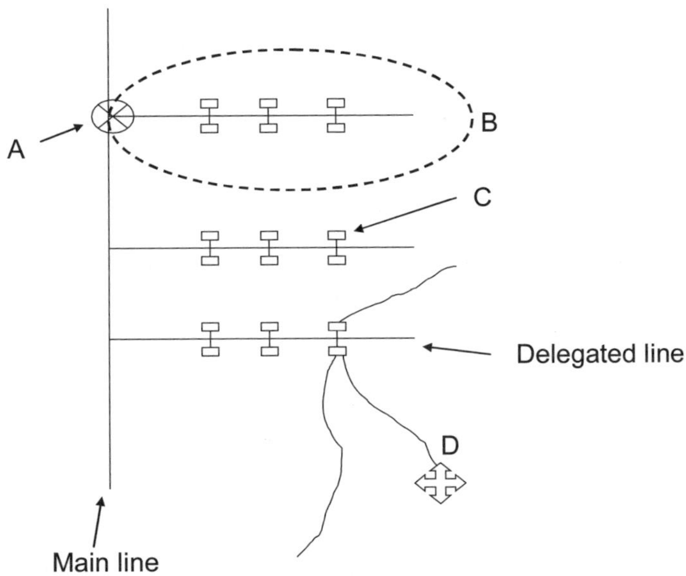

Figure 17.4 *The Nyalenda-delegated model*

Note: The bulk supplier builds the 'delegated lines' connected to the main line at 'A', where the supply is metered. Each delegated line is operated by a private entity, the 'master operator' 'B', who is responsible for management and payment to the bulk supplier. Each consumer connects to a meter belonging to the delegated line at 'C'. Households also resell water to neighbours 'D' at an official price.

In the Nyalenda model, transaction costs related to metering and, especially, collection of payments could be lowered since peer pressure within the community has instilled a culture of paying for services. In the other four delegated lines, the payment record has been very good and one of the master operators has already been able to expand the system using the internally generated surplus. The risk associated with insecure tenure is also somewhat controlled since the infrastructure put in place by KIWASCO is rather limited. Furthermore, the consumers themselves have a vested interest in it being kept in good order, both as consumers and as members of associations operating them (Adhiembo, pers comm, 2007). As shown by the Wamalanda case, the model is not safe from vandalism; but with the other lines these kinds of problems have not occurred (Ombogo and Morel, pers comm, 2007). It is also a clear-cut example of a nested approach, where the large-scale system has been combined with two subsidiary and local systems. The large-scale bulk supply (utility level) has been combined with the delegated line (community level), which is further combined with the regulated reselling (household level). The different institutions have also been formally nested; KIWASCO has a delegated management contract with the master operator, whose clients are regulated through the service agreement, and the reselling is regulated through the official price list.

Comparing the three models

All three models presented here – rural Sweden, Kampala and Kisumu – have evolved in their specific context of time, space and society. Nevertheless, it is of interest to compare similarities and distinctions, and to discuss in what way the different models could (theoretically at least) respond to the challenges of providing water to poor people in informal settlements.

We have argued in the foregoing that large-scale systems in their current form are unsuitable for service distribution in peripheral areas because of the relatively high cost of infrastructure and transaction costs leading to unfavourable incentives for the provider. With regards to infrastructure costs for distribution, all three models try to lower this by either outsourcing that cost to the consumers (Sweden and, partly, Kisumu) or, like in Kampala, by putting in place a most rudimentary network. The Swedish model presupposed some resources being available locally and was initially focusing on relatively resource-strong areas. This applies also to the Kisumu case, where at least the connections have to be built by the consumers themselves. Consequently, the first delegated line built in 2005 was put in place in an area where the inhabitants were not so poor – clearly an analogy with the Swedish case (Ombogo and Morel, pers comm, 2007). The Kampala model should probably offer the lowest investment cost; but, on the other hand, access may be a problem since every consumer will have to transport the water back from the vending point to the household. A higher transport effort generally lowers the quantity consumed, which, in turn, reduces the health benefits (Kjellén and McGranahan, 1997).

Regarding the transaction costs, all three models also make significant strides towards lowering them. The Swedish example and Kisumu/Nyalenda have almost an identical set-up for lowering transaction costs. The bulk supplier outsources the commercial risk and operational risk to the local operator, and will simply collect payment from one bulk consumer according to a contract and a meter. If there are losses (leakage, theft, non-payment, etc.) in the distribution system, the costs for that will only fall on the operator of the distribution system. One difference, however, exists: the ownership of the local infrastructure. In Kisumu, the master operator only leases the delegated line, and when vandalism occurs, the cost is born by the KIWASCO and LVSWSB. Since the latter two can access capital from donors in grant form for similar purposes, it has perhaps been the natural solution. The ownership issue is, however, an important one. The unclear land tenure situation in many informal areas could render that solution less attractive, adding to the risk premium and the transaction costs. The overall financial viability of the Kisumu model is still to be evaluated; but KIWASCO has, anyway, seen this as 'good business' since the project has reduced the losses from illegal connections (Ombogo, pers comm, 2007). The low transaction costs within the local distribution network have been demonstrated by the good payment record and the absence of illegal connections (Adhiembo, 2007). The problems of vandalism in Wamalanda should be under control once the delegated line is operational, providing water at low cost and, thus, pushing the vendors in the 'water cartels' out of the market.

With regards to Kampala, in theory it should offer very low transaction costs because of the absence of middlemen and because metering and payment for consumption are done simultaneously. The drawbacks are that the system is a fully automated extension of the large-scale system and not a locally operated network. There may, therefore, be less peer pressure to pay for use and less incentives for local self-monitoring since losses will be carried by the NWSC. Consumers will have the same options as before to connect illegally or to try to tamper with the vending machines. Hence, the commercial risk and the risk of losses through theft and vandalism are internalized with the NWSC, which, in fact, adds to the transaction costs. To lower this risk, the NWSC's ambition is to involve the local community in managing the infrastructure (Appelblad, 2009). It is still too early to assess the actual outcome of this pre-paid system, whether it will be efficient in reducing transaction costs or not. But the key problem with the Kampala model is its limited physical coverage. Its Achilles heel is that it does not combine the qualities of the large-scale supply system with those of a small-scale distribution system.

None of the models presented above is perfect. They are compromises in relation to the many different objectives that they try to address and the realities that they are subject to. Their key features are summarized in Table 17.2. What is important is that these models may be vital instruments for going beyond the ordinary conference jargon of 'pro-poor service' and actually contribute to creating favourable incentives for pro-poor provision.

CONCLUSIONS

In this chapter, three objectives were outlined at the start. First, we have tried to demonstrate the importance of history for the current water supply situation in towns in East Africa. Whereas the colonial systems that were put in place in the first decades of the 1900s were dominated by a large-scale technological paradigm, this did not mean that they were meant to serve the majority of the population. After independence, these socio-technical systems

Table 17.2 *Characteristics of the three models presented with regards to the key barriers to provision in informal settlements*

	Investment cost for the bulk supplier	*Transaction cost (property rights)*	*Transaction cost (payment)*	*Note*
Sweden	Low (fully outsourced to community)	Low (fully delegated risk)	Low (one contract for bulk supplier)	Requires substantial local resources
Kampala	Low (simple network)	Probably high (theft, vandalism)	Low (no middlemen)	Low access
Kisumu	Medium (cost for delegated line)	Probably medium (vandalism)	Low (one contract for bulk supplier)	Involves lease of assets

were, to a large extent, conserved both in regards to technological choice and institutional set-up and organization. The result of this inertia has been an effective discrimination of the urban poor by the public service systems. In the reforms initiated during the 1990s, most emphasis has been placed on cost recovery, which, although necessary in the long run, is no guarantee for improved service to the poor predominantly living in informal settlements. Second, we have reviewed the shortcomings of these historically shaped large-scale systems in providing service to informal settlements. We have argued that large-scale systems (although efficient for producing large quantities of good-quality water in bulk supply) are generally unsuitable for distribution in informal areas due to the high cost of network expansion and due to high transaction costs. The transaction costs stem mainly from the mechanisms needed to recover costs from the users, and from the risks associated with property rights enforcement: a particular challenge in informal areas. Third, we have demonstrated – through real-life examples – that there is a whole range of options available for policy-makers and water providers for overcoming the shortcomings of the large-scale systems. In a 'nested systems approach', the large-scale system will form the backbone of the provision system but be combined with small distribution networks that are outsourced to a local operator. More often than not, the physical as well as the institutional arrangements will have to be tailor made to suit its specific context. The key – we believe – is to find a 'smart' set-up to share costs between bulk supplier and the community supplied, involving both the investment cost and the transaction costs. Such nested systems have the potential of offering incentives that are favourable for providing quality service to poor people in informal areas at a reasonable price and under a cost-recovery approach. This can bring about the change needed for making public provisions cater for the needs of all citizens and not just a few.

NOTE

1 The historical analysis presented in this chapter draws heavily on previous work by Ezekiel Nyangeri Nyanchaga, Jenny Appelblad and David Nilsson.

REFERENCES

Appelblad, J. (2009) 'Parting from the privatization path? Public service reform strategies in the Ugandan urban water sector', in *The Provision and Politics of Public Services – Urban Water Governance in Uganda*, Licentiate Thesis, Department of Human Geography, Stockholm University, Stockholm

Appelblad, J. and D. Nilsson (2007) 'From "all for some" to "some for all"? – A historic perspective on current attempts for pro-poor water provision in Kampala,

Uganda', Paper presented at the Fifth International Water History Association (IWHA) Conference, Tampere, Finland, 13–17 June

Ballance, T. and S. Trémolet (2005) *Private Sector Participation in Urban Water Supply in Sub-Saharan Africa*, KfW Bankengruppe, Frankfurt am Main, Germany

Bayliss, K. (2003) 'Utility privatization in sub-Saharan Africa: A case study of water', *The Journal of Modern African Studies*, vol 41, no 4, pp507–531

Bransby Williams, G. (1907) *Report on the Sanitation of Nairobi and Report on the Townships of Naivasha, Nakuru and Kisumu*, British East Africa Protectorate, Kenya National Archives, GP 363.7 BRI, Nairobi

British East Africa Protectorate (1915) *Nakuru District Commissioner, Annual Report 1914–1915*, Kenya National Archives, DC/NKU/1/1, Nairobi

Collignon B. and M. Vézina (2000) *Independent Water and Sanitation Providers in African Cities, Full Report of a Ten-Country Study*, World Bank, Washington, DC

Hukka, J., T. Katko and O. Seppälä (1992) 'Kenya water sector – A position paper', Tampere University of Technology, Tampere, Finland

Kaijser, A. (1986) *Stadens ljus: Etableringen av de första svenska gasverken*, Liber förlag, Malmö

Kaijser, A. (2003) 'Redirecting infrasystems towards sustainability', in A. Biel, B. Hansson and M. Mårtensson (eds) *Individual and Structural Determinants of Environmental Practice*, Ashgate Publishing Ltd, Hants, UK, pp152–179

Kjellén, M. (2006) *From Public Pipes to Private Hands: Water Access and Distribution in Dar es Salaam, Tanzania*, PhD thesis, Stockholm University, Sweden

Kjellén, M. and G. McGranahan (1997) *Urban Water – Towards Health and Sustainability*, Stockholm Environment Institute, Stockholm

Kungl, Vattenfallsstyrelsen (1932) *Förslag till Stadgar m. m. för elektriska distributions förerungar*, Uppsala

Miller, C. (1971) *The Lunatic Express*, Penguin Books, London

Modig, H. (1984) 'El på landsbygden', in *Vattenfall under 75 år*, Kungl Vattenfallsstyrelsen, Stockholm

Nilsson, D. (2006a) 'A heritage of unsustainability? Reviewing the origin of the large-scale water and sanitation system in Kampala, Uganda', *Environment and Urbanization*, vol 18, no 2, pp369–385

Nilsson, D. (2006b) *Water for a Few: A History of Urban Water and Sanitation in East Africa*, Licentiate thesis, Royal Institute of Technology, Stockholm, Sweden

Nilsson, D. and E. N. Nyangeri (2008) 'Pipes and politics: A century of change and continuity in Kenyan urban water supply', *Journal of Modern African Studies*, vol 46, no 1, pp133–158

Nilsson, D. and E. N. Nyangeri (forthcoming, 2009) 'East African water regimes', in J. W. Dellapenna and J. Gupta (eds) *The Evolution of the Law and Politics of Water*, Springer, Berlin

North, D. C. (2005) *Understanding the Process of Economic Change*, Princeton University Press, Princeton, NJ

NWSC (National Water and Sewerage Corporation) (2006) *Kampala Water and Sanitation Program – Water Supply and Sanitation Services for the Urban Poor: Draft Final Report*, Software component, September 2006, Kampala, Uganda

Obudho, R. A. and R. A. Obudho (1992) 'The colonial urban development through space and time 1895–1963', in W. R. Ochieng and R. M. Maxon (eds) *An Economic History of Kenya*, East African Educational Publishers Ltd, Nairobi, pp405–444

Olsson, S.-O. (1984) 'Elektrifieringen ur avnämarsynpunkt: Användningsområden och distributionssystem', in *När Eelektriciteten kom. Tekniska museets årsbok 1984*, Tekniska Museet, Stockholm

Ostrom, E. (1990) *Governing the Commons: The Evolution of Institutions for Collective Action*, Cambridge University Press, Cambridge

Pakenham, T. (1991) *The Scramble for Africa*, Abacus, London

Republic of Kenya (1969) *Development Plan 1970–1974*, Nairobi

Republic of Kenya (2007a) *The Pro-Poor Implementation Plan for Water Supply and Sanitation*, Ministry of Water and Irrigation, Nairobi

Republic of Kenya (2007b) *The National Water Services Strategy 2007–2015*, Ministry of Water and Irrigation, Nairobi

Summerton, J. (1994) 'The systems approach to technological change', in J. Summerton (ed) *Changing Large Technical Systems*, Westview Press, Boulder, CO

Tarr, J. A. (1988) 'Sewerage and the development of the networked city in the United States 1850–1930', in J. A. Tarr and G. Dupuy (eds) *Technology and the Rise of the Networked City on Europe and America*, Temple University Press, Philadelphia, pp159–185

Thompson, J., I. T. Porras, E. Wood, J. K. Tumwine, M. R. Mujwahuzi, M. Katui-Katua and N. Johnstone (2000) 'Waiting at the tap: Changes in urban water use in East Africa over three decades', *Environment and Urbanization*, vol 12, no 2, pp37–52

Tumwine, J. (2002) *Drawers of Water II: 30 Years of Change in Domestic Water Use and Environmental Health in East Africa – Uganda Country Study*, International Institute for Environment and Development, London

UN-Habitat (2003) *Water and Sanitation in the World's Cities: Local Actions for Global Goals*, Earthscan, London

WASREB (Water Services Regulatory Board) (2007) *Socially Responsible Commercialization*, WASREB, Nairobi

WHO (World Health Organization) (1963) *The Organization for Water Development in Kenya: Report of a World Health Organization Consultant Team*, WHO, Kenya National Archives, BY/138/43, Nairobi

World Bank (2006) *Getting Africa on Track to Meet the MDGs on Water and Sanitation: A Status Overview of Sixteen African Countries*, World Bank Water and Sanitation Programme, WSP–Africa, Nairobi

WUP (Water Utility Partnership) (2003) *Better Water and Sanitation for the Urban Poor*, WUP, Abidjan

Interviews

Martha Adhiembo (2007) Katuoro Residents' Self-Help Group, Kisumu, pers comm, Interview in Kisumu, 15 November, 2007

Patrick Ombogo (2007) Lake Victoria South Water Services Board, and Alain Morel (2007) World Bank's Water and Sanitation Programme, Nairobi, pers comm (joint interview), Kisumu, 15 November, 2007

Charles Ochuodho and Malaki Obiero (2007) Wamalanda Development Group, Kisumu and Salome Okumo, Kiwasco, pers comm (joint interview), Kisumu, 15 November 2007

18

The South Asian Experience: Financial Arrangements for Facilitating Local Participation in Water and Sanitation Services (WSS) Interventions in Poor Urban Areas – Lessons from Bangladesh and Nepal

Roldan Muradian, Bishwa Nath Tiwari, Abu Jafar Shamsuddin and Laia Domènech

Introduction

Meeting the Millennium Development Goals (MDGs) of halving, by 2015, the proportion of people without sustainable access to safe drinking water and sanitation services (WSS), as envisaged in target 10 of the MDGs, encompasses significant financial challenges. Even though there is no consensus about the actual financial implications for achieving the goal, it is clear that a vast amount of additional resources are needed (Mehta et al, 2005). Estimates for the total amount of funding required to achieve MDG target 10 range from between US$6.5 billion and US$75 billion per year. Table 18.1 shows data for the coverage of WSS in Bangladesh and Nepal in 2002, suggesting that great efforts will have to be devoted to enhance sanitation coverage, both in rural and urban areas in these two countries. Moreover, Bangladesh is particularly far from reaching the target for drinking water, and in both countries the chances of success are hampered by chronic water shortages in urban areas.

Financing water and sanitation interventions in developing countries is facing serious difficulties, in part due to:

Table 18.1 *Drinking water and sanitation coverage in Bangladesh and Nepal (2002)*

	Drinking water coverage (%)				*Sanitation coverage (%)*			
	Urban		*Rural*		*Urban*		*Rural*	
	Total access	*House connections*	*Total access*	*House connections*	*Total access*	*Sewer connections*	*Total access*	*Sewer connections*
Bangla-desh	86	26	72	0	75	6	39	0
Nepal	93	48	82	8	68	12	20	0

Source: ADB (2008)

- scarce local public resources;
- insufficient international aid;
- the fact that foreign private investment in the water sector represents a small share of total foreign direct investment (FDI) in developing countries; and
- the fact that provision of water and sanitation among the poor is considered to be very risky by private investors, both local and foreign.

As a matter of fact, the water and sanitation sector worldwide suffers from chronic underfunding (UNDP, 2006). These are systemic conditions largely or entirely external to the WSS sector, but that exert a crucial impact upon it.

This context, characterized by public inability and private reluctance to invest in WSS, constitutes a significant challenge for growing poor urban communities worldwide, but meeting the MDGs also constitutes a unique opportunity given the far-reaching benefits that these communities may reap from investments in WSS. It is well established that the burden of deficient WSS systems is disproportionately allocated to the poorest population in urban areas (Kayaga and Franceys, 2007), who end up paying water tariffs considerably higher than better-off households, and additionally are more prone to suffer from sanitation-related diseases due to the unhygienic and crowded conditions of slums.

Nonetheless, the evidence shows that poor people are, indeed, very willing to invest in safe water supply and sanitation services, and to actively participate in WSS interventions. Community participation may mean a substantial shift in the distribution of the cost burden of implementing WSS systems, and therefore it should increase the chances of their execution. Although examples of participatory approaches in the WSS sector can be found in different countries, and particularly during the second part of the 20th century, since the early 1990s there has been a rapid and widespread development of participatory strategies. This has happened, to some extent, in response to the flaws of the top-down and supply-driven approaches to WSS that have prevailed in previous decades (Gomez and Nakat, 2002).

For instance, community participation in drinking water schemes has been mainstreamed in Nepal. The 1999 Local Self-Governance Act allocated responsibility to local bodies and communities for the design and building of drinking water systems, including repair and maintenance. In Bangladesh, commendable progress in the provision of basic water supply and sanitation has been achieved. The government set up a policy framework in 1998 by adopting a National Policy for Safe Drinking Water Supply and Sanitation, although slum communities have been rather neglected by government action. However, several non-governmental organization (NGO)-led models for enhancing WSS coverage in slums areas have been developed, which have extensively relied on local participation to achieve their objectives. In Bangladesh, NGOs have played the role of intermediaries between slums communities and the state, which has been traditionally reluctant to provide WSS to illegal settlements and holds a hostile vision of slums (massive evictions are not uncommon).

Against this background, this chapter looks at the interface between public policy and management by focusing particularly on the role of local participation in WSS projects. Within this framework, we pay special attention to the systematization of lessons learned and the identification of best practices with regard to the use of financial instruments for promoting the participation of local communities in WSS interventions in poor urban areas. The following section briefly summarizes the methodology used for data-gathering. It is followed by the presentation of the main results. The chapter finishes with some concluding remarks.

METHODOLOGY

Lessons and insights were drawn from 20 case studies: 10 in Bangladesh and 10 in Nepal. Fieldwork was conducted from September 2006 to June 2007. The case studies consisted of WSS interventions selected on the basis of the following two criteria:

1 They had to be located in a poor urban area.
2 They had to be schemes from which lessons and insights may be learned with regard the role of local participation in the performance of WSS systems.

We gathered both primary and secondary information. Primary data was collected through interviews with key informants and stakeholders, including representatives from local government bodies, the participating NGOs and members of community water committees. In addition, in each locality we administered a survey to between 70 and 100 households, randomly chosen, in order to assess their socio-economic status, as well as their involvement in and perceptions of the WSS intervention taking place in their community. Tables 18.2 and 18.3 summarize some of the basic characteristics of the case studies from both countries.

LESSONS LEARNED: BEST PRACTICES AND OBSTACLES

This section summarizes the lessons learned from the case studies. It is organized in five main subsections, each addressing the key categories of financial/managerial tools associated with local participation that we have identified in the study. These are:

Table 18.2 *Brief description of the Nepalese case studies*

Location	*Operating agencies*	*Short description*
Ilam municipality East	Department of Water Supply and Sewerage (DWSS) / Municipality	Water supply system with 1811 private taps, 7 community taps and 2 public taps. The system has been managed by municipality for more than 10 years with progressive tariff structure.
Damak municipality East	DWSS / Water Supply Users Committee (WSUC)	Water supply system managed by the users' committee with 2697 private taps and 3 community taps. There is demand for new connections but the users committee is unable to expand the system.
Dhulikhel municipality Centre	WSUC	Water supply system with 1033 private taps and 27 public taps, which has been managed effectively by a users' committee for a long time. A successful penalty and incentive system ensures prompt payment.
Khokana VDC, Lalitpur Centre	Environment and Public Health Organization (ENPHO) / People Welfare Organization (PWO)	Successful ecological sanitation (ECOSAN) piloting intervention aiming at constructing 64 ECOSAN latrines in a semi-urban locality of Nepal. The incorporation of innovative strategies ensured active collaboration of the beneficiaries.
Aalok Hiti, Patan, Lalitpur Centre	WSUC	Innovative water supply project totally led by community people. Up to 200 households have been benefited by the smart initiative. Capital cost and operation and maintenance (O&M) is 100 per cent born by the community.
Khadipakha, Kathmandu Centre	Lumanti / WSUC	Water supply and sanitation intervention in a slum area of Kathmandu with high community contribution.
Khairenitar VDC, TanahuWest	WSUC	Small Town Water Supply and Sanitation Project supported by ADB and implemented by DWSS-GoN. Community bears 50 per cent of the capital cost and the O&M of the system. There is special provision for urban poor.
Dandatole, Butwal municipality West	Lumanti / WSUC	Water supply and sanitation intervention in which partnership of municipality and the implementing NGO was established. Application of cost-recovery principle with 80 per cent of the capital cost born by community. The performance of the users' committee is not satisfactory mainly due to political differences.
Shreenagar, Surkhet Mid-West	Nepal Water for Health (NEWAH) / WSUC	Drinking water, health and sanitation project that provided 10 community taps and 99 latrines. Community contributed labour and materials to cover the capital cost. Users' committee is responsible for managing the system. Flat monthly tariff of NRs.10.
Mahendranagar municipality, Dhangadhi Far-West	Nepal Water Supply Corporation (NWSC)	Water supply system with 1,313 private taps and 27 public taps managed by NWSC inefficiently. Community contribution is absent.

Table 18.3 *Brief description of the Bangladeshi case studies*

Location	*Operating agencies*	*Short description*
Baganbari, Mirpur, Dhaka	Dushtha Shasthya Kendra (DSK) / Committee	Water supply and sanitation intervention in a slum area of Dhaka. Capital cost-sharing strategy based on ability to pay for shared water supply and sanitation facilities. Legal water connection from Dhaka Water Supply and Sewerage Authority (DWASA) with NGO acting as a guarantor. Payment of monthly tariff.
Hazighona, Chittagong	DSK / Committee	Water supply and sanitation intervention in a slum area of Chittagong. Capital cost-sharing strategy based on ability to pay for shared water supply and sanitation facilities.
Ta Block Slum, Mirpur, Dhaka	Association for Realization of Basic Needs (ARBAN)/ Committee	Water supply and sanitation intervention in a slum area of Dhaka. Capital cost-sharing strategy based on ability to pay for shared water supply and sanitation facilities. Legal water connection from Dhaka Water Supply and Sewerage Authority (DWASA) with NGO acting as a guarantor. Payment of monthly tariff.
Ghuntighar Slum, Jurain, Dhaka	Population Services and Training Center (PSTC) / Committee	Water supply and sanitation intervention in a large slum area of Dhaka. Capital cost-sharing strategy based on ability to pay for shared water supply and sanitation facilities. Legal connection of DWASA with NGO acting as a guarantor. Payment of monthly tariff.
Old Zimkhana, Narauanganj	Prodipon / Committee	Water supply and sanitation intervention in a slum area of Narayanganj. Capital cost-sharing strategy based on ability to pay for shared water supply and sanitation facilities.
Sirajganj	United Nations-Habitat/ Local Government Engineering Department (LGED)/ Committee	Local Partnerships for Urban Poverty Alleviation Project (LPUPAP). Water supply and sanitation intervention in which capital cost of water supply and sanitation schemes was fully born by the project. O&M is done by community. The project incorporates promotion of micro-credit and saving groups.
IG Gate Slum Faridabad, Dhaka	Urban Development Centre (UDC) / UNICEF/ Department of Public Health and Engineering (DPHE) / Committee	Environmental Sanitation, Hygiene & Water Supply in Urban Slums & Fringes Project. Cost-sharing strategy based on well-being ranking for water and full subsidy for sanitation. Payment of monthly water tariff.
Sreepur, Gazipur	Dhaka Ahsania Mission (DAM) / Local Government Body (LGB) / community	Decentralised Total Sanitation Project (Dishari). Joint initiative of several international and national agencies. The main focus of the project is on capacity-building of local government bodies and community. Only the very poor obtain some financial support.
Bashbaria, Sitakundo, Chittagong	Village Education and Resource Center (VERC) / community	Water supply and sanitation intervention in a semi-urban area of Sitakundo. Capital cost-sharing strategy based on ability to pay for water supply. Sanitation is promoted through CLTS, in which community is intensely mobilized to achieve the goal of total sanitation. No financial support is given for latrine construction.
Takhtarpool Slum Bakulia, Chittagong	NGO Forum / Prottyashi	Water supply and sanitation intervention in a private slum of Chittagong. Capital cost-sharing strategy based on ability to pay for shared water supply and sanitation facilities. The landowner contributed a fraction of the capital cost.

1. incentives for facilitating cost sharing;
2. tools for promoting participation and prompt payment;
3. instruments for including the poor;
4. mechanisms for ensuring efficient use of resources and sustainability; and
5. synergies with income-generating and local development activities.

The section is complemented with a brief discussion of some equity considerations arising from the cases.

Incentives for facilitating cost sharing

In most of our case studies we found that poor urban communities are both willing and able to bear a considerable share of the overall capital cost of WSS interventions. Table 18.4 shows the share of community contributions to the capital costs in a sample of selected cases from Nepal. Several examples from Nepal and Bangladesh demonstrate that as far as there is flexibility with regard to the type of contribution that can be made, a significant share (between 40 and 50 per cent) of the overall capital costs may be covered by local communities. A key factor that facilitated this outcome was a multi-stakeholder coalition, involving donors, NGOs, grass-roots organizations and local governments.

Aalok Hiti is one of the most interesting and creative cases that we have analysed. This intervention was initiated in 2003, and was designed and implemented by a group of young dwellers from the community. The very high proportion of local contribution to the capital costs reached (87.4 per cent) was possible due to two critical factors:

1. a strong local leadership that was able to mobilize local human and financial resources; and
2. a truly bottom-up approach.

The system is based on a traditional stone spout that has supplied water to this community for more than 500 years. The novelty of the intervention has to do with new ways of storing and distributing water, which is gathered at night and distributed at specific hours during the day. The group collected an initial seed fund from the households interested in improving the system for water supply. Another source of funds was a loan at zero interest rate that was granted by some community members. Furthermore, Lalitpur municipality donated a reserve tank, which completed the full capital cost of the system. This case shows very well that what could be termed 'bottom-up creative thinking' may play a major role both for finding innovative solutions and mobilizing local resources in a very effective manner. Nonetheless, the fact that the system can provide water for a few hours per day also shows the limitations imposed by systemic conditions (in this case, local water scarcity, high population density and insufficient government support for water infrastructure).

It is noteworthy that in Dandatole, Nepal, the local contribution was as high as 80 per cent for water supply and 62 per cent for sanitation. Dandatole's residents participated actively in the design and construction stages of the project. The system now provides 24 hours of water supply through 11 community taps. This was achieved due to a pioneering

Table 18.4 *Share of community capital cost provision in selected case studies from Nepal*

Case study	*Community*	*Other major stakeholders*	*Financial instrument used for covering community's contribution*
Aalok hiti (Water supply)	87.4%	16.7% (municipality)	Direct contribution in cash by the community
Dandatole (Water supply and sanitation)	80% for water supply 62% for sanitation	20% for water 38% for sanitation (Lumanti/WaterAid Nepal/ municipality)	Combination of direct contribution (in cash and low-skilled labour) and provision of a five-year loan to the community at no interest cost by WaterAid Nepal
Khadipakha (Water supply and sanitation)	46.5% for sanitation	53.5% (Lumanti/WaterAid Nepal)	Direct contribution in cash and in kind (low-skilled labour) by the community
Khairenitar (Water supply and sanitation)	50% for water supply	50% (GON/ADB)	A combination of direct contribution from the community in cash and kind (20% before and during construction period) and a loan for 30% of the total capital cost provided to the community by the ADB (8% interest rate) to be returned within 15 years
Khokana (Water supply and sanitation)	0% for drinking water 30% for toilet construction 25% for sewerage	100% for water (NWSC) 70% for toilet (ENPHO/ WaterAid) 75% for sewerage (VDC/ DDC)	Direct community contribution in the form of cash, labour and materials
Shreenagar (Water supply and sanitation)	41%	59% (NEWAH/ Municipality/DHSP)	Direct community contribution in the form of cash, labour and materials The community provides a periodical fee to cover operational and maintenance cost

Notes: ADB = Asian Development Bank; DDC = District Development Committee; DHSP = District Health Strengthening Programme; ENPHO = Environment and Public Health Organization; GON = Government of Nepal; NEWAH = Nepal Water for Health; NWSC = Nepal Water Supply Corporation; VDC = Village Development Committee.

financial arrangement in Nepal, based on what they call the principle of 'high-cost recovery'. In order to finance WSS interventions, Lumanti Support Group for Shelter – an NGO – and the Butwal Municipality created a poverty fund. Lumanti's contribution was financed by WaterAid Nepal, the local representative of the international NGO dedicated to water and sanitation. The poverty fund provided 20 per cent of the total capital cost of the water supply intervention as a non-refundable grant, and the rest was covered with an interest-free loan to be paid by the community in five years. The financial arrangement for the construction of sanitary units (pan latrines) in this case was slightly different. The beneficiary households obtained a free pan, skilled advice and construction materials for building a latrine up to the substructure or pan level, but they had to bear the full cost of finishing the construction.

Users, who were responsible for the construction of the water supply project, made an efficient utilization of the available fund. As a result, they were able to achieve significantly more than what was initially planned within the original budget. However, the performance of the loan repayment has not been so satisfactory, and after four years (out of five), less than 70 per cent of the loan had been returned. This low repayment rate was caused by political conflicts inside the community. According to some dwellers, the 2005 election of the users' committee was carried out with irregularities, and most of the members of the newly elected committee belonged to the same political party. Some residents of the community, who are not supporters of that party, decided to withdraw their support for the committee and were unwilling to maintain their participation in the decision-making process under the new leadership. This conflict illustrates the limitations that user committees may face when their members participate only on a voluntary basis since the fact of relying on voluntary work often facilitates the control of the system by particular groups within the community. Internal political conflicts triggered by community-led water interventions have been also reported in Africa (Page, 2003) and may be considered as part of the limitations of the participatory approach.

In Khairenitar and Khadipakha, Nepal, the local contribution to the overall capital cost of the interventions was 50 and 46.5 per cent, respectively, covered by means of low-skilled labour, cash and materials. In these two cases, no loan was provided to the local community. In Khairenitar, funds were made available by the Government of Nepal (through an Asian Development Bank credit), and in Khadipakha the bulk of resources were provided by Lumanti (again through a donation of WaterAid Nepal).

Several models of local participation in WSS interventions have been developed in poor urban areas in Bangladesh. One of the better known cases is the model generated by the NGO Dushtha Shasthya Kendra (DSK), which has shown to be very successful and is currently being widely replicated in poor urban communities in Dhaka and Chittagong. This scheme is based on the belief that the urban poor can manage a formal water supply system, and the role of the participating NGOs is just to provide seed funds and guidance for the creation of an effective users' committee (Akbar et al, 2007).

Most of the cases studied in Bangladesh follow a flexible capital cost-sharing strategy, which was developed by WaterAid Bangladesh and applied to a large number of localities, in partnership with different NGOs. The model is based on the 'ability to pay' approach – that is, the contribution of households to capital cost recovery depends upon their socio-economic status. WaterAid Bangladesh provides guidelines for the definition of five poverty categories. The classification is adapted to the local situation by the respective community by using indicators developed by local dwellers in order to categorize the poverty status of households. These indicators include, for example, the number of meals taken in a

Table 18.5 *The 'ability to pay' approach: Contribution of different household categories to fund the water and sanitation services (WSS) intervention (Baganbari slum, Dhaka, Bangladesh)*

Category	*Number of households*	*Household contribution to total capital cost (%)*
A: not poor	51	43
B: poor but relatively better off	124	35
C: moderately poor household	23	14
D: hard-core poor	11	8
E: extremely poor	68	0

day, the type of furniture and home appliances owned, the capacity to celebrate different festivals, etc. Under this model, the community is fully responsible for the operational and maintenance costs of the system. Table 18.5 depicts the contribution of the different household categories in Baganbari Slum to the overall capital cost. It has been found, however, that the poorest of these communities have a very low level of participation in users' committees and in the actual management of the systems (Hanchett et al, 2003). Hence, a participatory approach does not automatically ensure the inclusion of vulnerable social groups and may actually reproduce the asymmetries of local power structures.

In Sreepur municipality, also in Bangladesh, an interesting example of multi-stakeholder partnership for enhancing local participation was developed. A coalition between Dhaka Ahsania Mission (DAM), Plan Bangladesh, WaterAid Bangladesh and the Water and Sanitation Programme (WSP) of the World Bank was developed to fund and implement a sanitation intervention. The project only provided funds for covering the cost of project staff and office set-up, as well as for organizing training courses at various levels, workshops, seminars and an awareness campaign. The community covered the full capital cost of the installation of sanitary latrines in households. In order to facilitate the process, the project introduced four different types of low-cost technological options for latrines, and people selected the appropriate option for them according to their payment capacity. This case is an excellent example of successful participatory WSS interventions, given that flexibility in the choice of sanitation technologies (tailored to the need of the poor) has been identified as a critical issue for extending sanitation services to deprived urban areas (Paterson et al, 2007).

Tools for promoting participation and prompt payment

In Ilam, Nepal, the municipality has designed a penalty system to discourage delayed tariff payments. In Damak, also in Nepal, a high rate of timely payment has been achieved through the combination of a strong penalty system with an incentives scheme. For example, consumers may benefit from a 2 per cent discount on the total billed amount if they pay their bills within a week. In addition, an award is given to the two consumers who have the best payment record throughout the year. Furthermore, in this locality, the water committee has decided to introduce an incentive to reward the collaboration of the members. A financial incentive of 100 Nepalese rupees (US$1.5) is provided every meeting that they attend. This strategy has proven to be effective in substantially enhancing participation in the committee's meetings. There are clear advantages of making the work in the water committees more professional and rewarded economically as a strategy to avoid the kind of malfunctioning described earlier in the case of Dandatole.

It is noteworthy that in Aalok Hiti, a good performance (in terms of timely payment) has also been achieved by means of a much less stringent penalty system than in Ilam or Damak due to the high levels of social control and sense of empowerment experienced by the community.

Instruments for including the poor

One of the major risks associated with projects intending to maximize local financial contributions is that the very poor of the community may become easily excluded due to their low payment capacity. In this respect, the intervention in Ilam is very revealing. In this locality, the water supply intervention was aimed at building private taps for the population. Initially, this strategy was combined with the provision of public taps for those who were unable to afford the private option. However, after realizing that water misuse and conflicts among people gathering to get water from public taps became common problems (as still happens in Dhilikhel, which keeps public taps), the municipality decided to convert public taps into community taps. Access was fully free for public taps, while the use of community taps was conditioned to the creation of a users' committee and the payment of a flat monthly rate per household plus a variable initial fee. The municipality has introduced five different rates for the initial fee based on households' economic status, which is measured according to the quality of the house (e.g. cemented, bricks, mud, etc.). The poorest households that compose the fourth and fifth categories are not required to pay the initial fee. Therefore, the conversion from public to community taps resulted in a better management of the system without causing exclusion. Community taps for the poor are also provided in Damak.

Tailoring subsidies and flexibility in relation to households' contributions are two key factors that may be used as inclusion mechanisms. For example, in the sanitation intervention in Khokana, Nepal, households' contributions to the total cost of building latrines varied according to their socio-economic status. The amount of subsidy provided, as well as the design and size of the toilet, is discussed between households and practitioners, so both aspects and the financial support required from the users are tailored to their needs and capabilities. Similarly, in the WSS intervention in Khadipakha, those households that could not contribute in cash were encouraged to contribute with extra labour during the implementation phase. In Khairenitar, some degree of flexibility was allowed in the actual forms of access to water by the poor. For instance, some of the poorest households were allowed to obtain water from their neighbours' tap, either by paying them a negotiated amount or even without payment, while other poor households built a joint private tap to share expenses.

In Dandatole, the community adopted another strategy: the extremely poor households are offered free water through a private tap. Here, those households that cannot afford to pay the water tariff can request to have the monthly fee waived by the general assembly. This was also the model adopted in Shreenagar, Nepal, where around 20 per cent of users are exempted from paying their monthly charge due to their extreme poverty. Likewise, in the sanitation intervention in the Sreepur municipality, Bangladesh, the local government provides sanitary latrines free of cost to the most deprived members of the community.

In Sitakundo (Chittagong), Bangladesh, flexibility in relation to the technological choices made available was a critical factor in including the poor: these households were allowed to install a low-cost latrine in the initial stages and gradually move into the higher price and better-quality options. Moreover, all of the case studies from Bangladesh follow the 'ability to pay' approach described above, where households pay their share of capital cost according to five poverty categories.

Mechanisms for ensuring efficient use of resources and sustainability

The formation of a users' committee is a common feature in almost all the cases studied. The performance and capacity of the users' committee is probably the most important factor determining the long-term sustainability of the system. Normally, final decisions regarding planning, managerial and financial arrangements are taken during the general assembly. Thus, reporting to the assembly becomes an activity of major importance in this kind of participatory project. The good practices that we have found in this regard include the careful reporting of information concerning financial accounts presented to the general assembly by the users' committee in Aalok Hiti. This way of reporting, together with a generally high level of social control, reduces transaction costs since in these circumstances external auditing is not needed. It is also interesting to note that in Khadipakha the community willingly contributed more than the initially agreed amount.

Besides good practices for saving resources and increasing transparency, we have also identified examples of penalty systems for discouraging inefficient use of water by users. For instance, in Ilam, the municipality has adopted an elaborate scheme to deal with water misuse and pollution. Penalties are also imposed on those users who intentionally damage the system, do not install a meter or evade payment by manipulating the meter.

We have also found cases affected by lack of transparency. For instance, in Sitakundo (Chittagong) and Old Zimkhana (Narayanganj Municipality), both in Bangladesh, most dwellers were not aware of the fact that money collected through the water tariff was not being reinvested in the water system, but was just kept in the bank. This is a common feature of several cases in Bangladesh. In addition, following the majority rule for taking important decisions in the assembly often has limitations. For example, in Dhulikhel, there is a high demand for the expansion of the system by the un-served population, who do not have the right to vote in the assembly. However, the general assembly has decided not to expand it since the currently served population holding the voting rights believe that the expansion would probably negatively affect the performance of the scheme. This problem has been a source of conflicts between the two groups.

Synergies with income-generating and local development activities

There also exist excellent examples where WSS interventions are part of a broader strategy to trigger organizational and development processes in the communities, and which are often associated with income-generation activities. For instance, in Aalok Hiti, even though the water intervention project was not focused on sanitation, the improvement of local hygienic conditions was seen as a key complement to the provision of water supply. This initiative has also contributed to strengthen local organizational capacities and to consolidate the leadership of young dwellers committed to the development of the community. The model is currently being replicated in neighbouring localities as it has been very successful in triggering multiplier effects and rising local self-esteem.

The WSS intervention in Sirajgonj, Bangladesh, is part of the Local Partnerships for Urban Poverty Alleviation Project, which was jointly developed by the Ministry of Local Government, Rural Development and Co-operatives and the United Nations Development Programme (UNDP). As part of its activities, the project encourages community groups to develop sustainable saving and credit operations. There are 893 saving groups around Sirajgonj Pourasava, and 96 per cent of these groups are composed entirely of women. Paradoxically, despite the integrated approach informing the scheme in this project, we found one of the lowest levels of local contribution. Among other reasons for this poor performance, there was no provision for capital cost sharing for the installation of hardware, and local dwellers became reluctant to contribute to operational and maintenance costs, which was largely due to the low degree of local ownership of the project.

The WSS intervention in IG Gate Slum (Dhaka) followed a similar integrated approach. As part of the activities of the project, micro-credit and saving schemes were also introduced, led by ARBAN, a local NGO. Currently, all savings group members are women. Contrary to the previous case, in this locality the management of the WSS intervention was carried out by the beneficiaries through different committees. The slum area was divided into three clusters to provide legal water connections according to the convenience of the beneficiaries, adopting a capital cost-sharing strategy. This case shows that it is possible to combine integrated and participatory approaches. The previous two cases rely heavily on inter-sector partnerships for mobilizing financial resources and know-how. The latter is a key issue since the lack of pro-poor targeting and low levels of involvement of non-state actors have been identified as serious constraints for achieving the MDGs in the WSS sector (Gutierrez, 2007).

Equity considerations

Slum dwellers who are obliged to buy water from informal local vendors (the typical situation before WSS interventions) are particularly vulnerable to pay excessive charges. Participatory interventions of the kind that we have studied tend to considerably reduce water fees; but even then slum dwellers usually continue to pay higher fees than their counterparts in wealthier neighbourhoods. This typically has two causes. First, in general, a capital cost-recovery strategy of the type implemented in the slums entails relatively higher water tariffs since a large proportion of the initial cost has to be repaid, while wealthier communities normally benefit from a higher proportion of subsidies. Second, in the cases from Bangladesh, the water supply company keeps charging higher fees to the NGO that works as guarantor for connecting community water points to the general distribution network. These features suggest that participatory approaches for WSS projects tend to face a trade-off between, on the one hand, increasing the feasibility of the intervention (through cost-sharing and creating a local sense of ownership) and, on the other, improving the equity in the distribution of the costs involved in delivering WSS to urban areas.

Nonetheless, in many situations a cost-sharing strategy may constitute a preferable and more feasible option compared to the privatization of the service (Hall and Lobina, 2007), since the private sector normally faces considerable disincentives to serve the urban poor and tends to aggravate equity problems (Bakker, 2007; Castro, 2007).

CONCLUSIONS

The systematization of factors that facilitate and inhibit participation becomes a key input for improving the performance of WSS schemes (Manikutty, 1998). We have tried to do so through an extensive assessment of case studies. We think that the high level of local contribution that has been achieved in several of the cases studied is quite remarkable. This shows that cost sharing in poor urban localities is both feasible and effective. We have found that flexibility and tailoring of the schemes to meet local needs and conditions have been two critical factors in enhancing local participation and contribution. This conclusion is in line with what has been found in similar situations in other parts of the world (Gleitsmann et al, 2007). We have also found that participatory WSS interventions may become a driving force of local development if they strengthen local capacities, consolidate leadership and trigger multiplier effects. The most interesting cases that we have come across are those where local dwellers took an active role, designed creative solutions on their own and provided a very large proportion of the required human and financial resources. Moreover, our cases reveal that inter-sector partnerships play a very significant role in enabling local contributions and preventing the exclusion of vulnerable groups. These findings clearly demonstrate that the interface between public policies and management is a key area for concern in WSS interventions in poor urban areas.

However, community participation is far from being an easy and smooth process. A participatory approach towards WSS interventions is by no means without limitations or drawbacks. We have seen that users' committees may be co-opted by vested interests of particular groups, and decisions following the majority rule may entail, in some circumstances, the exclusion of socially disadvantaged groups from the benefits of WSS interventions. In addition, the very poor tend to be excluded from participation in users' committees, and structural and external conditions may impose serious constraints on the performance of WSS interventions (e.g. the risk of eviction in slum areas in Bangladesh discourages local investments). Nevertheless, despite the challenges and problems that have been identified, the 20 case studies briefly analysed here shed a rather optimistic picture about the role that community participation may play in enhancing the financial feasibility and sustainability of WSS interventions in poor urban areas.

ACKNOWLEDGEMENTS

This research was funded by the Asia Pro Eco programme of the European Commission (grant contract ASIA/PRO ECO/BD/2005/109993). We are very grateful to all the members of the teams from the International Training Network Center, Bangladesh University of Engineering and Technology (ITN–BUET) and the Central Department of Economics (CEDECON) at Tribhuvan University who conducted and facilitated fieldwork in Bangladesh and Nepal respectively. We are also indebted to local stakeholders who provided very valuable information.

REFERENCES

ADB (Asian Development Bank) (2008) *MDGs in Central Asia and the Southern Caucasus*, www.adb.org/Documents/Events/2007/MDGs-in-Central-Asia/default.asp, accessed June 2008

Akbar, H., J. Minnery, B. van Horen and P. Smith (2007) 'Community water supply for the urban poor in developing countries: The case of Dhaka, Bangladesh', *Habitat International*, vol 31, no 1, pp24–35

Bakker, K. (2007) 'Trickle down? Private sector participation and the poor water supply debate in Jakarta, Indonesia', *Geoforum*, vol 38, no 5, pp855–868

Castro, J. E. (2007) 'Poverty and citizenship: Sociological perspectives on water services and public-private participation', *Geoforum*, vol 38, no 5, pp756–771

Gleitsmann, B., M. Kroma and T. Steenhuis (2007) 'Analysis of a rural water supply project in three communities in Mali: Participation and sustainability', *Natural Resources Forum*, vol 31, no 2, pp142–150

Gomez, J. and A. Nakat (2002) 'Community participation in water and sanitation', *Water International*, vol 27, no 3, pp343–353

Gutierrez, E. (2007) 'Delivering pro-poor water and sanitation services: The technical and political challenges in Malawi and Zambia', *Geoforum*, vol 38, no 5, pp886–900

Hall, D. and E. Lobina (2007) 'Profitability and the poor: Corporate strategies, innovation and sustainability', *Geoforum*, vol 38, pp772–785

Hanchett, S., S. Akhter and M. H. Khan (2003) 'Water sanitation and hygiene in Bangladeshi slums: And evaluation of the WaterAid-Bangladesh urban programme', *Environment and Urbanization*, vol 15, no 2, pp43–55

Kayaga, S. and R. Franceys (2007) 'Costs of urban utility water connections: Excessive burden to the poor', *Utilities Policy*, vol 15, no 4, pp270–277

Manikutty, S. (1998) 'Community participation: Lessons from experiences in five water and sanitation projects in India', *Development Policy Review*, vol 16, no 4, pp373–404

Mehta, M., T. Fugelsnes and K. Virjee (2005) 'Financing the Millennium Development Goals for water and sanitation: What will it take?', *Water Resources Development*, vol 21, no 2, pp239–252

Page, B. (2003) 'Communities as the agents of commodification: The Kumbo Water Authority of Northwest Cameroon', *Geoforum*, vol 34, no 4, pp483–498

Paterson, C., D. Mara and T. Curtis (2007) 'Pro-poor sanitation technologies', *Geoforum*, vol 38, no 5, pp901–907

UNDP (United Nations Development Programme) (2006) *Human Development Report 2006: Beyond Scarcity: Power, Poverty and the Global Water Crisis*, UNDP–Palgrave Macmillan, New York

19

Water and Sanitation Services in China: Current Problems and Potential Solutions

Jiane Zuo and Lili Gan

Introduction

This chapter provides a synthetic account of the history and current status of Chinese water and sanitation services (WSS), including a consideration of the main problems and potential solutions facing the sector, with a special focus on urban areas. We present, first, a brief background description of the situation of WSS in the country, dealing with basic information about coverage and a reference to the main institutional and legal structures in relation to the management of these services. Then we consider the case of drinking water supply and explore the situation in the Chinese capital as an example of the problems and challenges facing the country in this particular area. The last section focuses on wastewater collection and treatment.

The context of water and sanitation services in China

During recent years, with China's rapid economic development, the urbanization process has gradually begun to speed up. According to data from the Ministry of Housing and Urban–Rural Development (MHURD) for 2007, there were, in total, 655 cities in the Chinese mainland, with a total urban population of 375 million people (MHURD, 2008). The rate of urbanization rose from 21 per cent in 1982 to 43.9 per cent in 2006 (the global average rate was 40 per cent in 2006), and it is estimated to reach around 50 per cent in 2010. This rapid process of urbanization has brought tremendous pressure to the Chinese urban and rural water supply and wastewater treatment services. Table 19.1 presents the rates of coverage for urban and rural WSS at the national level.

Table 19.1 *Urban and rural water and sanitation coverage in China (2007)*

Population (millions)		*Water supply*		*Sanitation*	
		Percentage covered	*Percentage covered over total population*	*Percentage covered*	*Percentage covered over total population*
Urban	375	93.8	24.6	62.8	16.5
Rural	1056	44.7	33.0	22.6	16.7
Total	1431		57.6		33.2

Source: MHURD (2008)

The overall picture presented in Table 19.1 clearly shows the asymmetries in service coverage both between water supply and sanitation services, as well as between rural and urban areas. Although substantial work has been carried out to develop WSS in the country since the creation of the People's Republic of China in 1949, much remains to be done, particularly in rural areas where the bulk of the population is concentrated. However, the problems facing Chinese cities are also significant, as illustrated in Table 19.2, which shows a sample of the largest urban centres.

As Table 19.2 shows, some of the main metropolises and large cities have been very successful in expanding WSS coverage, especially within their core urban areas. However, still in 2007, slightly over half (56 per cent) of the wastewater produced received any treatment, while around 200 Chinese cities have no coverage for sanitation services. This situation represents a formidable challenge for the future development of China.

Table 19.2 *Water and sanitation coverage in Chinese cities (2007)*

City	*Population (millions)*	*Water supply coverage (%)*	*Sanitation coverage within jurisdiction (%)*	*Sanitation coverage within core urban area (%)*
Beijing	16.33	100	92	100
Shanghai	18.58	100	78	100
Chongqing	28.16	95	48	95
Tianjing	11.15	97	75	97
Guangzhou	9.94	97	76	100
Chengdu	12.21	99	70	85

Source: NBS (2008)

MANAGEMENT, TREATMENT AND SUPPLY OF DRINKING WATER IN CHINA

Water supply institutions

The institutional framework for the management of water supply in China is based on the central role played by the public sector and can be defined as a case of 'government-led management structures', which are the responsibility of several government ministries and levels of administration. This central role played by the public sector is the key institutional feature of Chinese WSS, and can be traced back to the foundation of the People's Republic of China in 1949. There was no room for private ownership and management at that time, and the responsibility for essential WSS became a government preserve. As a result, all investments were provided by the national government, and the public sector became responsible for the administrative organization of the services. Although since 1978 there have been several reforms and important changes have been introduced in the Chinese WSS sector, the core principle of 'government management' has been maintained, not least because of the need to guarantee the population's safety. For instance, although in some cities the government has allowed the introduction of private corporations as stockholders in the WSS utilities, and the management of the companies is now run on a commercial basis, the government retains majority control (over 50 per cent of the shares).

The current institutional organization for water supply in China results from the reforms implemented during the last 30 years, the last of which took place in March 2008. The Ministry of Health and the Ministry of Environmental Protection (previously the State Environmental Protection Administration) are in charge of water quality control and guaranteeing the safety of the water supplied, including in rural areas. The MHURD (previously the Ministry of Construction) is mainly responsible for urban water management, especially centralized water supply. The legal basis for this organizational structure is provided by the 1996 Administrative Measures for the Sanitary Supervision of Drinking Water passed by the national government. Figure 19.1 outlines the responsibilities and management strategy for drinking water in China.

To a large extent, this division of responsibility for the quality of water supply among different ministries has been inherited from the original system established during the period of Chinese Planned Economy. Although this system was efficient during that specific period of the country's history, it has become increasingly problematic in the face

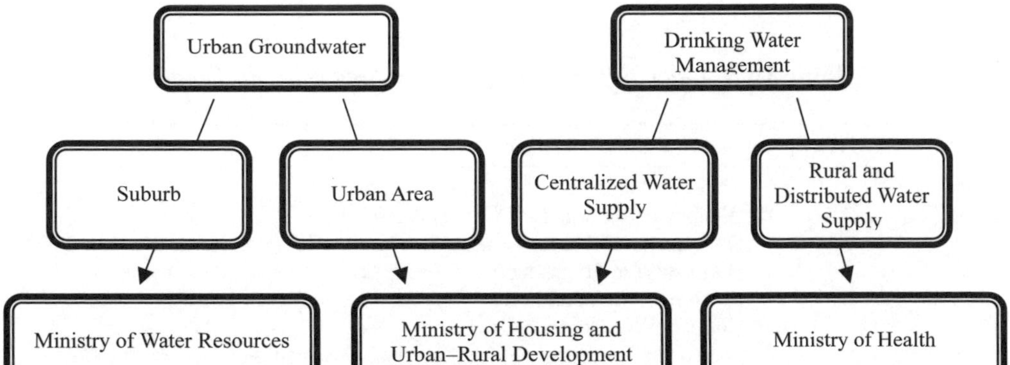

Figure 19.1 *Structures of responsibilities and management strategy for drinking water in China*

of the higher complexity of water quality management driven by the urbanization process and by China's rapid economic development. The bottom line is that the quality of water supply must meet the Sanitary Standard for Drinking Water (national standard), which is the lowest and most basic quality requirement. The parameter set for this indicator is low for a number of interconnected reasons: although the levels of development are very uneven throughout the country, the national standard is duly enforced nationwide (including smaller cities and villages) and, therefore, it is understood that the requirements of the standard will not be too high. In practice, compared to the situation in developed countries, the Chinese drinking water standards have fewer components and certain standards are not strictly monitored.

Besides the national standards, there are also special norms for specific industries such as bottled water production. In order to achieve different water quality control objectives for these special standards, local administrative departments can develop their own norms according to the conditions of local water resources, the needs of the population, the technical feasibility of water supply systems, or economic criteria. However, local norms must first meet the national standards and the specific water quality standards for the relevant industries. Thus, in fact, local water quality norms must be stricter than the national and industry-specific standards. These aspects are governed by several laws and regulations related to drinking water management, treatment and supply at different levels, which are listed in Table 19.3.

The most recent legal instrument listed in Table 19.3 also represents an essential and exciting change in Chinese WSS management: the 2006 edition of the *Sanitary Standards for Drinking Water* (GB 5749-2006), which replaced a 1985 norm (GB 5749-85) (Ministry of Health, 2006). The 2006 regulation is the first revision of drinking water quality standards in China, and all the changes that have been introduced are mainly a response to the impact of the country's economic development during the last 20 years and to the rising expectations of the population concerning the quality of the drinking water supply. As a result, the 2006 standards include a total of 106 parameters, compared

Table 19.3 *Laws and regulations for drinking water in China*

Year of promulgation	*Name of law or regulation*
1984	Water Pollution Prevention and Control Law
1985	Sanitary Standards for Drinking Water
1992	Development Plan for Urban Water Supply Technology for the Year 2000
1993	Regulation of Abstraction and Protection of Urban Groundwater
1993	Implementing Measures for the Water Licensing System
1994	Regulation of Urban Water Supply
1996	Administrative Measures for the Sanitary Supervision of Drinking Water
1997	Administrative Measures for the Supervision of the Water Licensing System
1999	Administrative Regulation of Urban Water Supply Quality
2000	National Water Law
2006	Sanitary Standards for Drinking Water

to only 35 in the 1985 edition. The most exciting transformation is the inclusion of four parameters to monitor pathogenic agents, including *Giardia* and *Cryptosporidium*. It is well known that these two pathogens are resistant to conventional disinfectants such as chlorine; but this is the first time that the 'two bugs' – as they are termed in the local technical jargon – have been included for monitoring in Chinese regulations and standards. This change also suggests that water treatment in China has entered a new era. Thus, besides the parameter for chlorine included in the 1985 edition, the 2006 norms include the monitoring of other disinfectants such as chloramines, ozone and chlorine dioxide, which are increasingly used in the country. However, the most significant increase in the number of water quality parameters concerns organic compounds, particularly persistent organic pollutants (POPs), pharmaceuticals and pesticides. In addition, the 2006 standards have set much stricter concentration limits for several parameters compared with the 1985 edition (Ministry of Health, 2006).

Nevertheless, even though the 2006 edition of water quality standards in China includes a similar number of parameters to the National Primary Drinking Water Regulations (NPDWRs) and the National Secondary Drinking Water Regulations (NSDWRs) of the US, there is still a large distance between Chinese and American standards when concentration limits are compared. In addition, the human health risks and the environmental impact of the chemical compounds listed in the standards have not been studied and properly understood yet, while in small cities and rural areas it is not possible to monitor and analyse certain specific compounds due to technical limitations. Therefore, there is still a long way to go to improve drinking water quality in China, a challenge facing the country over the next few decades.

Problems and potential solutions

Although great efforts have been made and significant improvements have been achieved in the management of drinking water treatment and supply in China, as discussed below, there are important problems that require serious consideration.

Duplication and overlapping of responsibilities

As mentioned earlier, the supervision of quality controls in urban water supply is in the hands of the Ministry of Health, while the management of water supply systems is the responsibility of the MHURD. Given that there are no clear definitions of 'supervisory' and 'management' roles in existing laws or regulations, often the departments in charge cannot tell the differences themselves. Therefore, there is much duplication of responsibilities between the different departments, while at the same time there are important administrative gaps in the system. A possible solution to this problem would be unifying the roles of 'supervision' and 'management' and concentrating the responsibility for both aspects in one single department.

Non-unified laws and regulations

The Regulation of Urban Water Supply was passed in 1994, and was shortly followed by the Administrative Measures for the Sanitary Supervision of Drinking Water, which were jointly issued by the MHURD and the Ministry of Health in 1996. In 1999, the MHURD also introduced the Administrative Regulation of Urban Water Supply Quality. In addition, the Ministry of Environmental Protection (MEP) is responsible for setting the quality standards for water sources. At the same time, local administrative departments that monitor urban development, health, environmental protection and water resources have their own regulations concerning the control of the quality of water supply and the protection of water resources in their own regions. As a result, in most places there are two regulatory systems for water quality control: the national and the local, operating simultaneously. This is compounded by the fact that the implementation of national laws and regulations is subject to the interpretation of different agents. As a result, local law enforcers often lack a proper understanding of what is required by the national regulations.

The search for solutions

Summarizing the above discussion, there are two systems of monitoring standards for urban water supply quality in China: the national health standards set by the Ministry of Health and the urban water supply standards set by the MHURD. In addition, there are quality standards for water sources set by the MEP. The overlapping and duplication between these institutional structures, the gaps between national and local norms, and the lack of clarity in the definition of roles continue to provoke significant problems. This is a major concern for the Chinese authorities and water experts, who are currently working on the development of a unified national standard for monitoring the quality of drinking water. This work follows current practice in the US on water quality standards, taking into account the specific conditions in China, and distinguishes two different classes or grades of standards: rural areas or small and medium-sized cities are required to conform to the class I standard, while large cities and megalopolises must comply with both class I and class II standards. Shanghai has been the first city to complete the unification of urban and rural water quality standards, and it is expected that between 2005 and 2010 the water quality of Shanghai will catch up with the current water quality standards of the US and the European Union.

Besides these problems facing the management of water supply systems, China needs to make further efforts on the conservation and protection of water sources. In the context of stricter drinking water quality standards already described, the monitoring of drinking water quality cannot be limited to the treatment process, but also requires controlling the quality of water sources, as problems arising at this end make the treatment processes much more complex and costly. A number of recent problems in China's drinking water supply systems, such as the Wuxi City drinking water crisis of 2007 caused by the blue-green algae bloom in Taihu Lake, have attracted the attention of Chinese academics and engineers and brought the importance of drinking water sources conservation to centre stage. As a result, non-point source water pollution control and monitoring the eutrophication of water bodies by nutrients have become crucial research topics in China.

The case of water supply in the Beijing metropolitan area

The city of Beijing, capital of the People's Republic of China, has a long history spanning over 3000 years, a period during which it was the capital of several Chinese dynasties. Today, Beijing has become a modern metropolis with a population of about 16 million people. However, the climatic characteristics of the region present a significant challenge for the city's water supply systems, as in recent years the average annual precipitation has been around 700mm and it is highly seasonal.

Water supply in Beijing

Beijing's first modern facility for drinking water supply was a water tower designed and constructed by the Germans in 1910, with a volume of $750m^3$. The first water treatment plant was built in 1937, which had a capacity of $50,000m^3$ per day (m^3/d) and used groundwater. A second water treatment plant was built in 1942, followed by six additional units built between then and 1982, all of them taking water from the shallow aquifer. The city's ninth treatment plant, and the first plant to treat surface water from reservoirs, was established in 1986, and today there are, in total, 19 water treatment plants in Beijing, with an 8000km long water supply network and a service area of $600km^2$. The water supply coverage for the 16 million inhabitants of Beijing is 100 per cent.

Currently, the construction and operation of all the drinking water treatment plants, as well as the maintenance of the pipe networks, are under the management of Beijing Waterworks Group, a state-owned holding company responding to the MHURD, the Ministry of Health and the Ministry of Water Resources (MWR). In 2006, the average per capita domestic water supply was 160 litres per day, and the total volume delivered reached 4 billion cubic metres per year, composed of around two thirds of groundwater and one third of surface water. This level of supply poses intense pressure on water sources, and it is estimated that the amount of over-abstraction of groundwater reached 200 million cubic metres in 2006.

The ninth water treatment plant. After a recent expansion of capacity, the ninth water treatment plant is now the largest drinking water plant in Asia, with a capacity of 1.5 million cubic metres per day, which accounts for 60 per cent of Beijing's daily water supply. The bulk of the raw water comes from the Miyun Reservoir (4 billion cubic metres), while a smaller amount is provided by the Huairou Reservoir (0.1 billion cubic metres), both located in the northern suburb of the city. The raw water from the Miyun Reservoir is channelled to the plant through an open aqueduct that is protected by specific regulations passed by the Beijing government in 1995 (Administrative Measures for the Protection of the Miyun Reservoir, the Huairou Reservoir and the Miyun–Beijing Aqueduct). The regulations prevent the location of industries near the channel. The raw water from Huairou Reservoir is conducted through a ductile iron pipe and three steel pipes.

The treatment processes within the ninth plant from the influent to the effluent are coagulation, sedimentation, sand filtration, activated carbon adsorption and disinfection. As in other plants in Beijing, the disinfection method used in the ninth plant is chlorination. Give that it is a low-cost solution, chlorination became a widely used disinfection method in China; but this situation is currently being reassessed for two main reasons. First, as already mentioned, with the stricter quality standards adopted, water quality is now also monitored for the incidence of *Giardia* and *Cryptosporidium*, which are resistant to chlorine; the authorities are in the search of disinfectants that are more effective to control these pathogens. Second, some water bodies in China are now facing the problem of trace contamination by POPs, pharmaceuticals, endocrine disruptors (EDs) and similar pollutants. These compounds are difficult to remove by coagulation and filtration, and it is known that their reaction with chlorine can produce carcinogenic disinfection by-products (DBPs). The human health risks caused by DBPs are not yet fully understood and their impact has already become a very important research topic in Chinese universities. However, despite the unknown risks of chlorine-related DBPs, many experts argue that new disinfection methods producing less or no DBPs should be substituted for chlorination. The feasibility of using neo-disinfectants such as chlorine dioxide, ozone, hydrogen peroxide and chloramines in China's water treatment plants is the subject of intensive research in the country. Furthermore, in addition to exploring the suitability of new chemicals, novel technologies are being introduced and tested, including pilot-scale experiments using ultraviolet (UV) radiation and electron beams, as well as other physical processes in water plants nationwide. So far, the primary disadvantage of the new methods compared to chlorination is the persistency of disinfection in the pipe networks. Thus, there is an increasing consensus that the solution may lie in the combination of chemical and physical methods.

Pollution risks in the distribution system. In spite of the fact that surface water disinfection in the ninth water plant is limited to chlorination, the quality of its water supply still meets the *Standards for Drinking Water Quality* set by the Chinese government because the plant also applies a process of activated carbon adsorption. However, the increasing water demand often requires that the plant also uses groundwater. When this happens, it faces the additional problem of treating the mixture of surface and groundwater. Unfortunately, the treatment plant was not originally designed to treat groundwater; consequently, a technical problem of compatibility between different salinities in the water becomes notable. Among other issues, the hardness of water causes scale and clogging in the network.

In addition, the urban water supply networks in Beijing are relatively old and have not been properly maintained, which means that there is a risk of secondary contamination during the distribution process; therefore, the quality of the tap water might be lower than that of the water delivered by the treatment plant. This is suggested by some research reports that found bacterial growth in the internal wall of water pipes, and metal erosion leading to metal concentrations in the water and leaking pipes. Currently, there is no system in place to prevent leaking given that the leaking monitoring system only detects changes in the pressure and in the water heads of the pipelines, which means that maintenance work always comes after leaks or bursts have already happened. This problem has also become a priority for the water authorities and scientists.

Beyond water treatment technologies. The above discussion on the problems facing the treatment of drinking water in Beijing has focused mainly on the search for technological innovations, which is a major concern. However, there is increasing awareness that there is a limit to what technology alone can achieve, and some experts now suggest that there is a need to change the very concept of water treatment. In particular, some scientists believe that more attention should be paid to the protection of water sources instead of the continued concern with developing more water treatment technologies. Facts speak louder. As already mentioned, the 2007 water bloom that broke out in Taihu Lake, the third largest freshwater lake in China, caused the failure of the water treatment plant in Wuxi City. The uncontrollable growth of blue-green algae (*Cyanophytes*) is due to the prolonged eutrophication of lake water, and the concentration of industries and farms around the Taihu Lake basin have been identified as the main culprits, particularly the dying and paper-making industries. The evidence provided by the government investigation into this incident showed that even if the factories have their own wastewater treatment facilities, a large amount of nitrogen and phosphorus is still discharged into the lake. However, the report also found that the volume of fertilizers used in agricultural farms around Taihu Lake is three to four times higher than the national average per unit area. Overdosed fertilization makes the soil even more sterile, while more fertilizers are applied to grow crops, and the vicious cycle goes on. In some respects, agricultural pollution is worse than industrial pollution, given that the irrigation drain water containing nutrient elements reaches the lake from the whole basin through rivers, streams and storm-water overflows, and this kind of non-point source pollution is by far more difficult to control than industrial pollution.

Although the task is daunting, we still believe that the protection of water sources must become a central strategy given that only by tackling pollution in this way can drinking water supply in China move to a safer and more efficient stage. There are successful examples of how this can be done – for instance, in Germany's North Rhine–Westphalia State. Thanks to the outstanding quality of management of water bodies in North Rhine–Westphalia, water treatment plants require simpler treatment processes, have low operational costs, and some plants do not even need to apply a disinfection process, which avoids the problems caused by DBPs.

MANAGEMENT, TREATMENT AND REUSE OF WASTEWATER IN CHINA

History and current status of wastewater in China

In China, the earliest modern urban drainage facility was built in the city of Shanghai, which is the second largest city, and the sewerage system was established 140 years ago. The first wastewater treatment plant (WWTP) was also built in Shanghai in 1921, known as the Northern District WWTP, and in 1949 its capacity was increased to $4700m^3/d$. Most large cities built sewerage systems during the early 20th century, which in the majority of cases use combined sewer systems. Since the 1980s, new cities and urban areas began to implement dual sewer systems, and some of the oldest cities also started to convert their combined sewer systems into dual systems.

According to data for March 2008, there are a total of 1321 municipal WWTPs in mainland China, with a designed treatment capacity of 80.43 million cubic metres per day. The sewerage network is 156,000km long. In 2006, 16.31 billion cubic metres of municipal sewage and industrial wastewater were treated nationwide, 13.04 billion cubic metres of which was municipal wastewater. The average municipal sewage treatment rate was 43.8 per cent, and among the provincial administrative regions, the highest rate (about 90 per cent) was achieved in Beijing, followed by Shanghai (about 78 per cent), while the lowest rate of treatment (10 per cent) was recorded in Guangxi Province.

In the Chinese capital, there are currently nine municipal WWTPs in operation, and four more are under construction, reconstruction or expansion. The operation of all these WWTPs is in the hands of the Beijing Drainage Group, a state-owned holding company that responds to the MHURD and the MEP. One of these plants, the Beijing Gaobeidian WWTP, is now the largest wastewater treatment plant in operation in mainland China, which has a capacity of 1 million cubic metres per day, treats 40 per cent of Beijing's wastewater and is run at full load all year around. Gaobeidian WWTP was built in two stages: the Phase I Project started in 1990 and was completed in 1993 with a capacity of 500,000m^3/d, while the Phase II Project was carried out between 1995 and 1999 to upgrade the facilities to their current capacity. The main treatment process employed in this plant is conventional activated sludge (CAS) without removal of biological nitrogen or phosphorus. There is also an anaerobic digestion process for sludge stabilization, which generates electricity that supplies about 20 per cent of the operational power demand of the plant. The average operational cost is about 0.4 China yuan renminbi (RNIB) per cubic metre (US$0.06 per cubic metre), which is charged in the drinking water fee.

Problems and challenges

As mentioned earlier, in most Chinese cities the sewer systems are combined, which often results in pollution incidents when the volume of sewage exceeds the WWTP's capacity and the untreated wastewater overflows into the receiving water bodies. At the same time, in almost all cases of dual sewer systems, the rainwater is released into the environment without any treatment, which causes non-point pollution. Moreover, existing plans for a geographic information system (GIS)-based urban sewerage information database have not yet materialized, and the activities of sewer leak detection and maintenance carried out by municipal utilities rely only on manual operations by the workers. This is compounded by the fact that urban development in most Chinese cities has been characterized by a lack of integrated planning and communication between the relevant government departments (municipal authorities, departments of communications, transportation and other administrative units, etc.), which has often resulted in multiple excavation of city roads, causing severe damage to the sewerage systems. Furthermore, in many cities the sewer networks are ageing and have been deteriorated by a number of factors, including normal wastage. The damage caused to the networks by these different factors result in the increasing intrusion of 'foreign water' into the sewer systems. In some areas of southern China where there is a high urban groundwater table, in the dry season the volume of 'foreign water' can be higher than the actual wastewater flow rate in the pipes, which dilutes the sewage and reduces the carbon component in the wastewater, but also significantly increases the hydraulic loading of WWTPs. In winter time, the very cold 'foreign water' often provokes a drastic reduction in the temperature of the wastewater, which causes operational problems in the treatment plants.

During the last five years, China has begun to solve some of the problems of the sewer systems. For instance, university researchers are developing an urban sewerage information database using GIS. They are using underground mobile cameras and other tools to collect information about the state of the systems, which will help to understand the operating conditions more comprehensively and, thus, help to plan and implement network maintenance more efficiently. At the same time, technologies and equipment for pipe repair and replacement have been introduced from advanced countries such as Germany in order to minimize the excavation of roads during network maintenance. In addition, in some dual systems, rainwater is already being collected and treated separately according to its quality. The small amount of dirty 'early storm water' is pumped to the WWTPs, while relatively clean 'late storm water' runs into the receiving water bodies after soil filtration. Some cities also have introduced technologies for storm-water treatment *in situ*. Pavements of sidewalks and open-air parking lots are hollowed out into a skeletonized structure, and laid on soil and gravel with storm-water-collecting pipes beneath. Instead of running directly into the sewerage system from the road surface, storm water percolates first through a sand filtration layer placed under the sidewalks and parking lots, and then into the pipe system. In this way, there is no special demand for land to build large soil filtration fields for storm water, and the soil and gravel beneath the pavement can be easily replaced when required.

In addition to this work towards solving the problems in sewer networks, China is also increasing the rate of municipal wastewater treatment and enhancing the quality of the treated effluent. As already mentioned, the average treatment rate for municipal wastewater is today less than 50 per cent, and some urban areas still have no sewer systems. As a result, the construction of WWTPs has been accelerated in all cities, and by March 2008 there were 771 WWTPs under construction, an average of more than one per city. Many existing WWTPs are also now under reconstruction and expansion in order to strengthen the protection of water sources and to meet higher discharge standards. For example, the *General Plan of Comprehensive Management of Water Environment of Taihu Lake Basin* passed in March 2008 imposed stricter requirements for the WWTPs operating in the lake basin. Until now, the plants were required to meet the class I-B or class II standards set in the *Standards for the Discharge of Pollutants for Municipal Wastewater Treatment Plants* (MEP, 2002); but now they have to meet class I standards. Table 19.4 lists the allowed concentration levels for major pollutants set in the 2002 standards (MEP, 2002).

From Table 19.4, it is clear that the class I-A standard calls for a stricter control of nutrient elements in the effluent. The conventional activated sludge (CAS) process can only reduce ammonia, but not total nitrogen (TN) or total phosphorus (TP). Therefore, the CAS process cannot meet the requirements of the new standards, and WWTPs using just CAS technology have to be upgraded to control the concentration of nutrient elements. New technologies are needed, such as anaerobic–oxic (A/O), anaerobic–anoxic–oxic (A/A/O), and chemical phosphorus removal treatment processes. For instance, Beijing

Table 19.4 *Allowed concentration limits for major pollutants (mg/l, daily average)*

Pollutant	*Class I*		*Class II*
	A	*B*	
Chemical oxygen demand (COD)	50	60	120
Biochemical oxygen demand (BOD_5)	10	20	30
Suspended solids (SS)	10	20	30
Total nitrogen (TN)	15	20	N/A
Ammonia-N (≥12°C)	5	8	25
Ammonia-N (<12°C)	8	15	30
Total phosphorus (TP)	0.5	1	3

Source: MEP (2002)

Gaobeidian WWTP is currently upgrading its technology to introduce A/O treatment for nitrogen and phosphorus removal and thus meet the class I-A standards. Similarly, in other parts of the country, such as in southern China, the efficient operation of WWTPs to meet stricter standards continues to pose enormous challenges. At the national level, by 2008 the process of technical upgrading and reconstruction of WWTPs has been completed in many cities, which are now better equipped with a wider range of wastewater treatment technologies. Table 19.5 illustrates the diversity of treatment processes currently applied in China.

Table 19.5 shows that the oxidation ditch technology is now very popular in China, which turns out to be an interesting development. The oxidation ditch process originated in The Netherlands and was designed for small-scale low hydraulic loading treatment; but in China, WWTPs equipped with this technology are built much bigger, thus losing the advantages of treating small wastewater flow rates associated with this process. For example, the Carrousel oxidation ditch process has been applied in Beijing Jiuxianqiao WWTP, which is now the second largest WWTP in Beijing with its 350,000m^3/d capacity, and treats the wastewater produced by 2 million people. Another phenomenon shown in Table 19.5 is that a large number of WWTPs still apply only primacy treatment processes, which can only remove part of the suspended solids from wastewater. These WWTPs serve small cities and towns, which means that the problem of organic pollution in wastewater is still unsolved in most small cities, not to mention the lack of control over nutrient elements required for tackling eutrophication.

In some cities with special water requirements, such as replenishment and recharge for surface water in recreation areas, some WWTPs need to achieve standards higher than class I-A, which requires the introduction of more advanced treatment processes for water reclamation. For instance, the Beijing Qinghe WWTP was originally equipped with A/A/O technology and had a capacity of 400,000m^3/d, but it has now been upgraded with an additional facility to treat a flow rate of 80,000m^3/d using ultra-filtration (UF), powdered activated carbon (PAC) adsorption, and reversed osmosis (RO) to recharge the lake water of the 2008 Beijing Olympic Games Park. Meanwhile, in order to ensure the water quality for scenery and recreation use in Beijing, Asia's largest membrane bioreactor (MBR) in

Table 19.5 *Diversity of treatment technologies in use in China's wastewater treatment plants*

Technology	*Percentage of WWTPs*
Oxidation ditch	24
Anaerobic–anoxic–oxic	17
Anaerobic–oxic	16
Primary treatment	15
Sequencing batch reactor	11
Conventional activated sludge	5
Adsorption biodegradation	5
Other	7

Source: MEP (2007)

operation was designed and built in Miyun City by Tsinghua University, with a capacity of 45,000m^3/d. Another 60,000m^3/d MBR project for water reclamation is now under construction in Beijing Beixiaohe WWTP, with the membrane modules manufactured by Siemens. It will provide 10,000m^3/d of reclaimed water to the Olympic Park and the remainder will be used in Taiyanggong Thermal Power Plant as cooling water. Since water reclamation has become increasingly necessary and accepted by the public, there has also been a growing need for new regulations and standards. In this connection, in 2006 the Ministry of Water Resources passed the *Standards of Reclaimed Water Quality* (SL 368-2006), which sets the norms for reclaimed water use for groundwater recharge, industry, agriculture, municipal needs, and scenery and recreation. In all of these uses, the standards set for reclaimed water are stricter than that of municipal WWTP discharge. As a result, the effluent of a WWTP cannot be reclaimed directly and technical upgrading has to be carried out for a WWTP to produce reclaimed water, which requires the introduction of advanced technologies, such as biological aeration filtration (BAF), UF or RO.

Besides the construction and upgrading of sewerage systems and WWTPs, China also faces urgent challenges in relation to the treatment of activated sludge in WWTPs. About 66 per cent of the sludge produced nationwide goes to landfill, while 17 per cent is used for composting. Very little sludge is processed through anaerobic digestion to recover energy. For example, among the nine WWTPs in Beijing, only Gaobeidian WWTP has anaerobic digesters for sludge stabilization. Owing to the crucial safety concerns surrounding sludge disposal practices, and in order to foster energy recovery, China is currently speeding up the expansion of sludge anaerobic digestion and incineration processes. It is estimated that the biogas power that could be generated from sludge anaerobic digestion can recover about 30 per cent of operating costs in Chinese WWTPs.

CONCLUSIONS

As the socialist system is now and will always be the founding stone of Chinese society, there is no doubt that the government will continue to play the essential roles in the investment and delivery of WSS. However, the far-reaching reforms and changes introduced in recent years suggest that there is a new trend in which water and sanitation systems are increasingly designed and built by private contractors and then transferred to government-controlled water utilities for management. In the foreseeable future, there is no possibility for privately owned water companies to have full control over WSS in mainland China's socialist system.

In this connection, in February 2006, China's State Council adopted the *National Mid- and Long-Term Science and Technology Plan* (2006–2020), which commits the country to achieve the goals of water pollution control and efficient management of water bodies over the next 15 years. The water component of the *Science and Technology Plan* has given Chinese scholars and engineers renewed energy to concentrate their work on improving the water environment. We believe that these efforts, focused on such tasks as protecting water sources, improving waste and storm-water treatment, and recovering energy from sludge digestion, among other key challenges, will be achieved in the near future.

The industry of water supply and wastewater treatment is now developing rapidly, corresponding to the high-speed development of the Chinese economy. However, what may be different in this process compared with the situation in many developed countries is that China is simultaneously experiencing the pressures of both quantitative and qualitative improvements in the provision of WSS. Although there is still a long way for China to catch up with Western countries in these respects, the country is set to achieve significant advances in the field of water supply and wastewater treatment in the near future, as well as in other important aspects of Chinese rapid development.

REFERENCES

MEP (Ministry of Environmental Protection) (2002) *Standards for the Discharge of Pollutants for Municipal Wastewater Treatment Plants*, MEP, Beijing

MEP (2007) *Statistical Report on the Environment in China for 2006*, MEP, Beijing

MHURD (Ministry of Housing and Urban–Rural Development) (2008) *Statistical Report of Urban and Rural Construction in China of 2007*, MHURD, Beijing

Ministry of Health (2006) *Sanitary Standards for Drinking Water*, Ministry of Health, Beijing

NBS (National Bureau of Statistics) (2008) *2007 Annual Statistical Bulletin of China*, NBS, Beijing

20

Water and Sanitation Policies in Brazil: Historical Inequalities and Institutional Change[1]

Léo Heller

Introduction

Brazil is the largest country in South America, with an area of 8.5 million square kilometres and an estimated population of 188 million inhabitants in 2007. According to the United Nations Human Development Index (HDI), the country belongs to the group of countries with high human development (HDI of 0.800 in 2005); but it has also been characterized by high levels of social inequality: data for 2007 show that Brazil has the 11th highest level of income concentration in the world (UNDP, 2007). From another perspective, the physical and demographic characteristics of the country – with extreme climate conditions, unequal distribution of surface water availability, and heterogeneity in the pattern of demographic occupation – often exacerbated by socio-economic, political and cultural constraints, act as systemic conditions that frame and even shape policies and actions in the water and sanitation services (WSS) sector.

In the last few decades, WSS in Brazil have experienced a singular historical course, in which the notion of 'water supply and sanitation' has been progressively replaced by the concept of 'basic sanitation' (BS), which encompasses a more comprehensive approach to policy interventions on human habitats, looking at the protection of public health and the environment. Basic sanitation includes the classical components of water supply and sanitation services, but also integrates the collection, treatment and disposal of solid wastes, storm-water drainage, and the control of vectors of transmittable diseases. Thus, although in this chapter I focus on WSS services as the main object of analysis, I examine these services in the context of the Brazilian tradition embodied in the concept of basic sanitation, especially in relation to the analysis of the sector's institutional evolution.

Another important assumption informing the chapter is that the adequate provision of WSS requires the integration of the technological dimension (involving the development and implementation of techniques for the design, planning, construction and operation of service units and systems) with the public policy dimension, which encompasses such issues as planning, assessment, institutional organization, and the provision of legal and administrative mechanisms for effective user participation. The chapter also adopts the concept of systemic conditions as an analytical tool to explore the constraints and opportunities derived from processes and factors mainly external to WSS, but that have influenced their historical evolution. I examine the challenges facing the pursuit of equality and universalization of services, focusing on how systemic conditions may enhance or hinder future progress. As a note of caution, due to the limits imposed by the length of the chapter, this analysis presents only a preliminary assessment grounded on work in progress involving case studies and comparative analyses of public policies and management models in the provision of basic sanitation in Brazil. Summing up, the chapter aims at contributing to the understanding of WSS policies in the context of the BS model implemented in Brazil, and argues for the adoption of approaches that integrate the technological and public policy dimensions in both analysis and practice.

Historical background: Improving services, worsening asymmetries

Other publications include a more comprehensive assessment of the historical phases characterizing the social, political, economic and cultural patterns of organization of basic sanitation services in Brazil elsewhere (see, for instance, Castro and Heller, 2006; Rezende and Heller, 2008). Here I adopted the same time frame used by Rezende and Heller (2008), whereby the course of WSS and BS policies in Brazil can be analysed in six stages:

1 early development and the implementation of the first sanitary actions (1500s to 1850s);
2 raising awareness about the interdependence of sanitary actions in a context characterized by an ambiguous relationship between public and private WSS companies (1850s to 1910s);
3 the consolidation of the national state as coordinator of sanitary policy (1910s to 1950s);
4 the reorientation of sanitary policies, their separation from health policy and the autonomy of WSS services (1950s to 1969);
5 the reorganization of sanitary policies during the military dictatorship (1970 to 2002); and
6 the institutional changes implemented during the Workers' Party national government (from 2003 to date).

In this chapter, I focus mainly on the last two stages, considering first the period of 1970 to 2002, and then address the developments since 2003 in the next section.

Over time, the Brazilian government's approach to WSS and the extent of its involvement in the organization and delivery of these services has gradually changed. These changes have usually been the product of factors external to the internal logic of the sector – for example, concerns associated with disease control and the necessity of improving sanitary conditions; the economic, political, social and cultural processes that influenced the country's development policies during each stage; or the understanding of the state's role that prevailed in each period.

It can be argued that the main features of current Brazilian WSS policies were laid during the 1970s, through the implementation of the National Basic Sanitation Plan (PLANASA) during the military dictatorship that started in 1964, which introduced far-reaching changes in the sector's organizational structure. Perhaps PLANASA's most important legacy was the reform implemented in the allocation of responsibility between the different levels of government for the provision of the services. In this regard, and especially during the military regime, the Brazilian federative structure has experienced a significant concentration of power in the hands of the provincial states to the detriment of the municipalities, and this process underpinned the transfer of responsibility for service management from local to provincial authorities. If until the 1970s the federal and provincial states had acted principally in the areas of technical assistance and funding of WSS projects, in the new policy environment created by PLANASA, the provincial states became the direct providers of services. This arrangement was institutionalized through the establishment of a contractual relationship between the municipalities and the state governments, where municipalities granted a concession to provide WSS in their jurisdictions to state-owned companies in a way not too dissimilar from the concession contracts currently used worldwide in cases involving private-sector participation.

This resulted in the creation of new state companies for water supply and sanitation (CESBs) in the 26 provincial states, in some cases by adapting existing structures. These companies became responsible for seeking funding, and developing and expanding WSS systems, as well as for the operation and maintenance of the services, while also gaining the right to collect the service fees. Another important feature of PLANASA was the implementation of a new source of public funding for WSS: the Employment Guarantee Fund (FGTS) gathered from workers' pension contributions, which was initially more reliable than the scattered funding sources that had prevailed before. In addition, PLANASA introduced new rules for financial management by establishing the compulsory self-sufficiency of the CESBs based on service fees, while allowing a policy of cross-subsidies within each company's jurisdiction.

Although the transition to democratically elected governments after 1985 led to a degree of change in institutional practices and in the forms of federal government activity, the model prevailing in the WSS sector was largely preserved. However, an important development took place with the enhancement of the role of municipalities, particularly after the new federal constitution passed in 1988. Although the constitutional reform did not bring about the changes required for effective decentralization, it resulted in greater political-administrative autonomy, larger budgets, and increased access to financing for the municipalities, causing new contradictions with the prevailing centralized model of WSS management inherited from PLANASA. Some of these contradictions are still prevalent.

From another angle, it is important to highlight some of the systemic conditions that determined the history of the WSS sector in Brazil, which was characterized by periods of advance and regression in the expansion of service access to the population. As already suggested, the shape of WSS policy in each historical period was often the result of factors associated more with the direction taken by the country's development than with the internal dynamics of the WSS sector. For instance, the progressive extension of access to WSS in the late 19th century was largely driven by the wider process of infrastructural development linked to the country's economic growth, such as the development of ports driven by the exports boom. Similarly, in later periods, the extension of access to WSS was the result of the urgent need to address public health concerns, such as the collective fear of epidemics in the cities that led to the expansion of WSS in urban and rural areas during the period of 1910 to 1930. The aid policies of the US during 1950 to 1960 and the rapid urbanization process that led to the implementation of PLANASA during the 1970s were also important drivers in the development of WSS. Moreover, the attempt by different Brazilian governments to improve the country's image in the international arena has also been an important factor, as it happened during the second half of the 19th century when the country's elite attempted to mirror European models of WSS. Perhaps a similar process is taking place today in Brazil because of the impact of the requirements brought about by international trade arrangements on the development of WSS.

Contrastingly, a number of factors largely external to WSS could also contribute to the explanation of the recurrent drawbacks in the expansion of the services and in the low priority awarded to these services in national policy decisions. These include:

- the use of basic sanitation as a source of political and economic power (e.g. the monopolization of municipal WSS by the CESBs controlled by provincial state governments since the 1970s, and policy pressures to introduce market-centred principles in the organization of WSS since the 1990s, etc.);
- the impact of economic crises, particularly those related to fiscal reforms and the external debt that significantly reduced funding for the sector (1980s and 1990s);
- the contradictions arising from the attempts to implement neoliberal reforms since the 1990s (e.g. expansion of private-sector participation in Brazil's WSS through concessions, service contracts, shareholding, etc.).

These few examples of factors external to WSS illustrate the remarkable influence of systemic conditions on the sector's evolution in Brazil. From an institutional viewpoint, while in developed countries publicly run WSS achieved full coverage during the 20th century, unfortunately in Brazil the state abandoned the principle of universality in the provision of essential services that had been adopted since the late 19th century. As described in the next section, by the 1990s, Brazil had become one of the most unequal countries in the world, and this was clearly reflected in the situation of inequality characterizing the access to WSS. During the 1990s, there were several attempts to introduce market-centred

WSS policies as a solution to overcome the sector's deficiencies; but these initiatives also failed to deliver the promised results. As a result, at the beginning of the 21st century, the institutional situation of basic sanitation in Brazil faced a number of crucial challenges, particularly the need for more consistent direction and clearer policy goals, financial and institutional stability, and a more effective articulation with the areas of public health, water resources management and urban planning. I consider these challenges later, but first examine the situation of WSS in the country.

THE CURRENT STATE OF ACCESS TO SERVICES

Although the Brazilian government keeps up-to-date and comprehensive data for WSS, the assessment of the actual situation of these services is often difficult because the data gathered gives priority to quantitative indicators to the neglect of more qualitative information. For instance, although the first step in any assessment is to quantify the proportion of the population included and excluded from access to the services, in the light of the high levels of income inequality characterizing the country we also need to measure the inter- and intra-regional asymmetries affecting their provision. In this regard, data from the Brazilian Institute for Geography and Statistics (IBGE, 2000a) enable an accurate assessment of the recent evolution in the coverage of water supply and sewerage.

Based on the IBGE data, water supply coverage in urban areas rose about 4.5 percentage points (from 87 to 91.4 per cent) during the period of 1991 to 2003, while, even more significantly, in rural areas it rose from 9.3 to 25 per cent, an increase of over 16 percentage points. This is certainly a positive trend, and yet these data show the difficulties facing the country to make further advances in universalizing urban water supply, given that the un-served population is predominantly located in peripheral and informal urban areas. The expansion of service to cover these sectors requires the adoption of specific WSS policies that need to be integrated with urban development strategies. Additionally, although water supply for the rural population has experienced significant improvements, the level of coverage is still very limited. As demonstrated by the IBGE data for the year 2000, around 12 million urban Brazilians lack access to safe water supply, while the figure rises to 22 million in the case of rural areas, a situation that is made worse by the growing demand for WSS fuelled by steady population growth. Furthermore, the breakdown of data for water supply coverage shows that there exist important regional asymmetries given that the municipalities affected by substandard coverage levels tend to be concentrated in the northern and north-eastern regions of the country, which are also the least developed.

However, there are important aspects that refer to the quality of the services, such as regularity, equity and affordability, or compliance with drinking water standards, which the IBGE findings do not reveal. In this regard, the quality of water supply very often does not meet the minimum requirements, which is compounded by inconsistencies in government policy concerning the financing of services and by the operational constraints faced by many utilities. Likewise, in relation to the quality of water provision, another IBGE survey shows that 22.5 per cent of the water treatment plants serving localities

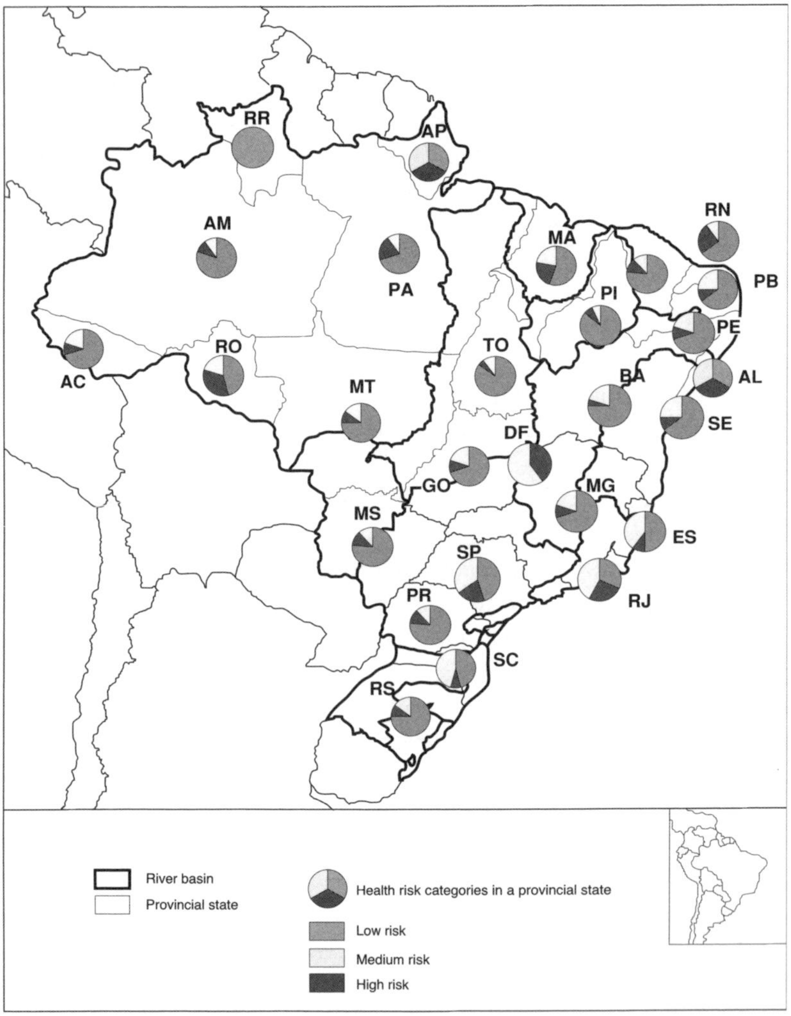

Figure 20.1 *Distribution of three levels of health risk due to drinking water quality in the provincial states and river basins of Brazil (2000)*

Source: Brazil (2000)

that rely on surface water are 'inadequate'. This is a crucial problem because in 32.3 per cent of these localities, the water sources are affected by pollution (IBGE, 2000b). In this connection, Figure 20.1 presents a map of health risks related to water supplies, based on the association between water sources, pollution levels and type of water treatment. The analysis demonstrates that the most critical risks are widely distributed across the country and that there is no clear concentration of water supply-related health risks, which are mainly the combined result of pollution of water sources and inadequate water treatment processes.

In the case of sewerage services, the data also suggest a clear pattern of service expansion during the period of 1991 to 2002, but there is a widening gap between urban and rural areas: while urban coverage was increased by 13 percentage points (from 62 to 75 per cent) in rural areas, the percentage of population served remained at 16 per cent despite an increase in the proportion of served population. In this regard, in the particular case of rural areas there is an ongoing debate over what would be the most appropriate technological solution for dispersed populations – whether on-site solutions or networked sewerage systems, which also presents the challenge of developing more adequate statistical parameters to measure the actual level of coverage. Incidentally, the historical absence of consistent policies specifically oriented towards rural areas is widely acknowledged in Brazil, which reinforces the precarious situation of rural sanitation.

As in the case of water supply, the quantitative indicators for sewerage coverage provide only partial information, as these do not take into account crucial aspects, such as the destination of effluents, which are often released untreated into the environment. This is significant since the provision of sewerage alone does not necessarily lead to real improvements in health and environmental conditions. Ambiguously, the construction of sewerage systems without provision of adequate interceptor lines and wastewater treatment plants, a situation that is very frequent in Brazil, may even aggravate the risks to human health and the environment. Thus, while wastewater directly infiltrates the subsoil in settlements not covered by networked sewerage, the introduction of sewerage without proper interception and treatment ultimately leads to the concentration of effluents in water bodies, which pollutes the urban environment and exposes the population to higher public health risks. Moreover, even the simple presence of interceptor lines and treatment plants may not guarantee the protection of human health and water quality in receiving water bodies given that treatment systems frequently have serious limitations in removing pathogens. On this issue, the *National Survey of Basic Sanitation* (PNSB) (IBGE, 2000b) reports that 4097 (42 per cent) out of the 9848 country districts are served by sewerage systems, but only 1383 (14 per cent of the total) have treatment plants. However, only 118 of these plants perform wastewater disinfection. Overall, of the total volume of wastewater collected in the country, only 35 per cent receives treatment before being released back into the environment. This means that around 9.4 million cubic metres of raw wastewater are discharged daily into water bodies across the country, and this figure only considers wastewater collected by the systems. Moreover, 3288 districts served by sewerage systems (80 per cent of the total) do not have interceptor lines, which suggests that, in all these cases, the water quality of urban streams may be seriously compromised.

Regarding both water supply and sewerage services, another important dimension of the analysis concerns the social asymmetries in service coverage. In addition to the previously mentioned inequality of access between urban and rural areas, the municipalities enjoying better WSS coverage are also those with the largest populations (see Figure 20.2) and the highest levels of development, as measured by the HDI (see Figure 20.3). The asymmetries in service access are also, unsurprisingly, linked to income-related inequalities (see Figure 20.4), as well as to the institutional characteristics of WSS utilities (see Figure 20.5). This disaggregation is important because it allows for a comparative analysis of the performance of the CESBs (state WSS companies) that have dominated the sector since

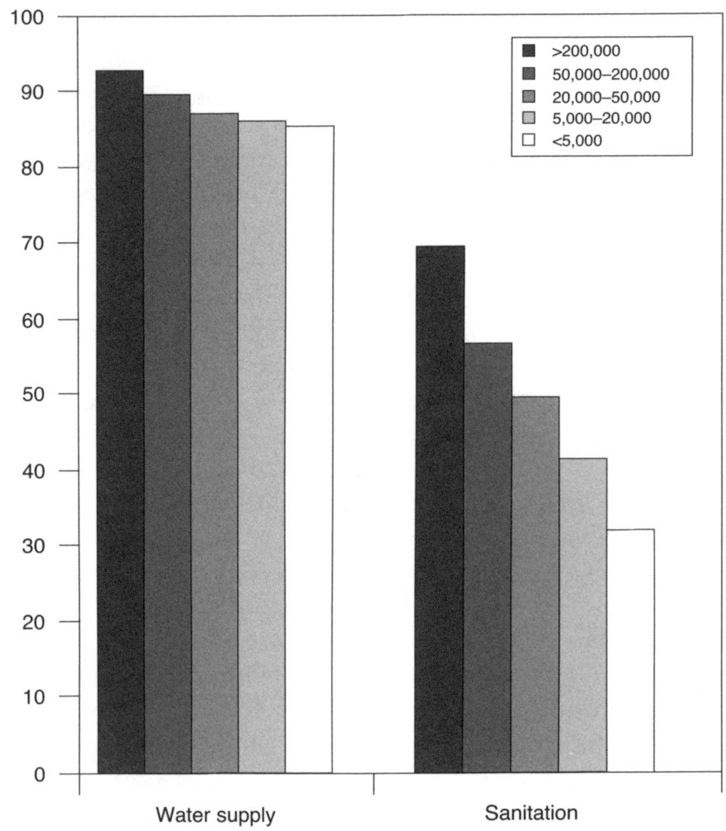

Figure 20.2 *Level of coverage of water and sanitation services (WSS) according to city size (inhabitants)*

Source: adapted from Rezende (2005)

the 1970s with that of municipal utilities, which shows that the effectiveness of both types of utilities has been similar (Rezende, 2005; Heller et al, 2006). These findings regarding the asymmetries in service delivery are consistent with similar analyses carried out in other developing or transitional economies, which highlight the role of such factors as the 'productionist' logic prevailing in the organization of WSS (Swyngedouw, 1995) or the impact of socio-demographic trends on inequalities characterizing the access to these services (McKee et al, 2006).

Regarding the latter, data on the performance of the CESBs indicate that after almost three decades, the model envisioned by PLANASA has not achieved the goal of universal coverage and that there exist significant inequalities in access to, and in the quality of, the services delivered between different municipalities. In addition, based on the evidence presented above and on complementary data (Rezende, 2005), it can be argued that PLANASA's model:

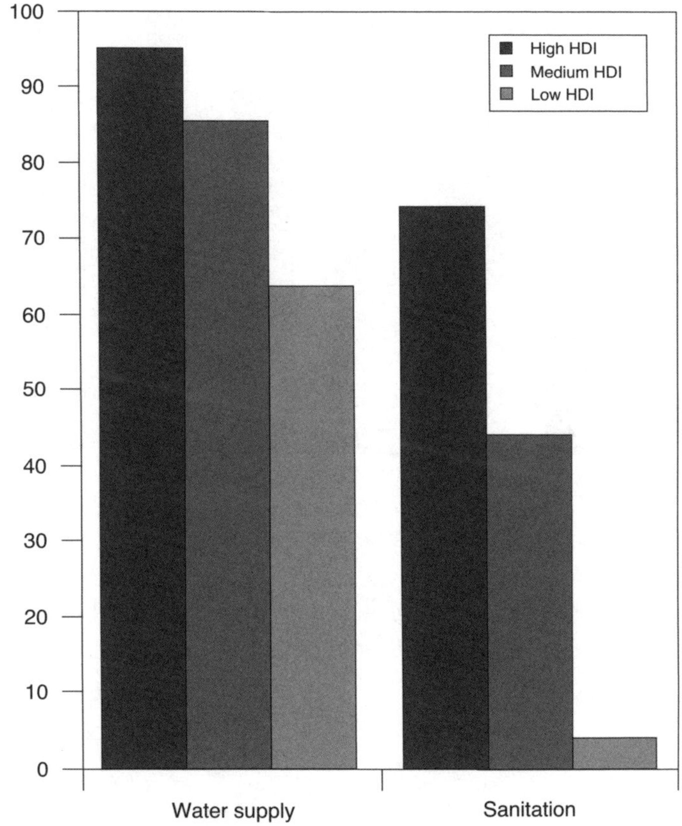

Figure 20.3 *Level of coverage of WSS according to position in the Human Development Index*

Source: adapted from Rezende (2005)

- privileged water supply to the detriment of sanitation (likely for reasons of economic-financial viability);
- was unsuccessful in establishing control over the WSS services in the municipalities of the most developed regions (perhaps because in these regions a greater proportion of municipalities exercised their autonomy and did not offer their services in concession to the CESBs);
- was unable to establish control over the WSS services of municipalities with the highest HDI, probably due to the fact that these were politically stronger and decided to run their own services, which were financially viable;
- in the case of sewage collection, clearly for reasons of economic-financial viability, gave less priority to municipalities with less than 20,000 people.

This evidence leads to the hypothesis that PLANASA's model could have even contributed to worsening pre-existing asymmetries in the access to safe WSS.

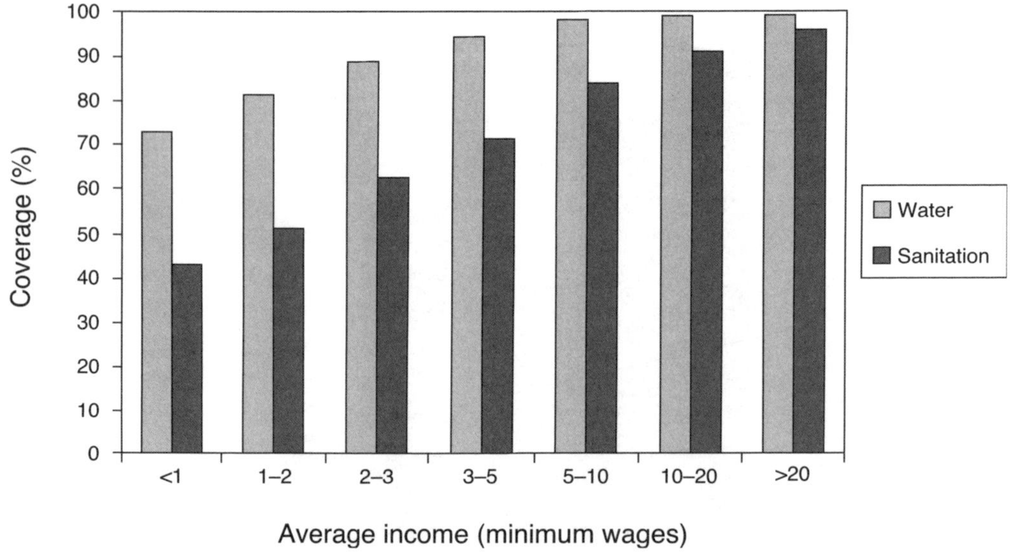

Figure 20.4 *Level of coverage of WSS according to family average income (minimum wage)*

Source: Costa (2003)

THE CURRENT CONTEXT: ADDRESSING SERVICE INEQUALITIES

With the arrival of President Luiz Inacio Lula da Silva's Workers' Party to the federal government in 2003, basic sanitation policies became the object of significant changes in the political, institutional and legal spheres. The most notorious development was the creation, in 2003, of the Ministry of Cities and its National Department of Environmental Sanitation (SNSA). The SNSA was entrusted with the mission of 'ensuring the fundamental human rights of access to potable water and to life in a healthy environment in the cities and the countryside, through universal access to water supply and sanitation, the collection and treatment of solid wastes, urban drainage and the control of vectors and reservoirs of transmittable diseases' (SNSA, 2006). The SNSA has responsibility for financing, assessment, implementation and establishing directives for basic sanitation at the federal level.

In addition to the SNSA, the government created the Council of Cities as a consultative and decision-making body with the purpose of 'proposing goals for the formulation and implementation of national urban development policy, as well as monitoring and assessing its execution' (SNSA, 2006). The council includes a Consulting Technical Committee for Environmental Sanitation, which has an advisory function, and the government has promoted the reproduction of similar structures at the state and municipal levels. These changes in the federal government represent a major departure from the past, when the

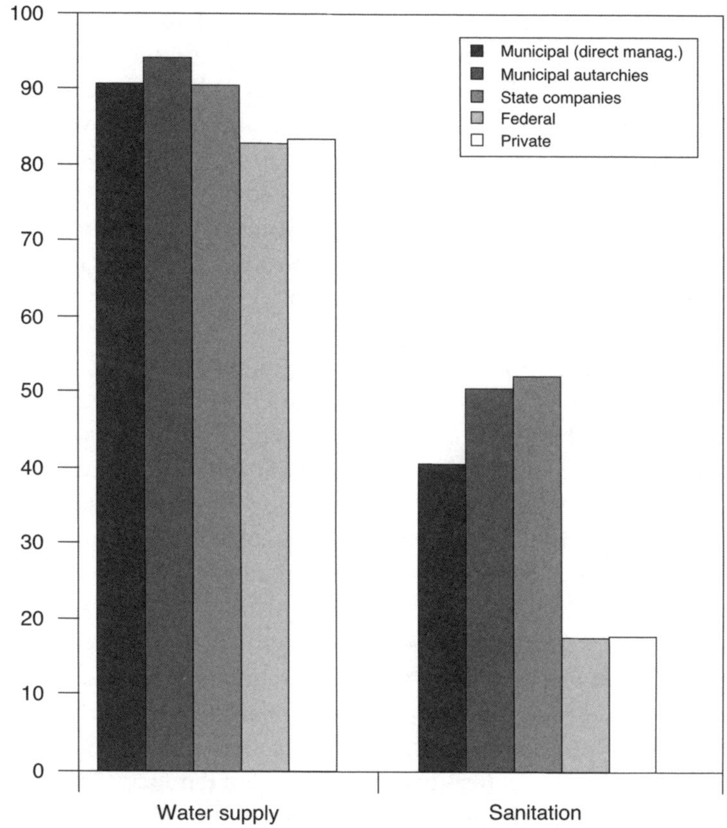

Figure 20.5 *Level of coverage of WSS according to type of utility management*

Source: adapted from Rezende (2005)

provincial states had the leadership in basic sanitation policy. Under the new arrangements, the national basic sanitation policy is informed by the deliberations of Conferences of Cities organized periodically around the country with the participation of a broad spectrum of representatives from the various areas of urban policy. Among other issues, the conferences have adopted the principle of universal access for basic sanitation, asking for a greater state commitment in delivering concrete solutions to the lack of safe basic sanitation services, especially for the most disadvantaged sectors of the population.

In addition to these initiatives, other federal departments also have an influential role in current basic sanitation policy. In particular, the National Water Agency (ANA) oversees the functioning of the river basin committees and has responsibility for supervising water abstractions, conservation, water planning and wastewater disposal (Brazil, 2006). The Ministry of the Environment is also responsible for implementing Agenda 21 in Brazil and has direct influence on a number of activities in the interface between basic sanitation and water management.

At the provincial level, there is often an absence of institutional arrangements regarding basic sanitation, and, in practice, the responsibility for these services has been in the hands of the state CESBs or municipal utilities. A crucial consequence of this situation is the limited scope for executive action available to the provincial governments in such issues as the universalization of WSS in their territories. This is the case because the CESBs only serve those cities where they have a concession to provide WSS, and do not have responsibility for extending coverage to the municipalities not included in their concession. Furthermore, although the federal constitution establishes that basic sanitation services are a municipal responsibility, this authority bestowed on local governments is rarely fully exercised. According to the prevailing system, the municipalities must either directly operate basic sanitation services, or grant a concession to third parties for their delivery, in which case the municipal government is responsible for regulating the concessionaire's activities. However, in practice, municipal authorities are not always fully aware of their rights and responsibilities regarding basic sanitation, and frequently they do not have the institutional, technical and financial capabilities for their exercise.

From another perspective, in comparative terms, basic sanitation policies in Brazil are laggards in the process of democratization experienced by the country since the mid-1980s. In this regard, while initiatives to implement mechanisms for democratic social control and popular participation rapidly attained legitimacy and efficiency in the areas of health, the environment, urban policy and water resources, in the case of basic sanitation services, the process has been significantly slower. Similarly, the enhancement of municipal authorities' capacity for local action during recent decades has fuelled a substantial process of decentralization in the areas of health and environmental policy. Contrastingly, there is a protracted deficit of democratization in basic sanitation policies, which could perhaps be explained by the resistance presented by interest groups and other power holders in the BS sector who refuse to become accountable and subject to democratic oversight and control. This strong resistance, mainly on the part of some CESBs, has been particularly evident during the debates and negotiations over proposals to reform the national legislation and introduce stricter regulation in the sector since the early 1990s.

In this connection, there have been several initiatives to establish a new legal and institutional framework for basic sanitation to replace PLANASA, such as a new law regulating the sector approved by parliament in 1993 to 1994, but then vetoed by President Fernando Henrique Cardoso in 1995 on grounds of 'economic reasons'. In fact, the absence of a well-defined legal and institutional framework has continued to hinder the design and implementation of basic sanitation policies with clearly defined objectives that are stable enough to resist the vagaries of the Brazilian political system. As a result, to a large extent, WSS in Brazil continue to be organized along the lines of PLANASA's main characteristics: the hegemony of the CESBs; the principles of financial self-sufficiency and cross-subsidization within the boundaries of provincial states; a lack of, or poor, accountability and democratic social control; and weak inter-sector linkages between basic sanitation, public health, water resources and urban planning, among other relevant sectors of activity.

Nevertheless, in recent years the government has passed two pieces of legislation that seek to lay the foundations for a new legal framework in the basic sanitation sector. First, the Law of Public Consortia (Law No 11,107), passed in 2005, that sets the ground rules for the creation of new public bodies for the delivery of basic sanitation services, including inter-municipal and municipal–provincial partnerships, or even more arrangements involving the federal government as a partner. Second is the 2007 National Sanitation Law (Law No 11,445), which is arguably the most important legislative innovation in the basic sanitation sector in decades and, as such, the first-ever federal law for WSS. This is a groundbreaking initiative that fills a historical gap in the sector's legislation after about 30 years of debate, and that for the first time in history makes possible the adoption of national guidelines for public policy and management in the basic sanitation sector. However, and although it is widely expected that this law will help to foster universalization and improvements in the quality of service delivered to the population, the process has been marked by ongoing confrontations. The first draft of this law was presented at the beginning of 2004 by the National Department of Environmental Sanitation and was subject to intense negotiations between the different actors involved, which included the federal government, the workers' unions, the state and municipal water utilities, the regulatory agencies, and the organized social movements with a direct interest in basic sanitation services. After the introduction of significant amendments to the original draft, congress finally passed the law in 2007.

Among the main principles worth highlighting, the National Sanitation Law has introduced important provisions aimed at fostering the long-sought democratization of the basic sanitation sector, including that:

- Access to a healthy environment is a public good, and the main objective of basic sanitation services, thus rejecting the reduction of these services to the category of economic goods.
- Basic sanitation services are public goods that are under state control and constitute a state duty.
- The state is committed to deliver basic sanitation as a universal service, with equity.
- Basic sanitation services constitute an integrated set of activities that must be regulated and subject to planning and monitoring.
- The delegation of basic sanitation services must be subject to clear guidelines, affirming the rights of the public authority to grant the concessions.
- Mechanisms to ensure participatory democratic control of basic sanitation services should be introduced, including providing the means for their implementation.
- The expansion of coverage to low-income populations must be promoted.

These recent innovations in the legislation have created the conditions for a new stage in Brazilian basic sanitation policies. In particular, the changes introduced require the reorganization of the prevailing arrangements existing between the federal, provincial and local levels of government and the providers of basic sanitation services. On the one hand, the new legislation leads to a redefinition of the institutional relationships between these actors, requiring the regularization of concession agreements, greater transparency and regulation of both public and private operators. On the other hand, it also establishes a more democratic framework for user participation, introducing a number of instruments for involving the citizens in the monitoring of basic sanitation policy and management. However, despite these significant advances in the legal framework, the actual process is far from perfect. It is not an easy task to foresee the future of the basic sanitation sector in Brazil, especially because history has shown that the external systemic conditions often have a more fundamental influence on its performance than the internal logic of the sector itself, which I briefly discuss next.

Systemic conditions and the future of basic sanitation in Brazil

We can identify various drivers that may favour or hinder the achievement of the long-overdue goals of universality and equity in the access to basic sanitation services in Brazil. Regarding the favouring drivers, we can mention the new legal instruments, the resulting institutional reform in the basic sanitation sector, and the disposition shown by President Lula's federal government to prioritize the allocation of financial resources to the improvement of BS services. In particular, the new legal framework may be instrumental in promoting the design and implementation of more effective management models. For instance, the creation of inter-municipal consortia between neighbouring cities can bring about economies of scale in the organization of basic sanitation services, while avoiding the problems posed by the existing mega-utilities (which the CESBs inherited from PLANASA) that tend to be oversized and largely distant from the users, both in geographical and organizational terms. Another important positive factor is the mechanism for tariff regulation introduced by the new legislation. According to the mechanism, the tariff structure must be transparent, covering the actual costs of each service (different from the current model in which each CESB has the same tariff for the entire area of the concessions) and providing the resources to invest in infrastructure maintenance and renewal, forbidding the use of any surplus to fund other purposes.

External factors could also play a role as beneficial drivers, such as the adoption of a national economic policy oriented towards the improvement of the country's position in the international arena, or the implementation of policies to improve the country's image in relation to key social indicators of international human development. In this regard, investment in basic sanitation can be a by-product of these broader policy objectives, as has happened at different moments of the country's history.

In contrast, there are also certain factors that could hinder the achievement of the cherished goals of equity and universality in BS services. In this regard, the protracted conflicts between the provincial and municipal levels of government over the jurisdiction of BS services constitute a major obstacle for progress. The provincial WSS utilities (the CESBs), with a track record of authoritarian top-down decision-making and operation, have been mostly unregulated, and have been largely free from democratic control and citizen scrutiny. The incentives for the CESBs to change their mode of operation continue to be minimal, and this situation has the potential to slow down improvements at all levels, most importantly in the expansion of service access. The balance of economic and political forces in the provincial governments, the real power holders in the CESBs, will continue to have a determining influence in the basic sanitation sector since they have the capacity to launch and sustain juridical battles against implementing the new legislation, as it is actually happening, and thus delaying much needed improvements.

Another important constraint, unsurprisingly, relates to the financing of service expansion and improvement. First, the relative financial bonanza experienced by the Brazilian basic sanitation sector since 2003 may not be sustainable since the currently favourable economic circumstances of the country may not last, which could slow down and even stop the investment flows required to meet the sector's targets. Second, even if the financial situation remains buoyant, the actual implementation of the investment programme in basic sanitation could still face significant delays for a number of reasons, including the gaps in technical expertise and leadership affecting the sector since the 1980s.

Regarding the factors internal to the sector, the asymmetric pattern of access to the services and of the quality of their provision could have an important influence in initiating improvements, as experienced in other countries such as Mexico (Castro, 2004) and South Africa (Beall et al, 2000).

Summing up, the history of Brazilian basic sanitation services suggests that social, economic-financial, political, cultural and other factors, mostly external to the logic of the sector, constitute fundamental constraints that may foster or hinder the success of public policy and management processes. A more rigorous and in-depth understanding of these factors is required in order to meet the goals of universality and equity in the delivery of these essential services. Finally, and yet importantly, as pointed out by other studies, institutional changes at various levels of water governance structures tend to be path dependent – that is, these changes are largely constrained or enabled by past institutional configurations (Kaivo-oja et al, 2004; Juuti and Katko, 2005; Livingston, 2005).

CONCLUSIONS: FUTURE CHALLENGES

This chapter has attempted a critical examination of the situation of basic sanitation services and policies in Brazil, introducing elements from public policy in what is conventionally a technical debate. It has argued that the most appropriate approach for analysing basic sanitation services, the forecasting of likely scenarios in the sector's future evolution, and the implementation of BS policies that may help to overcome current deficits must integrate the dimensions of management and public policy. In our perspective, adopting such an integrated approach is a fundamental requirement for achieving the goals of universal access to safe drinking water and sanitation services, urban waste collection, and storm-water control needed to deliver a better future for the Brazilian population.

This examination of the historical background and of the current scene for basic sanitation in Brazil helps to identify a number of challenges facing the project of universalization with equity in BS services:

- The historical evidence suggests that the advances in basic sanitation policy in Brazil have followed a pattern of gradual change largely influenced by factors and processes external to the sector, which have informed the changing role of the state in providing these services. Among other factors can be mentioned international influence in the development of contagious disease control since the late 19th century, transformations in the economic, political, social and cultural characteristics of the country, and, notably, changes in the prevailing understanding of the role of the public sector in relation to these services in each historical period.

- In the current circumstances, and considering the goals of progressively expanding access to high-quality basic sanitation services, these systemic influences must be incorporated within the analysis and taken into account in the design and implementation of public policy in the sector.
- From a public policy perspective, basic sanitation policies must also recognize the existing social asymmetries in service provision and seek to understand their structural determinants in order to develop well-targeted strategies to abate these social inequalities.
- However, any process of policy design and implementation in the basic sanitation sector must incorporate within the analysis the significant obstacles posed by the management model inherited from PLANASA, which is still dominant. This management model has historically bestowed political and economic power and control over BS services on the provincial WSS utilities (CESBs), which are likely to resist change and oppose the implementation of the new institutional and legal framework.
- Moreover, historically, political control over basic sanitation services has been the ambition of different interest groups, not least because it is seen as a strategic lever of power at the local and regional level. In the current situation, this has prompted ongoing political confrontations between political actors, particularly between provincial and municipal authorities, but also involving the federal state, and between the public sector, social movements and private interests. These social and political confrontations must also be integrated within the analysis, as well as within the design and implementation of public policies in the basic sanitation sector.

Note

1 This chapter is adapted and expanded from Heller (2007).

References

Beall, J., O. Crankshaw and S. Parnell (2000) 'Victims, villains and fixers: The urban environment and Johannesburg's poor', *Journal of Southern African Studies*, vol 26, pp833–855

Brazil (2000) *Lei no 9.984, de 17 de julho de 2000: Dispõe sobre a Criação da Agência Nacional de Águas – ANA, Entidade Federal de Implementação da Política Nacional de Recursos Hídricos e de Coordenação do Sistema Nacional de Gerenciamento de Recursos Hídricos*, Brazil

Brazil – Ministério do Meio Ambiente (2006) *Caderno setorial de recursos hídricos: Saneamento*, MMA, Brazil

Castro, J. E. (2004) 'Urban water and the politics of citizenship: The case of the Mexico City Metropolitan Area during the 1980s and 1990s', *Environment and Planning A*, vol 36, pp327–346

Castro, J. E. and L. Heller (2006) 'The historical development of water and sanitation in Brazil and Argentina', in J. Petri, T. Katko and H. Vuorinen (eds) *Environmental History of Water: Global Views on Community Water Supply and Sanitation*, IWA Publishing, London

Costa, F. J. L. (2003) *Estratégias de Gerenciamento dos Recursos Hídricos no Brasil: Áreas de Cooperação com o Banco Mundial*, Série Água Brasil, vol 1, Banco Mundial, Brazil

Heller, L. (2007) 'Basic sanitation in Brazil: Lessons from the past, opportunities from the present, challenges for the future', *Journal of Comparative Social Welfare*, vol 23, pp141–153

Heller, L., M. L. Coutinho and S. Mingoti (2006) 'Diferentes modelos de gestão de serviços de saneamento produzem os mesmos resultados? Um estudo comparativo em Minas Gerais com base em indicadores', *Engenharia Sanitária e Ambiental*, vol 11, pp325–336

IBGE (Instituto Brasileiro de Geografia e Estatística) (2000a) *Censo Demográfico*, IBGE, Brazil

IBGE (2000b) *Pesquisa Nacional de Saneamento Básico*, IBGE, Brazil

Juuti, P. and T. S. Katko (eds) (2005) *Water, Time and European Cities: History Matters for the Future*, Tampere University Press, Tampere, Finland, http://tampub.uta.fi/tup/951-44-6337-4.pdf, accessed 7 January 2008

Kaivo-oja, J. Y., T. S. Katko and O. Seppälä (2004) 'Seeking for convergence between history and futures research', *Futures, Journal of Policy, Planning and Futures Studies*, vol 36, pp527–547

Livingston, M. L. (2005) 'Evaluating changes in water institutions: Methodological issues at the micro and meso levels', *Water Policy*, vol 7, pp21–34

McKee, M., D. Balabanova, K. Akingbade, J. Pomerleau, A. Stickley, R. Rose and C. Haerpfer (2006) 'Access to water in the countries of the former Soviet Union', *Public Health*, vol 120, pp364–372

Rezende, S. C. (2005) *Utilização de instrumentos demográficos na análise da cobertura por redes de abastecimento de água e esgotamento sanitário no Brasil*, PhD thesis, Federal University of Minas Gerais, Brazil

Rezende, S. C. and L. Heller (2008) *Saneamento no Brasil: Políticas e Interfaces*, second edition, UFMG, Belo Horizonte, Brazil

SNSA (Secretaria Nacional de Saneamento Básico) (2006) 'Secretaria Nacional de Saneamento Básico', http://cidades.gov.br/index.php?option=content&task=section&id=17&menupid=215&menutp=saneamento, accessed 6 February 2007

Swyngedouw, E. A. (1995) 'The contradictions of urban water provision: A study of Guayaquil, Ecuador', *Third World Planning Review*, vol 17, no 4, pp387–405

UNDP (United Nations Development Programme) (2007) *Human Development Report 2007/2008 – Fighting Climate Change: Human Solidarity in a Divided World*, Palgrave Macmillan, New York

21

Challenges Facing the Universal Access of Water and Sanitation in Mexico

María Luisa Torregrosa Armentia and Blanca Jiménez Cisneros

Introduction

Despite efforts made by the Mexican government since the 1970s to expand water and sanitation services (WSS), coverage is still limited and of substandard quality considering the level of development of the country. WSS provision is also very unevenly distributed between regions and social groups. Overall, around 11 million people (almost 11 per cent of the population) have no access to water supply, while the number of people who lack access to safe water is substantially higher. However, even having access to water supply is often problematic given that the quality of the water delivered is poor. In the case of sewerage (termed sanitation in the official government reports), 15 million people lack access to any kind of service for the safe disposal of human excreta. Most of the un-served population is located in rural and peri-urban areas in the provincial states.

This unacceptable situation affecting WSS is taking place despite the overall economic strength of country, which according to recent reports has a per capita gross national product (GNP) of US$9803 (IMF, 2007). In an attempt to understand the reasons for this problem, the chapter analyses the relationship between the deficit in WSS and factors such as water availability, economic capacity, legal framework, institutional capacity and water policy. We close our discussion with a recommendation about what could be done in relation to each of these factors in order to ensure a safer water supply and greater coverage of sanitation services in Mexico.

THE CONTEXT OF MEXICO'S WATER AND SANITATION POLICY

Mexico has 2 million square kilometres and 103 million inhabitants. The population growth rate is 1.3 per cent per year with most people (78 per cent) living in urban areas. The country is rich in fresh renewable water resources with a mean pluvial precipitation of 772mm per year, which is relatively high compared to the world average of 225mm (CNA, 2005). National water availability is 4400m^3 per capita per year (pcpy) (CNA, 2007). However, at a regional level, water availability varies considerably, from 14,291m^3 pcpy in the south to 2044m^3 pcpy in the north, east and central parts of the country. The intensity of water use is also highly diverse: while the mean national value is 17 per cent, it ranges from less than 1 per cent in the south to about 154 per cent in the central area, where Mexico City is located (CNA, 2007). Moreover, water distribution is inversely related to the location of the population and to the spatial distribution of the gross domestic product (GDP). The north, north-west and central parts of Mexico have only 32 per cent of the renewable water resources, but concentrate 77 per cent of the population and 86 per cent of the national GDP (CNA, 2005).

As already mentioned, Mexico has a per capita GNP of US$9803. However, income distribution is very unequal, with nearly 10 million Mexicans (around 9.7 per cent of the total population) accounting for 80 per cent of total GDP, while the most deprived 20 million people (around 19.4 per cent of the total) have only 2 per cent of the country's wealth (Tuirán-Gutiérrez, 2006). This inequity is also observed across the country. For example, the Federal District and the state of Mexico have a per capita GDP of US$2137 and US$931, respectively, while the states of Colima and Nayarit – which are the poorest – both have a per capita GDP of only US$49. The polarization of wealth distribution has become more marked over the last 20 years in the context of neoliberal economic globalization. The policies inspired by this model have boosted inequalities in the appropriation and distribution of wealth worldwide, resulting in higher levels of socio-economic exclusion. One of the early consequences of these policies was observed in Mexico during the 1980s when a fall in oil prices and rising interest rates triggered a default on Mexico's external debt, which had a major impact upon the national economy (Morera-Camacho, 2005). Like most countries, Mexico adapted to the new international context through the introduction of financial, economic and fiscal reforms. Thus, in a relatively short period of time, Mexico moved from being a closed, highly regulated and highly oil-dependent economy (SEGOB, 2001) to one that promoted liberalization, deregulation, decentralization and privatization of state enterprises. These reforms, which were underpinned by Mexico's entry into the North American Free Trade Agreement (NAFTA) in 1994, contributed to one of the most severe economic crises in the country's history, which lasted from December 1994 to December 1996.[1] In fact, despite the adoption of different palliative policies, during the last two decades, governments have been unable to stabilize the Mexican economy. As a result, the country has paid a very high social cost, with an alarming increase in poverty rates, marginalization, unemployment, the expansion of the informal economy and a marked worsening of social inequality patterns (Katz, 2002; Meyer, 2005; Székely, 2005).

In this context, there have also been significant policy changes in WSS, especially in relation to the financial sources for expansion and maintenance of the services' infrastructure. Mexico turns principally to the World Bank and the Inter-American Development Bank (IDB) for multilateral credit and to the Japanese Bank for International Cooperation (JBIC) for bilateral credit. It is important to note that one of the conditions on which these loans are granted is the decentralization of utilities and the incorporation of private-sector participation in the administration of potable water, sewerage and sanitation services.[2]

THE SITUATION OF WATER SUPPLY AND SANITATION SERVICES

The most recent official reports show that the levels of coverage for water and sanitation services are 89.2 per cent for water supply and 85.6 per cent for sanitation (CNA, 2007). But these values hide the differences in provision between urban and rural areas. In urban centres, 95 per cent of the population is covered, while in rural areas, the figure drops to 70.7 per cent. Concerning sanitation, urban coverage is 94.5 per cent, while in rural areas it is only 57.5 per cent (CNA, 2007). Additionally, these data do not show the existing differences in the quality of the service provided (such as the intermittence of the service or the use of malfunctioning latrines) or the inequalities in access between well-off and poor neighbourhoods, or between urban and peri-urban areas. These inequalities are significant, as shown by the fact that of the total population reported as having sanitation coverage, only 67.6 per cent are actually connected to a sewer system, 15.9 per cent have latrines (most of them malfunctioning) and 2.1 per cent discharge their waste directly to the soil or to surface water bodies.

From another angle, it is useful to compare the situation of WSS in Mexico with other countries with similar economic conditions.[3] In this regard, Mexico has a lower level of coverage for WSS than countries such as Argentina, Chile, Costa Rica, Uruguay and Turkey, which not only have a lower GDP per capita, but some of which also spend a higher percentage of their national income on international debt payments. In other words, while Mexico has the highest GDP per capita in Latin America, the country ranks fourth among Latin American countries in the level of coverage for water supply services, and drops to the tenth position for sanitation coverage and to the thirteenth for rural sanitation (CNA, 2006).[4]

Furthermore, although some may argue that the differences in coverage could be caused by external systemic conditions, such as hydrological and climatic constraints on Mexico's water availability, as shown in Figure 21.2, Mexico has an even lower level of coverage for WSS than countries such as Cuba or Israel, where water availability is comparatively lower.

Wastewater treatment and drinking water quality

As explained earlier, data for sanitation services in Mexico only refer to the collection of wastewater from households and do not consider wastewater treatment and its safe

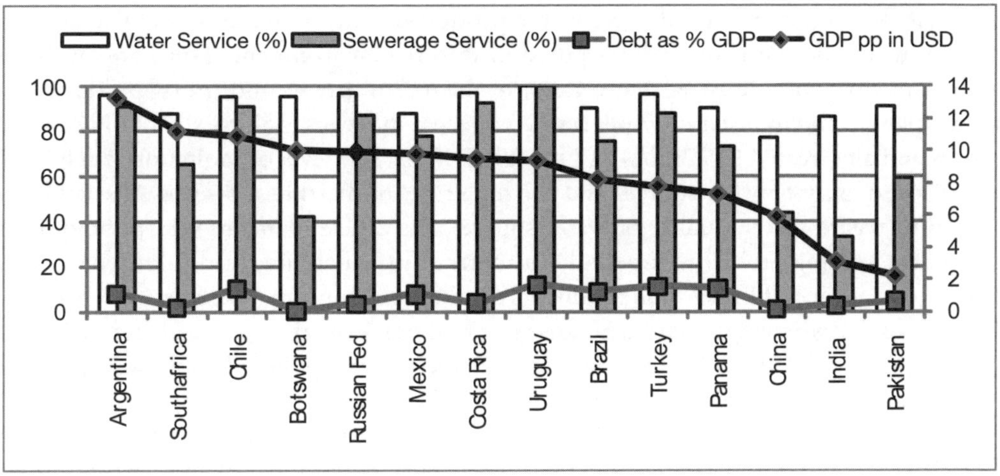

Figure 21.1 *Water and sanitation services, gross domestic product (GDP) and debt as a percentage of GDP*

Source: UNDP (2006) and WRI (2007)

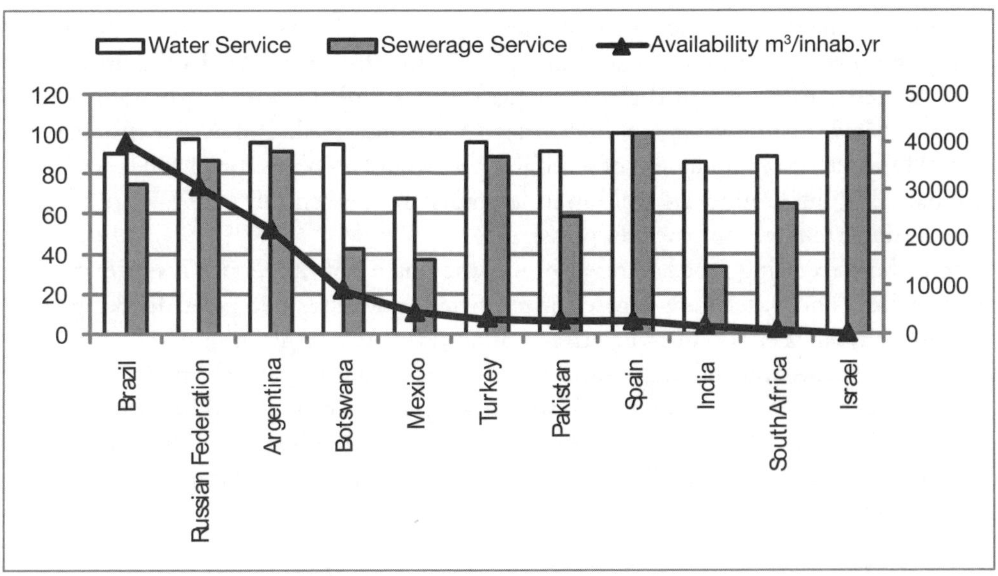

Figure 21.2 *Water and sanitation services and water availability in selected countries*

Source: UNDP (2006) and WRI (2007)

disposal. In fact, of the total wastewater produced, only a fraction is collected and treated, while the rest is unsafely disposed of into the environment. This happens both with municipal and industrial wastewater: of the $412m^3$ per second (m^3/s) produced by municipalities and industries, 87 per cent is collected in sewers ($358m^3/s$), but only 21 per cent is actually treated ($86.2m^3/s$) (CNA, 2005). This low level of sanitation is reflected in the low quality of water sources and the high rates of diarrhoeal diseases observed.

With regard to the quality of water sources, the official data are very poor and do not reflect existing problems, especially in terms of its suitability for drinking purposes, understood as the fulfilment of the country's drinking water norm (Mexican Government, 2000). This is a worrying situation because due to the lack of safe municipal and industrial wastewater disposal, water sources are becoming increasingly polluted, as shown by several studies reporting pollution problems in water sources across the country (Alcocer et al, 1998; Carrillo and Drever, 1998; Pacheco et al, 2000; Armienta and Rodríguez, 2002; Escolero et al, 2002; Rodríguez et al, 2002; Foster et al, 2004; Jimenez and Chávez, 2004; Carrillo-Rivera et al, 2006). In spite of this, water treatment processes have remained unchanged for more than 40 to 50 years and are mostly limited to chlorination (CNA, 2005).[5] However, the pollution of Mexico's water sources includes the presence of toxic and non-toxic organic matter, nitrates, metals and fluorides, as well as a wide variety of microorganisms resistant to chlorine (Espinosa-García et al, 2002; Mazari-Hiriart et al, 2002). Nevertheless, faced with the risks of distributing water that has not been disinfected, adding chlorine is always preferable. Still, it is estimated that only between 85 per cent (according to the Mexican Ministry of Health) and 95 per cent (according to the National Water Commission) of the water distributed is chlorinated (CNA, 2005; Secretaría de Salud, 2005).

In cities, the water service is highly heterogeneous. Networked water is distributed mainly to downtown businesses and medium- and high-income residential areas. Most other domestic users obtain their water from water tankers, which provide a limited amount of low-quality water at a high price (Torregrosa et al, 2002). For instance, the Federal District[6] has a population of 8 million people and accounts for 21.8 per cent of the national GDP, but around 1.1 million inhabitants (14 per cent of the total) are supplied through water tankers (Jiménez, in press).

For those receiving piped water at home, the quality of the service is not necessarily much better. The water flow is often intermittent and does not meet drinking water standards (Jiménez et al, 2002; Ezcurra et al, 2006).[7] As a result, households have their own storage tanks and indoor pumping systems in order to gain permanent access to water, but need to buy safe water to drink from vendors or have a domestic system to potabilize the water. This increases the cost of water by 30 to 50 per cent over the official price. For a four-member family living on a minimum wage, drinking water represents 11 to 21 per cent of their income.[8]

Water services have an impact upon health and the environment

From a health perspective, Mexico is considered a country in transition, with diseases typical of underdeveloped countries (e.g. diarrhoeas), but also, increasingly, diseases typical of the developed world (Gutierrez et al, 1994). According to official data, the incidence of diarrhoeas is declining due to an increase in chlorination of water supply from 55 to 95 per cent between 1990 and 1992. But despite this improvement, these infections are still considered the main causes of infant mortality in the country (NIPH, 2002). Moreover, the health achievements have been uneven at the local level and they have been especially limited in areas where the coverage of WSS is low.

The lack of safe wastewater treatment and disposal affects the quality of water sources. According to an international survey applied to 122 countries, Mexico ranks 102nd for its low water-quality sources (UN, 2003). According to this survey, Mexico ranks below countries with similar income levels (such as Uruguay, Costa Rica, the Russian Federation, Botswana and Chile), countries with much lower income levels (e.g. Cuba, Pakistan, Guatemala, China, Venezuela, Turkey and Panama) and even below countries with lower water availability (e.g. Turkey or South Africa).

WATER MANAGEMENT AND PUBLIC POLICIES

Drinking water quality is regulated through a norm dating from 1994. This norm considers 40 parameters for assessing potability; but, in practice, at most only one parameter is measured regularly by water utilities. Information on water quality is frequently kept confidential. Therefore, the enforcement of this norm is quite weak.

In order to control water pollution, since 1973 the government has passed different regulatory frameworks, with modest results in practice. A thorough analysis of the situation led, in 1996, to the development of new norms specifically designed to address the needs and problems observed, which considered:

- the reasons why previous regulations could not be enforced;
- the fact that around two thirds of the wastewater produced was released on soil (reused for agricultural irrigation) rather than emptied into surface water bodies;
- the need to coordinate the analysis of the parameters used to control the conditions of wastewater discharges with those used to monitor the quality of water sources (Jiménez, 2005).

Under the new legal framework, in eight years, wastewater treatment was increased by 300 per cent compared with the situation observed under previous norms. Unfortunately, the new regulation has not been fully enforced due to a lack of political will to implement complementary actions, such as training, access to sound financial programmes, and the set-up of political agreements between provincial and local governments from different political parties. One example was the situation of Mexico City, which produces 20 per cent of the country's total wastewater and accounts for 21 per cent of national GDP. The city already had a wastewater treatment programme, including financial support to implement it, but could not build the required infrastructure due to the lack of agreement with the federal and local governments, which are from different political parties.

Wastewater treatment technology has not been a problem either. Prior to the 1990s, the most popular treatment systems were stabilization ponds. Since then, several pond systems that were malfunctioning due to deficient designs and a lack of maintenance (more than 50 per cent were out of service) were replaced. The new systems built use activated sludge (when the effluent is to be discharged into surface water bodies) and advanced primary treatment (when the effluent is to be discharged into soils for agricultural irrigation). Because the latter was an economic option for fulfilling the NOM 001 to reuse wastewater for agricultural irrigation, several plants were built using this technology (Mexican Government, 1997; Jiménez, 2005). Therefore, the combination of the new regulatory framework with new technological options resulted in a notable increase in treated wastewater.

The process of political and administrative transformation

The management of WSS has undergone far-reaching transformations during the last two decades in an attempt to change the previous model based on 'administrative rationalism', highly concentrated at the federal level (Castro, 2005), to more decentralized forms that require local management and private and social participation, although these are incipient. These new characteristics were promoted by international financial institutions (mainly the World Bank and the IDB) as part of their policies to transfer functions and services from the federal government to provincial and municipal authorities, and from the public to the private sector (Dourojeanni et al, 2002; Tortejada et al, 2004; Torregrosa, 2006). These changes also introduced modifications in the organization of public institutions and prompted a diversity of responses from different sectors of society that try to adapt to new conditions.

One of the most important institutional changes occurred in late 1976, when the functions and resources of the Ministry of Hydraulic Resources (SRH) and the Infrastructure Commission of the Ministry of Health (SSA), which was in charge of constructing water works in rural areas, were transferred to the Ministry of Human Settlements and Public Works (SAHOP). The new Ministry of Agriculture and Hydraulic Resources (SARH) preserved only the responsibility for major hydraulic works. This change created an artificial division between what was called the Bulk Water Supply Works that remained the responsibility of SARH, and the water infrastructure works dealing with WSS (Torregrosa and Vargas, 1994; Carabias and Landa, 2005).

During the 1980s, the decentralization of the country's WSS was accelerated. The president transferred the management of water services from the federal government to provincial administrations, which, in some cases, passed them on to municipalities. In 1983, Article 115 of the Mexican Constitution was reformed, transferring the responsibility for WSS from the federation to the municipalities. In 1986, the Water Rights Federal Law was reformed to introduce fees for water usage in order to promote efficient water use and to improve the payment of water services. These fees are annually updated, and in 1991 the law was reviewed to also include fees for wastewater discharges.

However, the economic resources collected through the water service fees are often used for funding activities unrelated to the water sector, which is partly caused by the lack of institutional capacity in the provinces and municipalities. This situation, along with the lack of a well-defined policy for water tariffs owing to the political use of the water fees, among other problems, has led to the de-capitalization of water systems and the increasing need for government subsidies. Unfortunately, subsidies do not always result in benefits for the poor sectors that they are supposed to help (Belausteguigoitia and Rivera, 1992; Bonilla-Castañeda, 1997). An example is the Federal District where, due to the deficiencies of the water supply system, most of the poor are not served by the networked water systems. In fact, they have to buy small volumes of poor-quality water at a much higher price than the middle- and high-income classes, whose water supply is subsidized.[9] Similar situations have been observed in most of the country's cities (Quiñones, 2001; Montesillo-Cedillo, 2004).

Another important step in the reorganization of WSS took place in 1982, when the federal and municipal water responsibilities were moved from the SARH to the Ministry of Urban Development and the Environment (SEDUE), which further decided to decentralize the construction of water works. SEDUE retained only the function of technical agent in order to be eligible for obtaining international loans; the SARH retained the management of bulk water supply works, in which several states were involved.

The latest stage of the reorganization began with the creation in 1989 of the National Water Commission (CNA). All public water management functions were allocated to the CNA; but at the same time, the old project to decentralize nearly 300 provincial and municipal water utilities was reactivated. Simultaneously, important legal reforms were introduced to modernize the water sector (Torregrosa et al, 2004; Carabias and Landa, 2005). In 1992, the crucial Article 27 of the Mexican constitution was reformed, fundamentally changing the rules governing the ownership of land and water in the country. A regulatory law to enforce the changes came into effect in 1994. The main objectives of the 1992 constitutional reforms were:

- promoting the integral management of water, with strong user participation;
- sustainable water management;
- consolidating the CNA as the only federal authority with responsibility for water management;
- granting legal certainty to water users and private companies;[10]
- making water use more efficient and rational; and
- promoting stronger private participation in the water sector.

In this regard, the reforms introduced the possibility of private participation[11] in the construction and operation of WSS works through concessions and other instruments. However, as already shown by the relatively poor performance of Mexico in relation to WSS coverage, the implementation of the new law had a rather low impact on the improvement of these services, while the evidence suggests that it has contributed to the observed increase in social conflicts surrounding water resources and water services in recent years (de Alba and Kloster, 2007).

One of the main challenges ahead is to improve the situation in urban areas, where there is intense social pressure for water services. To meet this demand, the government has followed different strategies. At a federal level, the CNA began to promote integrated water resources management (IWRM) along the lines of the international initiatives on the topic. In recognition of the centrality of social participation in the implementation of IWRM policies, the CNA created the Water Citizen Movement (Castro et al, 2004; Tortejada et al, 2004), which seeks to foster the development of a 'new water culture' and raise environmental awareness among the population. The Water Citizen Movement has produced a number of initiatives, such as a national water council and provincial citizen water councils, which stem from longstanding water-user associations and NGOs. These councils contribute to the design of specific programmes (e.g. environmental education) and liaise with differing degrees of coordination with the CNA, the Ministry of Environment and Natural Resources (SEMARNAT), the Ministry of Public Education (SEP) and the Ministry of Health (SSA), among others. However, in practice, this process has had limited impact upon improving access to WSS, or, more broadly, upon enhancing the living conditions of the population. To some extent, the lack of impact of this type of citizen involvement may be explained by the fact that these groups have developed very close ties with the government, which tends to reproduce its traditional top-down approach to water policy. For instance, the government defines who can participate in the different councils, which severely restricts the representation of actual water users and of the water needs of wider social groups.

Mexico is formally committed to fulfilling the WSS targets of the Millennium Development Goals. However, it remains to be seen if the government will show the political will required to make the relevant decisions in time. For instance, although according to national and international experts, the key for meeting the WSS goals is the investment of fiscal resources, during the last two decades Mexican water policy has been influenced by the argument that the bulk of the investments needed should come from the private sector and from the users themselves. Yet, recent international experience shows that private investors do not have the capacity to extend the coverage of WSS to the poorest sectors of society. The experience of developing countries such as Bolivia, Argentina and Kenya, but also Mexico, amongst others, has shown that private water companies do not invest in developing or replacing infrastructure, and that they tend to rely on highly subsidized public funding or loans. In fact, this has been the case in developed countries too, as the history of the private water companies in France demonstrates (see Chapters 11 and 14 in this volume). This is understandable because the main rationale of private water companies is not to provide a public service or to extend coverage to the poor who cannot afford to pay the full cost of commercialized WSS, but rather to make a profit for their shareholders (Castro, 2007). In this regard, although Mexico's position in the world economy suggests that the country's economic conditions would support rapid improvements in the provision of essential services, significant changes will be needed in Mexican WSS policy if the country expects to meet the internationally agreed targets for these services.

CONCLUSIONS

In Mexico, the government has historically been responsible for providing WSS, covering different functions that range from planning and implementation of policies to direct management. The lack of participation by other social actors in the organization of these services, especially those that have played important roles in other countries, such as local authorities, cooperatives and NGOs, may be an important factor in explaining the delays experienced by Mexico in the universalization of WSS. Moreover, the participation of unregulated non-governmental actors, particularly private water vendors, has contributed to the low-quality standards characterizing water supply services in the country. More recently, the government has promoted private participation in the sector, which has been mainly limited to service contracts (e.g. metering and billing) and a few long-term integral concessions, to which we could add the creation of a commercial but municipally owned company. Nevertheless, in Mexico, the best examples of efficient WSS management and capability for universalizing WSS come from public and not private utilities. This is notably the case with the WSS companies of Nuevo León and Tijuana, among others, which have not only proven to be more efficient than private operators, but also effective at covering the needs of the most deprived sectors of the population, who are precisely the main target of the MDGs.

In this connection, the decentralization of WSS that started in the 1980s was carried out in conjunction with a series of administrative, policy and legal reforms focused on the promotion of the private sector. However, the overall process and, in particular, the concession of water utilities to private operators were characterized by contradiction in the context of a very weak institutional framework. Concessions, for instance, were granted to private companies in the absence of regulatory mechanisms or an appropriate legal framework to monitor the performance of the private operators. This situation prevailed even when the first problems with the concessions were observed, such as lack of compliance with contractual agreements, a problem that has not yet been properly addressed.

Decentralization also had unpredictable impacts upon WSS, owing to the political context. In this regard, WSS have ranked very low in the government agenda, not least because the political reforms that, among other significant outcomes, brought to an end 80 years of uninterrupted single-party rule in the year 2000 are still very incipient and fragile. One crucial aspect has been the absence of any meaningful political representation in congress that could offer an effective counterbalance to the central government. This happens despite the emerging power of provincial governors, who have become stronger thanks to the decentralization of the federal functions and budget, but that with very few exceptions have lacked the institutional conditions to make WSS a priority.

These problems have been compounded by that fact that decentralization came up against severe obstacles in the form of long-established political mechanisms based on clientelistic relations, where public services had been traditionally obtained and maintained in exchange for political favours. This has hampered the plans for developing self-sustainable WSS with clearly established user rights and responsibilities as foreseen in the 1992 Water Law and related legal and administrative reforms. In practice, the traditional corporative-clientelistic model of service provision, where economic payment for basic services was not as important as payment through political support (Ziccardi, 1998), continues to be entrenched in the Mexican political system. Unfortunately, this way of managing water resources and access to WSS is still one of the key elements of the 'clientelistic' game, characterizing the forms of social participation promoted in Mexico's water sector (Castro et al, 2004).

In this context, achieving universal access to WSS will not be an easy task. We have observed that the decentralization of policy and management of water resources and WSS and the introduction of private-sector participation have not contributed to meeting the water needs of the most deprived population both in urban and rural areas. We argue that making safe water and effective sanitation available to all sectors of Mexican society can only be accomplished by adopting an integrated and multidimensional approach in which WSS provision is organized through a diversity of arrangements and options that take account of the different local conditions. Such strategies have proven to be successful in many developed and developing countries, and range from municipal ownership and management with outsourcing of non-core operations to improved public–private cooperation, regional (supra-municipal) cooperation, cooperative models, and community ownership and management, etc. (see, for instance, Chapters 9, 10, 13 and 18 in this volume).

Nevertheless, we maintain that the government continues to have the main responsibility in ensuring the country's achievement of the MDGs, which implies reducing by half the proportion of the Mexican population who lack access to safe WSS by 2015. The government should adopt an integrated approach involving the coordination of the different actors involved (all levels of government, water users, academics and professional experts, practitioners, politicians, etc.) to design a medium- and long-term plan for WSS provision in Mexico in accordance with the needs and conditions specific to each region and locality. We recognize that the sustainable and universal provision of WSS is not an easy task: this challenge, therefore, demands the close interaction of scientists to develop an interdisciplinary framework that brings together the technological, socio-cultural, economic-financial and political factors of WSS policy and management. We also recognize the need for social and political leaders who can drive the required transformations effectively.

NOTES

1 The crisis hit bottom during those two years. In January 1997, the growth rate began to pick up; but even to date, employment, productivity and wage levels have not recovered to pre-crisis levels (Puyana, 2007).

2 As an example of the above, 2 billion pesos (around US$2 million) were given to the Programme for the Modernization of Water and Sanitation Utilities (PROMAGUA) in 2001. The programme was implemented by the National Works and Public Services Bank (BANOBRAS) and is aimed basically at supporting water utilities that serve localities with more than 50,000 residents, which represent over 50 per cent of the country's population.

According to the plans, the CNA will provide sunk investments to the operators who sign up to the programme, and the funding would be allocated to a short-term (three-year) investment programme to increase efficiency. The share of the funding that the federal government would contribute to the programme is determined by the overall efficiency of the operator, which is ascertained by the *Diagnostic Study and Integral Planning* report required to participate in the programme. Only those utilities that demonstrated an overall efficiency coefficient above 45 per cent received sunk investments for developing new sources of water supply and extending coverage. Crucially, PROMAGUA gave more facilities for accessing funding to those WSS utilities that incorporated some form of private participation. However, after six years of operation, the demand for the programme's loans has been very low.

3 Traditionally, water services at the international level are described by means of two indicators, one referring to the water supply and another to 'sanitation'. However, in practice, the meaning of these two indicators varies for different countries. Water supply refers to the percentage of the population receiving water:

- in their house or at a 200m distance;
- continuously or intermittently; and
- that may be potable or not (i.e. may or may not fulfil the drinking water norms).

In turn, 'sanitation' refers to the collection of wastewater from households by a network that in some cases may or may not end in a sewerage system. Normally, developing countries do not include wastewater treatment under 'sanitation'. Despite these variations, considering the universality of these two parameters, we will use them throughout this chapter.

4 Although macro-economic indicators show a positive trend in the Mexican economy during this period, the benefits have not been evenly distributed. Thus, the top tenth decile of the population concentrates half of Mexican household income, while 40 per cent of the households receive only 6 per cent. Also, 70 per cent of the country's municipalities have a sizeable share of the population living in conditions of marginalization.

5 With the exception of some specific localized sites where problems caused by iron, manganese and arsenic were detected several decades ago and where more complex processes are used as a result.

6 Considering only the Federal District and not the whole metropolitan area, where the figures are much worse.

7 Independent studies carried out by academic institutions revealed that even though a high percentage of water is chlorinated, the water received by the households fulfils the requirements of residual chlorine and microbiological content in only 30 per cent of the cases (Jiménez et al, 2002).

8 The minimum wage is 43 Mexican pesos (US$4.10).

9 In the Federal District, the minimum price of water supplied by private vendors is 10 pesos (around US$0.93) per cubic metre, while the public water tariff is 2 pesos (around US$0.19) per cubic metre.

10 Through the creation of a Public Water Rights Registry (REPDA).

11 In Mexico, privatization does not refer to the sale of water-sector assets or the total transfer of service provision and its responsibilities from the government to private actors. Private participation is being promoted only through time-limited contracts. This includes several modalities, ranging from the partial management of WSS to the participation of private companies in sharing a certain degree of financial risk. In the most advanced schemes, the private companies participate in build–operate–transfer (BOT) activities (Barocio and Saavedra, 2004).

References

Alcocer, J., L. E. Marín and E. Escobar (1998) 'Geochemical evaluation of five cenotes for use as potential drinking water supplies in northeastern Yucatan, Mexico', *Hydrogeology Journal*, vol 6, pp293–301

Armienta, M. A. and R. Rodríguez (2002) 'Metals and metalloids. Case of study: Pollution by arsenic in the groundwater of Zimapan, Hidalgo; Environmental problematic and methodological approach', in B. Jiménez and L. Marín (eds) *Water Viewed from Academy*, Mexican Academy of Sciences, Mexico City, pp79–98 (in Spanish)

Barocio, R. and S. J. Saavedra (2004) 'Private participation in water and sanitation services in Mexico', in B. Jiménez and L. Marín (eds) *Water Viewed from Academy*, Mexican Academy of Sciences, Mexico City, pp289–316 (in Spanish)

Belausteguigoitia, J. C. and J. M. Rivera (1992) 'Water fees as an element of rational water allocation', in R. Samaniego (ed) *Essays on the Economy of Mexico City*, Pórtico de la Ciudad de México, Mexico City, p405 (in Spanish)

Bonilla-Castañeda, J. (1997) 'The market economy and Mexico's future', www.javierbonilla.com/1997/1997_01_20.htm, accessed May 2008

Carabias, J. and R. Landa (2005) 'Water, environment and society: Towards the integral management of water resources', in J. Collado and P. Martínez (eds) *National Autonomous University of Mexico (UNAM)*, El Colegio de México and Gonzalo Río Arronte Foundation, Mexico City, p221 (in Spanish)

Carrillo, A. and J. I. Drever (1998) 'Environmental assessment of the potential for arsenic leaching into groundwater from mine wastes in Baja California Sur, Mexico', *Geofísica Internacional*, vol 37, pp35–39

Carrillo-Rivera, J., A. Cardona and T. Herat (2006) 'Induction of deep thermal water to superficial zones: Aguascalientes, Mexico', in B. Jiménez and L. Marín (eds) *Water Viewed from Academy*, Mexican Academy of Sciences, Mexico City, pp137–158 (in Spanish)

Castro, J. E. (ed) (2005) *Final Report, PRINWASS Project*, University of Oxford, Oxford, www.prinwass.org, accessed May 2008

Castro, J. E. (2007) 'Poverty and citizenship: Sociological perspectives on water services and public-private participation', *Geoforum, Special Issue on Pro-poor water. The Privatization and Global Poverty Debate*, vol 38, no, pp756–771

Castro, J. E., K. Kloster and M. L. Torregrosa (2004) 'Citizenship and governance in Mexico: Conflict and social participation around water management', in B. Jiménez and L. Marín (eds) *Water Viewed from Academy*, Mexican Academy of Sciences, Mexico City, pp339–369 (in Spanish)

CNA (Comisión Nacional del Agua) (2005) *Water Statistics in Mexico 2005*, CNA, Mexico City (in Spanish)

CNA (2006) *Water Statistics in Mexico 2006*, CNA, Mexico City (in Spanish)

CNA (2007) *Water Statistics in Mexico 2007*, CNA, Mexico City (in Spanish)

de Alba, F. and K. Kloster (2007) 'Water in the city of Mexico and the political fragmentation factor', *Perfiles Latinoamericanos*, vol 29, January–June, pp137–159 (in Spanish)

Dourojeanni, A., A. Jouravlev and G. Chávez (2002) 'Management water at basin level: Theory and practice', *Natural Resources and Infrastructure Series*, no 47, United Nations Economic Commission for Latin America and the Caribbean (ECLAC), Santiago de Chile (in Spanish)

Escolero, O. A., L. E. Marín, B. Steinich, A. J. Pacheco, S. A. Cabrera and J. Alcocer (2002) 'Strategy for the protection of karstic aquifers: The Merida Yucatan, Mexico example', *Water Resources Management*, vol 16, no 5, pp351–367

Espinosa-García, A., M. Mazari-Hiriart, A. Arias and S. Ponce de León (2002) 'Presence of enteric viruses in water in Southern Mexico City', in *Proceedings of the International Conference: Water and Wastewater, Perspectives of Developing Countries*, Indian Institute of Technology Delhi-International Water Association, Nueva Delhi, 11–13 December

Ezcurra, E., M. Mazari, I. Pisanty and A. Guillermo-Aguilar (2006) *The Basin of Mexico*, Fondo de Cultura Económica, Mexico City (in Spanish)

Foster, S., H. Garduño, A. Tuinhof, K. Kemper and M. Nanni (2004) *Urban Wastewater as Groundwater Recharge: Evaluating and Managing the Risks and Benefits*, Briefing note 12, World Bank, Washington, DC

Gutierrez, G. H., H. Guiscafré, H. Reyes, R. Pérez, R. Vega and P. Tome (1994) 'Reduction of mortality caused by acute diarrhea diseases. Experiences from a research-action program', *Salud Pública de México*, vol 36, no 2, pp168–179

IMF (International Monetary Fund) (2007) *World Economic Outlook Database*, IMF, Washington, DC, www.imf.org/external/pubs/ft/weo/2007/02/weodata/WEOOct2007all.xls, accessed May 2008

Jiménez, B. (2005) 'Treatment technology and standards for agricultural wastewater reuse: A case study in Mexico', *Irrigation and Drainage*, vol 54, no 1, pp23–35

Jiménez, B. (in press) 'Management of water in Mexico City', in L. Mays (ed) *Integrated Urban Water Management in Arid and Semi-Arid Regions around the World*, UNESCO, Paris

Jiménez, B. and A. Chávez (2004) 'Quality assessment of an aquifer recharged with wastewater for its potential use as drinking source: "El Mezquital Valley" case', *Water Science and Technology*, vol 50, no 2, pp269–273

Jiménez, B., A. Chávez, S. Lucario and C. Maya (2002) *How to Improve Water Services in Quantity and Quality in the Coyoacan Delegation*, National Autonomous University of Mexico, University Programme for Studies of the City, Mexico City (in Spanish)

Katz, I. (2002) 'Inflation, growth, poverty and inequality in Mexico', *Gaceta Económica*, year 7, pp154–174 (in Spanish)

Mazari-Hiriart, M., Y. López-Vidal, S. Ponce de León, J. J. Calva-Mercado and F. Rojo-Callejas (2002) 'Significance of water quality indicators: A case study in Mexico City', in *Proceedings of the International Conference: Water and Wastewater, Perspectives of Developing Countries*, Indian Institute of Technology Delhi-International Water Association, Nueva Delhi, 11–13 December

Meyer, L. (2005) 'Poverty in Mexico: Approximation to the great historic problem', *Comercio Exterior*, vol 5, no 8 (in Spanish)

Mexican Government (1997) 'NOM-001-ECOL-1996: Mexican Official Norm to set the maximum allowable limits to discharge wastewater to national water and goods', Official Diary of the Federation, Mexico City, 6 January 1997 (in Spanish)

Mexican Government (2000) 'NOM-127-SSA1-1994: Official Mexican Norm. Environmental health, water for use and human consumption, allowable limits and treatment needs for potabilization', *Official Diary of the Federation*, Mexico City, 22 November 2000 (in Spanish)

Montesillo-Cedillo, J. L. (2004) 'Tariff structure: Economical analysis of tariff structure of potable water service in the Distrito Federal', in V. Libreros-Muñoz, M. C. Martínez-Omaña, R. I. López-Hernández, J. L. Montesillo-Cedillo, A. Ortiz-Rendón and A. M. Quiñónez-Castillo (eds) *Water Management in the Distrito Federal: Challenges and Proposals*, UNAM, PUEC, Legislative Assembly of DF, II Legislatura, Mexico, p199 (in Spanish)

Morera-Camacho, C. (2005) 'The new economic power and the origins of injustice in Mexico: Sources of the great mobilization for democracy', *Memory in Monthly Political and Cultural Magazine*, no 197, July, http://memoria.com.mx/node/602 (in Spanish), accessed June 2008

NIPH (National Institute of Public Health) (2002) *Health Atlas 2002*, NIPH, Mexico City, (in Spanish)

Pacheco, A. J., S. A. Cabrera and L. E. Marín (2000) 'Bacteriological contamination assessment in the karstic aquifer of Yucatan, Mexico', *Geofísica Internacional*, vol 39, no 3, pp285–291

Puyana, A. (2007) 'The project and the realities', Paper presented at the International Workshop on The Modern Market and Traditional Politics? A workshop to explore the economics and politics of the last two decades in Latin America, Oxford University, Oxford, 4–5 June

Quiñones, A. (2001) *Roll of Tariff as a Mediator Factor of the Relation Population-Environment: The Case of Water Consumption in Delegación Milpa Alta*, MA thesis, Latin Amercian Faculty of Social Sciences (FLACSO), Mexico City (in Spanish)

Rodríguez, R., A. Armienta, J. Berlin and J. A. Mejía (2002) 'As and Pb groundwater pollution of the Salamanca Aquifer system: Origin, mobilization and restoration alternatives', *Red Book IAHS*, vol 275, pp561–565

Secretaría de Salud (2005) *A First Report on the Indicators and Available Measurements of the Hydric Diseases Transmission in the Country*, Secretaría de Salud, Mexico City, www.salud.gob.mx/ssa_app/noticias/datos/2004-02-23_804.html, accessed May 2008 (in Spanish)

SEGOB (Secretaría de Gobernación) (2001) *National Development Plan, 2001–2006*, SEGOB, Mexico City (in Spanish)

Székely, M. (2005) 'Poverty, inequality and macro-economic environment in Mexico', *ICE*, no 821, March–April, pp125–142

Torregrosa, M. L. (2006) 'Integral, decentralized and participatory water management in Mexico: A construction process', in *Water-First Part*, Science and Technology Council of Guanajuato, *Gaceta Electrónica*, vol 16, no 28, p51 (in Spanish)

Torregrosa, M. L. and S. Vargas (1994) *Restructuration Policy of Water Services for Agricultural and Urban Use*, Research report, Mexican Institute of Water Technology, Subcoordination of Social Participation, Jiutepec, Morelos, Mexico (in Spanish)

Torregrosa, M. L., F. Saavedra and E. Padilla (2002) *Elements for the Construction of the D2 Baseline Report*, Preliminary report, PRINWASS Research Project, University of Oxford, Oxford, www.prinwass.org (in Spanish)

Torregrosa, M. L., C. Arteaga and K. Kloster (2004) *Opportunities, Constraints and Room for Improvement of the Current Framework Regulating Water Supply and Sanitation in Peri-Urban Areas: The Case of Mexico City*, Final report of the Service Provision Governance in the Peri-Urban Interface of Metropolitan Areas project, University College London, Development Planning Unit, London

Tortejada, C., V. Gerrero and R. Sandocal (2004) *Towards an Integral Management of Water in Mexico: Challenges and Alternatives*, Third World Centre for Water Management, A. C., Porrúa, and Cámara de Diputados, Mexico City (in Spanish)

Tuirán-Gutiérrez, A. (2006) *Inequality in the Distribution of the Monetary Income in Mexico*, Final report, National Population Council (CONAPO), Mexico City

UN (United Nations) (2003) *Water for People, Water for Life*, United Nations World Water Development Report, UNESCO and Berghahn Books, Barcelona

UNDP (United Nations Development Programme) (2006) *Human Development Report 2006: Beyond Scarcity – Power, Poverty and the Global Water Crisis*, UNDP, New York

WRI (World Resources Institute) (2007) *Earth Trends: Water Resources and Freshwater Ecosystems – Country Profiles*, http://earthtrends.wri.org/country_profiles/index.php?theme=2, accessed May 2008

Ziccardi, A. (1998) *Gobernability and Citizen Participation in Capital City*, Institute of Social Research, UNAM, Porrúa, Mexico, p240 (in Spanish)

Index